Children's Well-Being: Indicators and Research

Volume 23

This series focuses on the subject of measurements and indicators of children's well being and their usage, within multiple domains and in diverse cultures. More specifically, the series seeks to present measures and data resources, analysis of data, exploration of theoretical issues, and information about the status of children, as well as the implementation of this information in policy and practice. By doing so it aims to explore how child indicators can be used to improve the development and the well being of children.

With an international perspective the series will provide a unique applied perspective, by bringing in a variety of analytical models, varied perspectives, and a variety of social policy regimes.

Children's Well-Being: Indicators and Research will be unique and exclusive in the field of measures and indicators of children's lives and will be a source of high quality, policy impact and rigorous scientific papers.

More information about this series at http://www.springer.com/series/8162

Donell Holloway • Michele Willson •
Karen Murcia • Catherine Archer •
Francesca Stocco
Editors

Young Children's Rights in a Digital World

Play, Design and Practice

Springer

Editors
Donell Holloway
School of Arts and Humanities
Edith Cowan University
Perth, WA, Australia

Michele Willson
School of Media Culture and Creative Arts
Faculty of Humanities
Curtin University
Perth, WA, Australia

Karen Murcia
School of Education
Curtin University
Perth, WA, Australia

Catherine Archer
Strategic Communication, School of Arts
Business, Law and Social Sciences
Murdoch University
Perth, WA, Australia

Francesca Stocco
School of Arts and Humanities
Edith Cowan University
Perth, WA, Australia

ISSN 1879-5196 ISSN 1879-520X (electronic)
Children's Well-Being: Indicators and Research
ISBN 978-3-030-65915-8 ISBN 978-3-030-65916-5 (eBook)
https://doi.org/10.1007/978-3-030-65916-5

This Springer imprint is published by the registered company Springer Nature Switzerland AG.
The registered company address is: Gewerbestrasse 11, 6330 Cham, Switzerland

Acknowledgements

This book arose out of a Digitising Early Childhood Conference held in Perth, Western Australia, in 2018, which was supported by an Australian Research Council (ARC) Discover Grant, *Toddlers and tablets: Exploring the risks and benefits 0–5s face online* (DP150104734), and a pre-conference in Manchester (UK), which was also supported in part by an ARC grant (DP180103922), *The Internet of toys: Benefits and risks of connected toys for children*. Many of the authors whose chapters form part of this collection presented at these conferences. Importantly, however, this book also includes chapters from members of diverse networks and bodies such as DigiLitEY and EU Kids Online, as well as academic contacts gathered from a wider call for papers.

The conferences, research networks, and academic collaborations bring together in this volume collaborating authors from across wide-ranging interdisciplinary backgrounds including cybersecurity, marketing, media/communications and cultural studies, commerce, engineering, early childhood education, ethics, and physiotherapy fused with psychology. This dynamic and interdisciplinary mix provides a wealth of research expertise and experience underpinning children's rights to participation, provision, and protection in a digital age, which fosters children's autonomy and freedom to explore play opportunities.

The editorial team for this volume consists of researchers drawn from across the four main universities in Western Australia and from a number of disciplinary backgrounds. Our team especially appreciates the resilience and persistence of our chapter authors, who finalised their chapters throughout the first half of 2020, as the COVID-19 pandemic took hold. Despite experiencing lockdown restrictions, working from home (often combined with childcare and home-schooling), and looking after vulnerable loved ones, these authors continued to edit and format their chapters and meet deadlines.

The editors gratefully acknowledge the vital assistance of Linda Jaunzems. Her attention to detail, organisational skills, and deep experience of the editing process helped enormously. They are also indebted to Springer for its editorial support and especially to Ameena Jaafar for her valuable advice and insight throughout the editing and publishing process.

Finally, the anonymous referees who provided our chapter authors valuable feedback are thankfully acknowledged. Their thoughtful comments and suggestions are much appreciated.

Contents

List of Figures

List of Tables

Chapter 1
Introduction

Karen Murcia, Michele Willson, Catherine Archer, Francesca Stocco, and Donell Holloway

In 2014, the UN's Committee on the Rights of the Child met to discuss 'Digital Media and Children's Rights'. The resulting Digital Rights Framework (Livingstone & Bulger, 2014) reconfigures how we understand children's needs, agency and vulnerability to harm in today's digital world. The rights framework also implies the roles and responsibilities for a variety of social actors including the state, families, educators within schools, not-for-profit and commercial entities, researchers and the children themselves.

This book foregrounds research which is centred on young children's rights in a digital world. It gathers current research from around the globe focussing on young children's rights as agential citizens to the provision of and participation on digital devices and with content—as well as their right to protection from harm (Livingstone, 2007). Implicit within this book is the acknowledgement that children of various ages, abilities, socio-economic and geographic backgrounds should have equal access and experiences with new digital technologies and content alongside adult support, and guidance to enhance these experiences.

The book is divided into four parts: The Early Childhood Home, Pedagogy Approaches and Challenges, Connected Toys at Home and School and Privacy and Protection. These parts scaffold the discussion ranging from considering the early childhood home, parental concerns and practices, connected toys, pedagogical

K. Murcia (✉) · M. Willson
Curtin University, Perth, WA, Australia
e-mail: karen.murcia@curtin.edu.au; m.willson@curtin.edu.au

C. Archer
Murdoch University, Perth, WA, Australia
e-mail: c.archer@murdoch.edu.au

F. Stocco · D. Holloway
Edith Cowan University, Perth, WA, Australia
e-mail: fastocco@our.ecu.edu.au; donell.holloway@ecu.edu.au

D. Holloway et al. (eds.), *Young Children's Rights in a Digital World*, Children's Well-Being: Indicators and Research 23,
https://doi.org/10.1007/978-3-030-65916-5_1

approaches, privacy and data analytic challenges that present some obstacles for children in which research seeks to overcome. This book has an international focus with contributors and their diverse research projects drawn from across the globe. This international range enables similarities and differences across particular locales to be brought to the forefront. It becomes quickly apparent that parents, educators and regulators around the world are grappling with how to better support children's digital engagement to maximise affordances and opportunities in innovative ways, while being mindful of the need to minimise potential harm. Each chapter offers more focussed examination on a particular issue, question or concern, the collection as a whole also offers the reader the potential for a multifaceted reflection on the challenges and opportunities for our very young people within an increasingly interconnected, and digitally mediated landscape.

Part I: The Early Childhood Home

Part I provides research findings from a variety of family-based studies internationally. The authors discuss children's digital technology use, parental attitudes and beliefs about screen use in the home, parental mediation strategies, family communication practices, and the social implications of screen use in the family home (such as reducing or enhancing interactivity between family members). In all cases, parents are endeavouring to protect their children from harm, while at the same time providing them with the digital opportunities to participate, learn and play. However, they are often caught in a myriad of contradictory messaging and paradigms from media, education providers, health professionals, family and the community alike that can be difficult to reconcile and to navigate.

In Chap. 2, *The tablet is my best friend!": Practices and perceptions of young children and their parents*, Patricia Dias and Rita Brito base their chapter on the interviews and activities conducted with 25 Portuguese families, including the parents and their children. The authors sought to discover how families are adopting digital technologies in their home. They selected a varied sample, considering the gender of the child, family composition and socio-economic status, to obtain a variety of narratives on the phenomenon being studied. Discrepancies were identified in the discourses of parents and children regarding the perception of dangers and benefits (parents value learning and place more emphasis on risks, while children enjoy fun, and enjoyment), the digital skills of children (children are more autonomous and capable than parents think), and the criteria for setting rules (parents tend to be restrictive of screen-time, but not of content and activities).

In Chap. 3, *Family mediating practices and ideologies: Spanish and Portuguese parents of children under three and digital media in homes*, Mitsuko Matsumoto, David Poveda, Ana Jorge, Raquel Pacheco, Vítor Tomé, Cristina Aliagas and Marta Morgade explore how parents accompany their very young (under 3 years old) children's inclusions into 'digital society' by examining their mediating practices and ideologies. The authors draw on diverse data (observations/video-recordings and interviews with parents at home) from case studies of five children from middle-class families from Spain and Portugal, collected in 2017, following the protocol developed for *A Day in the Digital Lives of 0–3 Year-Olds* (Gillen et al., 2019). The authors problematise the complex relationship between parental beliefs, self-perceptions and their actual practices regarding the place of digital technologies in children's lives. They do so by examining mediation as an emergent process in which family members co-create interactional ecologies (Kyratzis & Johnson, 2017; Erickson, 1996), and by seeing mediation as a set of strategies within family routines (Livingstone, 2007). In terms of the digital ecosystem, they analyse mediation at the levels of the digital media ecology/environment in the home, the actual digital media activities and mediation practices, and the parents' broader media ideologies and beliefs on technologies (Gershon, 2010), finding, as in the previous level, contradictions between the various stakeholders.

Portugal is also featured in Chap. 4, *Digital parenting and transnational migration: Cultural and emotional drives for digital media use*, with a focus on the country's history as being marked by flows of immigrants and emigration. Authors Teresa Castro and Cristina Ponte explore the niche of the transnational families with the two research questions: how are late modern transnational families incorporating and making sense of communication tools in their everyday parenting chores, and secondly, families' interactions and how they are regulating children's digital provision and protection? The authors selected six different families (in Portugal and England) in terms of cultural and socio-economic characteristics, as part of an ongoing qualitative longitudinal study (iTec Families). The chapter builds on the families' testimonials, perceptions and practices to stimulate the debate around children and technology in the domestic context, and on parental guidance and mediation from a socio-constructivist standpoint. They investigate whether the adoption and appropriation of digital technology is a relevant element in shaping families' daily lives, and whether digital artefacts and their perceived affordances help foster connectedness.

In Chap. 5, *Children's rights to 'good' and 'bad' screen time: Parental narratives of how children do family online*, authors Helga Sadowksi and Lina Eklund draw on interview data with six extended Swedish families spread over 18 households, and investigate how parents and grandparents manage, relate to, and assist in children's digital family work. For many families, digitally connected technology has become an ingrained and indispensable part of family life. Smartphones and other digital communication technologies help families to keep up with each other during school/work hours; organising family gatherings can be coordinated in chat-groups created, specifically to facilitate time for family catch ups; and keeping up with grandparents overseas has become much more vivid thanks to video-telephony. However, integrating this technology into families' everyday lives means new (micro) tasks, new opportunities, but also new troubles. In their qualitative analysis, the authors demonstrate how particular roles and tasks are assigned to and expected of the children. They find that the children are often put into a paradoxical position: On the one hand, they are understood as digital natives "by default", who embrace digital technologies and for whom communicating online is automatically fun and easy; On the other hand, they are positioned as victims of omnipresent digitalisation who need to be protected against a backdrop of an idealised 'natural childhood'.

Meanwhile, in an Australian context, authors Will Balmford, Larissa Hjorth and Ingrid Richardson in Chap. 6, *Taking over the living room: Children's mobile mediaplay in domestic space*, explore scenarios of use concerning the play practices of children on mobile devices, and the subsequent domestic tensions that arise out of such practices. In a domestic media landscape, mobile media has increasingly featured in the contested and contradictory practices of the home (Morley, 2000). Mobile media constantly blur distinctions between public and private, work and leisure. The chapter draws on fieldwork from the Games of Being Mobile (GoBM) Australian Research Council (ARC) project—a 3-year, nationwide study of Australian mobile gaming practices—to unpack how these tensions manifest within the domestic environment. In order to best examine these tensions, the chapter initially reviews key literature on the relation between domestic space and media practices, before providing an overview of the GoBM project. The discussion component of the chapter uses two detailed key case studies to unpack how domestic usage of mobile devices by younger children are challenging earlier, and more traditional models of media use in the home. The phenomena the authors explore in this chapter highlights the shifting needs and agency of children in a digital world, a key focus of this volume. Furthermore, the various familial tensions are indicative of the difficult balancing act parents face between ensuring online safety, data privacy, security and protection, as well as encouraging digital literacy for their children that has been touched on in some way in all of the preceding chapters.

In Chap. 7 by Maureen Mauk, *Think of the parents: Parental controls in digital TV and family implications,* the author uses a combination of discourse analysis and discursive interface analysis to interrogate Netflix's algorithmic affordances, against their claims of easing parent/subscriber burdens with its 'informed viewing' parental control tools. Traditionally, media regulations are often framed as being put in place 'for the children' with the expressed intent to 'protect society's most vulnerable'. This framing often fails to consider the needs of parents who are playing the role of familial gatekeeper, while contending with exigent public scripts on children's media consumption. The chapter argues that to adequately consider the algorithmic design and practice of children's participation in the digital space, it is also necessary to take into account the implied roles and responsibilities of families as they navigate their children's interactions with media content. Given the children's digital content industry boom coupled with the "psychological space" (Jordan, 2016) that content mediation requires of parents, this chapter examines the cultural implications of parental controls beyond regulatory execution, to consider its affordances. Proposing a shift to the approach for platform control away from government standards and towards better utility of the affordances of algorithmically driven content, this chapter offers a new framework and potential solutions that give families more control, creating new allowances in time and mental space 'for the parents'.

Then in Chap. 8, by Kylie Stevenson, Lelia Green, Donell Holloway and Kelly Jaunzems, *Screening language acquisition skills in a mediated childhood,* the authors used an observational ethnographic case study approach, informed by play-based research methods, to explore a young child's media and play practices within a bilingual home. The 2-year-old child, Lavinia, was an ardent fan of *Peppa Pig* and, during the in-home observation and interview with her mother, researchers observed her playing *Peppa Pig* in Mandarin on an iPad in parallel with the same episode in English in streaming video mode on the television. The researchers watched Lavinia set up this entire system of media retrieval and replay. Lavinia effectively created a personal tutorial to practice her Mandarin–English bilingual comprehension using *Peppa Pig* in a self-directed manner. In this chapter, the authors have shared the importance of multiliteracies to this family's everyday life, addressing how Lavinia has engaged creatively with technology and culture even at a very young age.

Part II: Pedagogy Approaches and Challenges

Part II explores the pedagogical actions and roles taken by educators to the development of children's digital competencies. Authors in this section report on learning design analysis, and the pedagogical decisions made by educators in response to changing curricula, parental expectations and children's learning needs as they develop as active creators with digital technologies.

Increased access to digital technologies is argued, by Judith Dinham, in Chap. 9, *Media Arts in early childhood: A framework for developing young children's creative participation in the digital world,* to have democratised mass-media communication and made it possible for everyone to create and communicate to a wider audience. This chapter explores the potential for young children to be active creators, by adopting multimodal and multi-medial practices, as part of a participatory culture. The nature and role of media arts practices in contemporary early childhood learning is presented from a participatory culture perspective. The author offers a model for conceptualising the way media arts can be a meaningful part of the daily early childhood learning program and provides a framework for reimagining children's learning, and redefines the role of the contemporary early childhood educator within this personalised learning journey. It also situates the educator in terms of five roles that are enacted daily in a dynamic and synergetic relationship: artist, researcher, designer, co-constructor, and advocate. The author suggests that this model could assist educators charting their own professional growth in a participatory culture, and also serve to frame research about contemporary educational practices.

The focus on teachers' roles in learning environments infused with digital technologies is further explored by author Vicki Schriever in Chap. 10, *The impact of digital technologies on the role of the early childhood teacher.* The author highlights significant and contemporary challenges facing early childhood teachers as they manage the emergence of curricula expectations regarding children's engagement with digital technologies that are integrated into early learning environments. The author shares the findings from a grounded theory (GT) investigation concerning how early childhood teachers understood and managed their changing roles regarding information communication technologies (ICT) in the context of their kindergarten. Semi-structured, in-depth interviews were undertaken with 19 practising early childhood teachers, employed at different kindergarten settings across a regional area of Queensland, Australia. The author exposes the distinct ways early childhood teachers perceive their role, reveals the significant impact ICT has on the role of the early childhood teacher, and uncovers the actions taken by early childhood teachers to manage digital technologies within their kindergarten.

In Chap. 11, *Bridging Communities: Developing digital literacies and introducing digital technologies in the Montessori Early Childhood Education classroom,* challenges and opportunities are explored in the development of authentic methods that are responsive to Montessori pedagogy, for developing children's digital literacies. The authors, Sharon Davies, Samantha Owen, and Sarah Iles share a richly descriptive account of their action research project, initiated by 'River' Montessori School in response to a change in the local authority's mandated curriculum. In this chapter, the authors discuss their whole school project and the professional learning actions implemented that supported teachers' experiments, including possible strategies for digital technologies implementation. The contention made in this chapter is that while there was no agreed approach or consensus reached for implementation of digital technologies in the international Montessori community, the heart of the struggle lay with the River community, as there was no consensus around attitudes to use which could guide implementation. The authors explain how ultimately the identification of a shared language resolved the internal contradiction in the school as educators were confident that the approach with digital technologies was consistent with Montessori pedagogy, and this encouraged a channel of communication between educators, children, and families and—finally—a common approach.

Chapter 12 turns to focus on the voice of children and their right to playful digital learning in the early years. Authors Helle Hovgaard Jørgensen and Helle Marie Skovbjerg, in their chapter *Understanding the mutuality of play and media literacy in young children: An ethnographic investigation of pre-primary school children's perspectives on media literacy as seen through the lens of play,* propose that in a digital world, children's play and their implicit right to play extends to play involving digital technologies. The authors aim to link children's right to play to a broad and socio-culturally inspired stance of media literacy that involves dimensions of play. The authors revisit the concept of media literacy to frame and understand the nature of play with digital technologies. The importance placed on children's agency and play is based on a long-term field study of children aged 5–6 years in two Danish public pre-primary schools. Throughout this chapter, the importance of the children's perspective is recognised as they actively participate in digital play. The authors share the voice and ideas of children as they explain how to construct a game, and in so doing, provide a new perspective on media literacy from the child's perspective.

Next, author Jo Li Marie-Joelle Tay gives consideration to the learning design of tasks and the cognitive load they place on children as they play and learn. In Chap. 13, *Digital technologies and children: Does more digital interactivity make for better learning?* the author examines the concept of interactivity in relation to learning experience, technology, and cognitive load. The author argues that it is unclear whether more digital interactivity actually makes for better learning. She discusses the limitations of working memory and argues that children can only process a small number of different items at any one time. Working memory becomes problematic when there is high element interactivity in a given task, and the interactions between many elements must be learned at the same time. This simultaneous interactivity results in a high cognitive load (Sweller, 1994). In this chapter, a model of interactivity is presented and used by the author to analyse three different mathematic Learning Experience Scenarios (LES) which are: a printed worksheet, Mathletics (an online learning mathematics website), and Minecraft (a popular multi-player online game). The concept of cognitive load is used to interpret the results of visually mapping the dimensions of interactivity in each task. The author suggests that when used together, cognitive load theory and the innovative model of interactivity that are presented in this chapter could provide a lens for critiquing the suitability of learning tasks for children.

In Chap. 14, authors Catia Malaquias and Katie Ellis take a social and human rights approach to disability, as they draw attention to the disruptions caused to education systems and inclusive practices by the COVID 19 pandemic. Quantitative and qualitative research undertaken by Children and Young People with Disability Australia during COVID-19 identified a lack of appropriate materials, and social support. This chapter draws on this evidence and the broader context of digitisation articulated within disability media to highlight the persistence of disabling attitudes in preventing the realisation of human rights such as inclusive education.

Part III: Connected Toys at Home and at School

Implicit in Part III, is recognition of young children's right to play and to learn through play. This section of the book focuses on the design features of devices, software applications and learning environments with their roles in influencing and enhancing children's engagement in playful learning experiences that are aligned with their developmental needs. In all of these chapters, children are viewed as competent contributors in digital environments, where they are ideally supported to be creative participants and active learners.

In Chap. 15, *Internet of toys and forms of play early education: A longitudinal study of preschoolers' toy-based learning experiences*, Pirita Ihamäk and Katriina Heljakka engaged with 20 Finnish preschool children and their teachers to understand children's play and their educators' understandings of this play, in the newly emerging hybrid and connected context of digitally connected and enabled toys. The authors raise questions about how to best bridge the gap between pedagogical understandings of play with connected toys, and young children's experiences with digital technologies, digital play, the digital dimension of toys and their relation to popular culture. This 6-month study examines how educators have adapted two IoToys: Fisher-Price's Junior Smart Toy Bear and Wonder Workshop's Dash Robot as part of early education in the preschool environment, and how these children have experienced toy-based learning.

Continuing the exploration of children's educational and connected play, in Chap. 16, *Digital play objects as part of preschool children's imaginative play*, Kristín Dýrfjörð and Anna Elísa Hreiðarsdóttir explore how children play with digital play objects they have not encountered before in an environment they are contextualised in, with support from recognised materials. The connection between children's imaginative play and coding devices, in particular, a little beetle Blu-Bot and the coding blocks Cubelets are examined. They investigate children's relationships with both each other and the devices to explore the creativity, emotions and problem solving that appear during play. Part of the chapter's intent is to explore how the culture of children's local community permeates and becomes a significant part of their play.

In Chap. 17, *Co-creating hybrid toys as an approach to understand children's needs in play experience*, authors, Tamara Pinos Cisneros, Felipe Escobar Vega, Ben Kröse, Ben Schouten and Geke Ludden suggest the use of hybrid toys as an innovative way to engage children in personalised healthcare in the delivery of physical therapy. However, they asserted that there was a requirement to first understand the needs of children in their digital-physical play experience to be able to effectively design these toys. The aims of this explorative study were to identify the needs of children in their play experience and to examine co-creation workshops as a mean to do that. Finally, the authors highlight the importance of involving children in the development of new smart technologies for play, by empowering children to design their own smart toys.

Part III also includes research involving the development of hybrid toys to enable the diagnosis of developmental delay and to facilitate the delivery of physical therapy for children with disabilities. In Chap. 18, *Assessing developmental difficulties in children through connected Smart Toys*, authors Diego Rivera, Maria Luisa Martin-Ruiz, Luis Cruz-Piris, Kevin van der Meulen, Antonio García, Cristina Serrano García, Susel Fernández, Bernardo Alarcos, and Juan R. Velasco, consider how the use of playing activities and tools have been proven to be a powerful method for the assessment and monitoring of children's psychomotor development, from a psychology perspective. According to the authors, the traditional manual measurement techniques used in psychology limit the accuracy of the results, and also restricts the range of people who can carry out the tests (as they should be experienced specialists). The improvement of sensor-based technology and the rise of the Internet of Things (IoT) paradigm, they suggest, provide an interesting opportunity for the design of platforms, devices, and methods which could improve assessment outcomes. The authors propose the design and development of an IoT-based platform composed of smart toys designed specifically for the assessment of children. As part of their research, they have designed and built sets of sensor-embedded toys mimicking some of the most used tools in the current validated psychology scales.

In Chap. 19, *Young children learning to code: A digital technologies framework for the early years*, Karen Murcia explores the role of digital technologies in early childhood education and how young children's experiences with them are increasingly discussed and negotiated in a learning centre. Specifically, the author used a grounded theory (GT) approach to identify and document how educators see and understand learning opportunities for young children, as they played with tangible coding technologies (TCT's) or 'robots'. The 6-month action research project involved four early childhood educators working in a University's Early Years Centre that provided long daycare and a kindergarten program. In collaboration with the educators, the author investigated and compared how children learned through play with two types of TCT's: Bee-Bots and Cubetto. It was evident that educator's technological pedagogical content knowledge was key to the quality of children's learning experiences with the digital technologies. An outcome of this research was an innovative early year's digital technologies (DigiTech) framework that positioned young children's digital literacy as a multi-faceted construct that sits at the intersection of computational thinking, digital technology skills and social-emotional capabilities. The participating educators reported that the framework assisted them to see and respond to learning opportunities for the children as they played with the tangible coding technologies.

Part IV: Privacy and Protection

In Part IV, a slew of privacy concerns are also raised as children's data is collected and as private corporations offer tech to education and health providers in potentially problematic alliances. National and international, commercial and educational aims and objectives, policies and practices may not easily align or be managed in ways that ensure the complete protection against data privacy and security risks, and the agency of the child in these experiences. This section considers children's rights to privacy and protection through research about the normalisation of parental, commercial and state surveillance practices (including big analytics and predictive algorithms), and their complex interactions that can impinge upon children's rights.

Chapter 20, *Researching representations of children and childhood on Instagram: Ethical and methodological considerations,* focusses on the agentic rights of the child as digital citizens (DEEWR, 2009; Robinson & Jones-Diaz, 2016). However, the authors, Madeleine Dobson and Jenny Jay suggest that children's voices are silenced as a result of the surveillance involved. Sharenting has become a 'normalised' part of the parenting repertoire resulting in a high level of engagement with 'influencer' parents, and 'microcelebrities' who disseminate posts of children's activities on social media, curating an idealised representation of their children's and families' lives (Blum-Ross & Livingstone, 2017; Leaver, 2017). Children's images and content can be commodified through sponsored posts that portray children in an idealised fashion. Predictive algorithms analyse posts to construct future content relating to children, including blogs, images, and videos. Sharenting can result in a number of accuracy and privacy-related consequences that warrant protection for children's images, such as the misrepresentation of marginalised groups (Choi & Lewallen, 2017). This is in contradiction to the principle that children from different socio-economic backgrounds should have equal experiences with digital technologies, and on social media platforms. The chapter argues that parental media literacies should be further supported by acquiring digital skills and competencies to effectively manage the dissemination of their children's images in the public arena, such as by providing children with filtering tools, that have settings to restrict viewers (Molina et al., 2010). A multi-phase case study approach (each a month-long of data collection) was used to identify how the child was conveyed on Instagram posts by popular influencer parents, children's brands (collected in accordance with Australia's fair dealing principles), and high-profile celebrity parents. The analytical matrix analyses the different visual, discursive and creative aspects within the photograph/video/story, and any captions, tags and comments from the Instagram posts. Prospective influencer parents who use Instagram were invited to participate in interviews to capture their perspectives of the platform, and if they involved their children in creating posts. The results of images portrayed on Instagram were analysed in comparison to the traditional pedagogical images of capable and competent children that early childhood educators have constructed.

Chapter 21, *The 'sharent' trap: Parenting in the digital age and a child's right to privacy* explores the popular practice of sharenting,—particularly by mumpreneur influencers, and those parents who are also categorised as 'micro-microcelebrities'. The authors, Anna Potter and Renee Barnes, argue that sharenting involves a conflict of interest. This is because parents (as primary caretakers) have a vested interest in their role to protect their children's right to privacy. However, they also share their children's data and information online in ways that reduce their children's autonomy to control their own personal data. In order to examine this tension, the literature reviews parents' enthusiastic attitudes towards posting about their children, positioning them as part of an audience's social media gaze; however, suggests that this is in contrast to the protectionist view of children who make up a special audience of media as vulnerable minors, and who have developing competencies (Buckingham et al., 1999; Lemish, 2007). Parents are responsible for curating posts of their children, without asking for their consent. An online survey (N = 613) of Australian parents with children aged 0–13 years was used to explore their social media (Facebook and Instagram) usage, views on children's privacy and their implementations used to protect their children's privacy online. The majority of parents at 58% reported that children earned the inherent right to privacy from birth, and 74% of parents believe children's permission should be asked before posting content on either Facebook or Instagram to protect their right to privacy. The results demonstrated that 27% of parents believe posting is very risky, and the majority of parents manage their privacy using social media privacy settings; with 64% of parents aged 45–54 years using the custom lists function when sharing post's containing their children's content. The disconnect between parental beliefs and attitudes about their children's rights to privacy, recalibrates the relationship between children, parents and the media.

Chapter 22, *Santa's little helper and star of Instagram, Elf on the Shelf: Gendered labour, normalising surveillance and digitising a childhood phenomenon*, by

Catherine Archer and Tama Leaver situates the Elf on the Shelf (EotS) as a popular interactive and worldwide cultural phenomenon that sits on the shelves of families' homes and reports back to Santa about children's behaviour during the festive period. Despite being a popular children's Christmas toy, the EoTS has earnt a contentious reputation as a hyper-surveillance device that is used to listen in and 'spy on' children's activities, both by parents through intimate surveillance (Leaver, 2017), and by commercial and state actors who commodify children's data into monetary value contributing towards surveillance capitalism (Zuboff, 2015). The data was downloaded from Instagram that resulted after searching for #elfontheshelf2018 in June 2019. This data identified the top nine posts that were ranked highly by algorithms, was analysed thematically, through visual narrative analysis and social semiotics. This data was investigated as to whether mumbloggers support or challenge the notion that the EoTS toy normalises surveillance by commercialising children's data. The EoTS Instagram posts often generated traffic by directing followers to mother's blogs or professional profiles, and there were also different projections, cultural significance and characterisations of the Elf.

Chapter 23, *Digital predictions: Children's futures, opportunities and obstacles*, by Michele Willson discusses the increasing role played by predictive algorithms (employing artificial intelligence techniques such as pattern recognition, big data analytics, machine learning and behavioural modelling), to shape children's futures. The author suggests predictive algorithmic decisions can influence and impact children's future opportunities in education, health, commercial and social spheres. Intimate children's data (Leaver, 2017), physiological and behavioural, is tracked by and manipulated through algorithmic processes from conception onwards. Commercial and state data is collected, analysed and aggregated, for children who are positioned as vulnerable subjects in need of protection. The ability of insightful commercial and educational data to perceive and predict an imagined future for children is enticing, often occurring via engagement with third-party commercial providers, that offer digital learning activities. Furthermore, data that is extracted from biometric devices such as children's wearables can have impacts in creating foreseeable learning and development outcomes, by engaging diverse academic bodies within psychology, health, and neuroscience disciplines. The results of pre-emptive decision-making based on of predictive algorithmic outcomes for children is initiated by the data collected from diverse sources, intents and perspectives. This can result in perverse outcomes that undermine or thwart the child's and parental agency. Once again, this is in contradiction to the idea that digital technologies can enable all children from different 'walks of life' to seek and fairly access equal opportunities.

Chapter 24, *Research Ethics and Digitising Early Childhood* by Madeleine Dobson, Karen Murcia, Kim Gifkins and Donell Holloway investigated ethical issues which children and researchers face within an environment of digital technologies. Researchers, in collaboration with parents, make pragmatic interpretations using their knowledge of ethical principles, and evaluate different approaches to make impactful ethical decisions regarding their work with children and digital technologies. Four vignettes are presented in this chapter from researchers' practice, and the ethical process used in each was analysed through the 'Digital Child' Ethical framework, developed from the current National Statement on Ethical Conduct in Human Research, the EECERA Codes for Ethical Conduct and the Guidelines for Early Childhood Australia. As an example, and to illustrate, the first vignette concerned the *negotiation of initial consent* and *anonymity* by using open and clear information to communicate respectfully with prospective participants regarding extracting and anonymising data from families' Instagram posts to analyse the construction of childhood. A pragmatic decision was reached by researchers to use crystallisation techniques to blur any facial identifying features of children, and adding captions to convey emotions, that protected the privacy and identity of the children. Significantly, the vignettes presented and analysed in this chapter highlighted that researchers should: consider the negotiation of ongoing consent through regular progress updates; seek creative ways to maintain anonymity; and plan ethical considerations when disseminating research outcomes within appropriate contexts to various stakeholders which could include: participants, researchers and a general community audience. The Digital Child Ethical Framework aims to provide researcher's with guidance for resolving ethical dilemmas that are frequently experienced when researching with children in a digital environment.

This book acknowledges that very young children's internet use, and that of their parents and educators, carries with it a variety of opportunities and risks for children. The book seeks to find a balance between children's rights to provision, participation and protection;—a balance that does not diminish young children's rights to play and learning in a digital world. With respect to the opportunities and risks online, it is evident throughout the book that children's play practices with digital media, their screen use, smartphones and mobile media play, and, on a broader level, sharenting on social media platforms, create many conflicting messages and domestic tensions for parents, educators, regulators, and children. On one hand, there is the perception that these devices and platforms present the opportunities for children to foster innovative play and develop important digital literacies; however, on the other hand, these same devices and platforms can open up experiences and access to others which could inflict harm upon vulnerable children. The harm may be implicit, such as screen overuse, or mobile media blurring the boundaries between different aspects of life (public, private, work and leisure) beyond children's play practices. We are reminded by the research shared throughout the book that children's content on social media platforms could be mined by algorithmic artificial intelligence practices, which enable the commodification of their data, targeted marketing schemes and remarketing to parents on social media. Shifts are seen to occur regarding children's agency in the digital world, as algorithmic assemblages can potentially provide and predict opportunities and future content for children. The question then is how do we protect children's vulnerability and provide parents with more control on content-mediated platforms.

Authors throughout this book recognise that online digital platforms, connected toys and digital tools provide tremendous opportunities and challenges for all who surround and support children in their educational and life journeys. In this book, pedagogical, social and parental understandings and practices are shared that could maximise learning and contribute to positive social and emotional opportunities for children, while minimising risks. Digital technology and its use can provide innovative and engaging ways to inform pedagogy, enhance the development of children who have learning difficulties, increase digital literacy in society, and to facilitate increased social engagement as we navigate diverse socio-economic and cultural contexts. In order to achieve the vision created by the authors, approaches are sought that respect children's agency and that grapple with the tensions, contradictions and desires of the various social actors involved. Within the covers of this book we have aimed to achieve a greater understanding of our children's needs, rights to agency and vulnerability to harm in today's digital world.

References

Blum-Ross, A., & Livingstone, S. (2017). "Sharenting," parent blogging, and the boundaries of the digital self. *Popular Communication, 15*(2), 110–125. https://doi.org/10.1080/15405702.2016.1223300.

Buckingham, D., Davies, H., Jones, K., & Kelley, P. (1999). Public service goes to market: British children's television in transition. *Media International Australia incorporating Culture and Policy, 93*(1), 65–76.

Choi, G. Y., & Lewallen, J. (2017). "Say Instagram, kids!": Examining sharenting and children's digital representations on Instagram. *Howard Journal of Communications, 29*(2), 144–164. https://doi.org/10.1080/10646175.2017.1327380.

Department of Education, Employment, and Workplace Relations [DEEWR]. (2009). *Belonging, being, and becoming: The early years learning framework for Australia*. https://www.education.gov.au/early-years-learning-framework.

Erickson, F. (1996). Going for the zone: The social and cognitive ecology of teacher-student interaction in classroom conversations. In D. Hicks (Ed.), *Discourse, learning, and schooling* (pp. 29–62). New York: Cambridge University Press.

Gershon, I. (2010). Media ideologies: An introduction. *Journal of Linguistic Anthropology, 20*(2), 283–293.

Gillen, J., Matsumoto, M., Aliagas, C., Bar Lev, Y., Clark, A., Flewitt, R., ... & Pacheco, R. (2019). *A day in the digital lives of children aged 0-3*. Full report: DigiLitEY ISCH COST Action 1410 Working Group 1: Digital Literacy in Homes and Communities. http://www.research.lancs.ac.uk/portal/en/publications/-(19b42af9-7828-4950-afca-69fdce62702e).html.

Jordan, A. B. (2016). Presidential address: Digital media use and the experience(s) of childhood: Reflections across the generations. *Journal of Communication, 66*(6), 879–887. https://doi.org/10.1111/jcom.12265.

Kyratzis, A., & Johnson, S. (2017). Multimodal and multilingual resources in children's framing of situated learning activities: An introduction. *Linguistics and Education, 41*, 1–6.

Leaver, T. (2017). Intimate surveillance: Normalizing parental monitoring and mediation of infants online. *Social Media + Society, 3*(2). https://doi.org/10.1177/2056305117707192.

Lemish, D. (2007). *Children and television: A global perspective*. Blackwell.

Livingstone, S. (2007). Strategies of parental regulation in the media-rich home. *Computers in Human Behavior, 23*(2), 920–941.

Livingstone, S., & Bulger, M. (2014). A global research agenda for children's rights in the digital age. *Journal of Children and Media, 8*(4), 317–335. https://doi.org/10.1080/17482798.2014.961496.

Marsh, J. (2006). Emergent media literacy: Digital animation in early childhood. *Language and Education, 20*(6), 493–506. https://doi.org/10.2167/le6660.0.

Molina, M., Oltra-Gutierrez, J. V., & Sarabdeen, J. (2010). An exploratory study on the privacy of children's images in Spain's most widely used social network sites (Tuenti and Facebook). *International Review of Law, Computers & Technology, 24*(3), 277–285. https://doi.org/10.1080/13600869.2010.523953.
Morley, D. (2000). *Home territories: Media, mobility and identity*. Routledge.
Robinson, K., & Jones-Diaz, C. (2016). *Diversity and difference in childhood: Issues for theory and practice* (2nd ed.). Open University Press.
Sweller, J. (1994). Cognitive load theory, learning difficulty, and instructional design. *Learning and Instruction, 4*(4), 295–312. https://doi.org/10.1016/0959-4752(94)90003-5.
Zuboff, S. (2015). Big other: Surveillance capitalism and the prospects of an information civilization. *Journal of Information Technology, 30*(1), 75–89.

Karen Murcia is an Associate Professor in the School of Education, Curtin University. Her research interests include scientific literacy for citizenship, creativity with digital technologies and STEM education. She is a researcher with the Australian National Centre of Excellence for the Digital Child, Chair of Child Australia and a Scitech Discovery Centre Board Member.

Michele Willson is Professor of Internet Studies at Curtin University, Perth in Western Australia. Her research explores the intersection of technology and the everyday with a particular focus on algorithms, digital childhood, social games and online sociality more generally.

Catherine Archer is a Senior Lecturer and researcher specialising in social media and strategic communication at Murdoch University. Her current research interests include social media, focusing on families' use of social media, influencer relations and ethics, and the blurring of lines between media, marketing, public relations and communication.

Francesca Stocco is PhD candidate at Edith Cowan University researching connected toys, with particular reference to the commercial governance of these toys. She is currently analysing selected privacy policies (PPs), terms and conditions (T&Cs) and frequently asked questions (FAQs) related the connected toys.

Donell Holloway is Senior Research Fellow at Edith Cowan University, Australia. She is a Chief Investigator on three Australian Research Council grants, the ARC Centre of Excellence for the Digital Child, The Internet of Toys: Benefits and risks of connected toys for children, and Toddlers and Tablets: Exploring the risks and benefits 0-5s face online.

Part I
The Early Childhood Home

Chapter 2
"The Tablet Is My Best Friend!": Practices and Perceptions of Young Children and Their Parents

Patricia Dias and Rita Brito

Introduction

The society we live in is profoundly shaped by the integration of digital technologies in our daily lives. Children are being born in homes populated with computers, smartphones and tablets, and coming into contact with these devices at an increasingly younger age (Kucirkova, 2011; Plowman, 2015). This new domestic media landscape poses challenges to parents, who mediate access to devices and content for young children (Dias et al., 2016). Perceiving both risks and opportunities in the digital environment, parents struggle to reconcile the main axis of the Digital Rights Framework established by the UN in 2014—protection, provision and participation.

Our study provides an in-depth look at the digital practices of families with young children, using a qualitative approach. It aims to shed some light on how parental mediation styles are coping with the need to protect young children from online risks, in a way that still provides them skills and allows them to participate in the opportunities that the digital environment also holds.

P. Dias (✉)
Universidade Católica Portuguesa, Lisbon, Portugal
e-mail: pdias@ucp.pt

R. Brito
ISEC Lisboa, Lisbon, Portugal

D. Holloway et al. (eds.), *Young Children's Rights in a Digital World*, Children's Well-Being: Indicators and Research 23,
https://doi.org/10.1007/978-3-030-65916-5_2

Young Children and the Tablet

Why Young Children Prefer the Tablet

There is a recent body of academic literature stating that children have access to digital media very early (usually under 2 years old) and are spending increasing time interacting with such media (ChildWise, 2016; Kabali et al., 2015; Lauricella et al., 2015). Studies conducted in Europe, the US and Australia concur that the tablet is currently the favourite digital device of children under 8 years old, and that their favourite activities are playing games and watching videos (cartoons and music) on YouTube (Chaudron et al., 2018; CommonSense Media, 2013; Luscombe, 2015; Plowman, 2015; OfCom, 2015)—unfortunately there is a lack of research on this topic in other contexts. Although the choice of activities is very similar—games and videos—it is also strongly gendered. Chaudron et al. (2015) found that the choice of apps tends to reflect boys' and girls' fictional universes of superheroes and princesses, respectively.

How Children and Parents Perceive Digital Media

The perceptions and attitudes of children under 8 years old towards digital devices are generally positive (McKenney & Voogt, 2010). Contrasting with the parents' view, who are very interested in supporting learning and the development of their children, young children perceive the tablet as a multifaceted toy that offers entertainment. They enjoy being able of autonomously exploring their interests (e.g. searching on Google about their favourite animal) and learning (e.g. developing skills to go to the next level on a game), but their main motivation for engaging in this kind of activities is 'having fun', they don't acknowledge learning or being empowered as benefits afforded by the tablet (Chaudron et al., 2015).

Concerning parents, Plowman et al. (2008) discuss the 'technologisation' of childhood, claiming that most parents regard this process as positive and believe that digital technologies are important tools for their children's professional future. Negative perceptions of digital media are usually related to lack of digital literacy and also previous negative experiences with older siblings (Nikken & de Haan, 2015).

Concerning the perception of online risks, parents are quite fearful of negative consequences of excessive screen time, such as hyperactivity, decrease of attention span, eyesight problems, and underdevelopment of social and physical skills, and also of contact with strangers. For addressing the first problem, parents monitor and restrict screen-time (Nikken & de Haan, 2015). Concerning the second, the parents of children under 8 years old believe they are not yet exposed to contact with strangers, as they do not know how to read and write, and they do not interact with other people on social media (Chaudron et al., 2015). As a consequence, many parents have not addressed the topics of online safety and media literacy with their children yet (Chaudron et al., 2018).

Previous research has, however, identified some risks that most parents are unaware: the main one is being exposed to age inappropriate content, mostly on YouTube, by following recommendations, and also due to in-app advertising; another important risk is data collection and further intrusive or unethical commercial exploitation of such data (Dias & Brito, 2018).

What Is the Role Played by Parents

Several studies agree that parents play a very important role in the adoption, practices, perceptions and attitudes of young children towards digital media (Bittman et al., 2011; Kucirnova & Sakr, 2015; Livingstone & Third, 2016; Plowman et al., 2008).

Parents are the first mediators of digital experiences, as they decide what is available in the home, and what children are allowed to use (Bittman et al., 2011; Kucirnova & Sakr, 2015). Also, at this age, children learn a lot by imitation. Lauricella et al. (2015) described a 'mirror effect' according to which the practices of children are correlated to those of the parents. This happens regarding time of use, favourite activities, and attitudes towards each device. In addition, when young children have doubts or difficulties, older siblings and parents are the ones they ask for help, guidance and support (Chaudron et al., 2015; Kucirnova & Sakr, 2015).

The concept of parental mediation refers to the role played by parents as mediators of the children's engagement with media, thus shaping their practices and perceptions. Recent research has focused on the parental mediation of digital media, which can be thought as an option between two extremes: restrictive (with younger children, parents tend to be more worried about time of use than content); and enabling (Livingstone & Third, 2016; Morentin et al., 2014; Nikken & Janz, 2014). One of the most consensual proposals is the model of Valcke et al. (2010), which explains parental mediation by correlating it with parenting styles. The authors consider two axes regarding internet use in the home, control and warmth. The crossing of the axes results in a matrix of four parental mediation styles: authoritative (the most common), authoritarian, permissive and *laissez-faire*.

According to Goh et al. (2015), parents are generally permissive regarding digital media use by young children, but Chaudron et al. (2015) found some rules which were relatively common in the homes of young children: not being allowed to play games before finishing homework, not being allowed to play for long periods of time, and not being allowed to play before bedtime.

Research states that co-using (or co-playing) of the device is the most beneficial for children's development (Livingstone et al., 2017; Nikken & Jansz, 2006), safety and general wellbeing, and this practice is commonly found in households in northern Europe, the US and Australia. In southern Europe, it is more common for parents to leave children exploring the tablet autonomously, keeping them entertained while the parents are busy with house chores or work (Genc, 2014; Kucirnova, 2011). This 'babysitting' function of the tablet is quite common in all contexts, a useful resource for parents when they need to keep the children entertained in public spaces, such as restaurants or waiting rooms (Chaudron et al., 2018).

The literature on parental mediation of digital media explores factors that may influence the parental mediation style. Several authors agree on the importance of sociodemographic features of the parents such as age, gender, income and education (Goh et al., 2015; Valcke et al., 2010). Others highlight variables related to internet usage, digital skills and perceptions and attitudes towards digital technologies (Bittman et al., 2011; Connell et al., 2015; Nikken & Janz, 2014; Valcke et al., 2010)—positive perceptions are associated with enabling mediation and negative perceptions with restrictive mediation.

Our research sets out to explore the parental mediation of the tablet in a particular context—Portugal—and factors that may bear a stronger weight on the adoption of certain parental mediation actions.

Methodology

Research Questions

This chapter presents partial results from an exploratory research project that aimed to gain in-depth knowledge about the digital practices of families with young children. It sets out to explore if the perceptions of parents and children about digital media influence the parental mediation style.

Research Design

This study follows an interpretivist approach and a qualitative method.

We used a purposive design of the sample, aiming to get a wider range of narratives about the use of the tablet in the home by young children and their families (Lindlof & Taylor, 2002). This is adequate for an exploratory study that does not intend to be representative.

For collecting data, the method adopted was visits to families, including a diversity of techniques: semi-structured in-depth interviews; age-appropriate games and activities; and a 'digital tour'. This diversity of activities was very important to motivate the participation of young children. The study followed the protocol developed for an EU-level wider research project (Chaudron et al., 2015). Initially, we interviewed the whole family together, using a game with stickers in order to get to know the daily routines of the children; then, parents and children were simultaneously but separately interviewed by each of the researchers. The interview to the parents was standard, but the interview to the child (and siblings if it was the case) was enriched with different activities, such as a card game about their favourite activities, identifying apps from a standard set, and promoting a 'digital tour' in which children showed us their digital devices, their favourite places for using them, what they like to do and how they do it. The data was registered in audio and in photograph, as well as in participant observation notes made by the researchers.

For the analysis, we used thematic analysis following Braun and Clarke (2006) and also qualitative comparative analysis (QCA) following Boyatzis (1998) and Ragin and Rihoux (2004). For classifying the frequency of digital media use of parents and children, we applied the criteria followed by Livingstone et al. (2015) in the EU kids online project, and for classifying the parental mediation styles of the families studied, we used the model by Valcke et al. (2010), also used in the European Commission Project Young Children and Digital Technologies (Chaudron et al., 2015).

Participants

Our sample includes 25 Portuguese families, with (at least) one child between 3 and 8 years old who uses at least one digital technology, as least once a week. We opted for a purposive sample in order to obtain variety of variables such as gender of the child; family composition (both parents *versus* mono-parental; without *versus* with siblings; older *versus* younger siblings), and sociodemographic features such as education and income of the parents. These characteristics are presented in Table 2.1. The interviews took place between June and November 2015.

Table 2.1 Socio-demographic information about the participating families

Variables considered for obtaining a diverse sample	Description of the sample
Age of child	0–2 year olds: 3 3–5 year olds: 10 6–8 year olds: 13
Gender of child	Girls: 10 Boys: 16
Family composition	Two parents: 20 One parent: 5 No siblings: 8 Siblings: 17
Education of parents	Graduate: 13 Undergraduate: 12
Income of parents	Low: 13 Medium: 9 High: 3

To contextualise the description of our sample, it is important to mention that in 2018, 34% of the Portuguese households was composed by parents with at least one child, and 11% were single-parent families (Pordata, 2018a). Concerning education, only 18, 7% of the Portuguese over 25 years old have completed Higher Education in 2018 (Pordata, 2018b). Regarding income, the minimum wage in Portugal is currently 635 euros, one of the lowest in the EU, which leads to a high number of families with low income.

Findings and Discussion

Digital Practices of Young Children

Our results corroborate previous research, identifying the tablet as the favourite digital device of young children. In fact, they often prefer the tablet to other devices (even to television), other toys, and other activities (even outdoors). A few mothers recognise the tablet as their children's favourite 'toy', but one goes beyond that and states that the tablet is her daughter's "*new best friend*" (Family 9, Mother).

Children like their tablet because "*it has more games*" (Family 25, 5 year-old boy) and "*is bigger*" [than the smartphone] (Family 17, 5 year-old girl), so they can "*see better what's inside*" (Family 12, 3 year-old girl). They use the tablet mainly to play games, in many cases replacing the console, and in others television, as they can watch similar content on YouTube. Preferences concerning games are strongly gendered. Girls prefer casual games, such as make up, dressing dolls and taking care of virtual pets. Boys prefer action and adventure games with their favourite characters from movies and cartoons. They also like sports games, mostly soccer, or building games such as Lego and Minecraft. Watching videos on YouTube is the second most frequent activity mentioned by children. All children like music videos, but the 0–5 year-olds enjoy watching cartoons, while the 6–8 year-olds have expanded their preferences: girls enjoy make up videos and boys like tutorials about console games.

Among the parents of the 0–5 year-olds, the tablet is becoming the new '*baby-sitter*'—as Family 8 mother puts it, *"I can't be always on top of everything, or I wouldn't do anything else"*. The children usually play with the tablet on their own, and they reveal relatively sophisticated digital skills. They autonomously experiment and are very savvy in achieving their goals. They get tired of the games very easily, and they are keen to explore the App Store or Google Play searching for new and more attractive games.

> Family 9, mother: "She is always looking for something new, she sees one game, plays it a dozen times, and then it holds no secrets for her anymore, she gets bored and wants to play another game."

Six–eight year-olds who already know how to read and write are able to perform more complex and accurate searches on YouTube and Google, while 0–5 year-olds rely on following suggestions the appear in the search box, or related contend suggested by YouTube, App Store and Google Play. They also ask older siblings or parents to write what they want to search for.

Perceptions of Children and Parents Towards Digital Media

We found a few contradictions between the perceptions of the parents and the perceptions and observed digital practices of the children in our sample.

Firstly, the parents in our sample reveal mixed perceptions, acknowledging digital media as important for learning and for the future of their children, but also fearing some risks, mainly negative consequences of excessive use such as sleeping disorders, addiction, obesity and lack of social interaction. Young children are enthusiastic about digital devices, preferring the tablet, and perceive it as a source of entertainment and fun.

Secondly, although parents fear online risks, they believe that young children are protected from most of them because their digital practices are limited. For example, because they do not use social media, they are protected from meeting strangers online. However, when questioned about negative experiences with the tablet, children reported negative experiences with YouTube: one of the girls had already watched a pornographic video with older friends at school (Family 3, 7 year-old girl) and "*didn't like it, it was disgusting*", and two boys reported feeling scared and having nightmares due to videos they watched on YouTube (Family 1, 6 year-old boy, and Family 5, 6 year-old boy). Also, we watched one of the boys (Family 1, 6 year-old boy) playing an online multiplayer game where another player tried to chat with him, but he didn't pay attention to it because it was in English and he didn't understand.

Thirdly, most parents have shown some concern about the types of games installed by the children and the contents they may come across, but they do not monitor the children's activities, they focus instead in limiting time of use. Most children use YouTube without supervision, following suggestions and easily coming across inappropriate content. For instance, one mother claims to be alert about the tablet activities of her daughter, "*I watch, I listen, I end up controlling* [what her daughter does with the tablet]", yet she admits that "*most of the times she* [the daughter] *ends up using the tablet when I am not available, so we do not keep each other company.*" (Family 10, mother). Another mother claimed that her children "*don't really ask to install violent games*" (Family 4, mother), nonetheless the children told us that their favourite games are "*games with shootings, GTA.*" (Family 4, 6 year-old boy), where "*we have to get guns to go to the next level and the goal is stealing cars*" (Family 4, 7 year-old boy).

Parental Mediation

Consistently with previous research, instead of adopting a consistent mediation style, the parents in our sample combine restrictive and enabling actions (Livingstone et al., 2017; Ponte et al., 2017). Parents are generally permissive and let children explore the digital devices autonomously, often resorting to the tablet as 'babysitter', but there are also some rules in most families. Usually, such rules are enforced by the parents following negative behaviour of the children or undesired effects of their digital practices. Such restrictions apply mostly to time of use and not to content or practices.

> Family 8, mother: "he stops interacting with his sister, with me. We don't spend much time together. Sometimes I am here in the living room with him, and he does not pay any attention to me, he just wants to focus on it [games]."

> Family 2, father: "whenever they play before bedtime, they end up having nightmares during the night, and then on the next morning they do not want to get up, and when they go to school they are unable to pay attention."

There are other rules, such as withdrawing the tablet if the children misbehave, or having to ask for permission to install new apps. Some of the children are also forbidden to install violent games, but some end up installing them anyway, and the parents delete them when they find out. Also, the perceptions about what actually is a 'violent' game were also very different from parent to parent: about half, mostly the mothers, had a strict sense of violence, including physical and verbal actions, while the other half, mostly fathers, found fighting and guns tolerable to some point, and only described as violent audio-visual content and games where these situations were very frequent and intense (e.g. Grand Theft Auto, Call of Duty).

There is an effort by parents in directing children to using educational apps, but most children dislike them. Motivating the children to use them usually requires the participation of the parents and co-using, and they claim not having time to engage in such practices regularly.

> Family 9, mother: "I try that she ends up choosing games that will make her learn something. Those for dressing up dolls, I also let her play, but they are not the most fun or important."

> Interviewer: And when you play with her, do you try to direct her towards the games you find best?

> Family 9, mother: "Yes, the others she plays on her own."

Two mothers of girls mentioned the only cases of educational apps. They had apps for learning English and Math, but the children explained that they are far from being their favourites—*"I don't really like maths."*, says Family 9, 7 year-old girl. These apps were installed by the parents or suggested by teachers. However, her mother told us that she had recently found an app for reading e-books *"and liked it."* (Family 9, mother). According to the girl, she enjoys reading books in the tablet because *"it is more fun."* The 6 year-old girl from Family 10, uses the website of

Table 2.2 Technology use by the children and parents and parental mediation style

Parental mediation style	Technology use by children	Technology use by parents
Laissez-faire: 6	Low: 0 Medium: 5 High: 1	Low: 3 Medium: 6 High: 2
Permissive: 6	Low: 1 Medium: 4 High: 1	Low: 3 Medium: 6 High: 3
Authoritarian: 6	Low: 3 Medium: 2 High: 2	Low: 2 Medium: 9 High: 0
Authoritative: 7	Low: 1 Medium: 3 High: 3	Low: 3 Medium: 5 High: 5

Virtual School to use digital versions of her syllabus, and the mother claims that is her only pedagogical activity with the tablet.

Despite the rules that we found in each home, all the parents told us that they believe more in dialogue and information when it comes to protecting their children from online dangers, than in filters or prohibitions, which can be contoured. Nonetheless, some parents shared how they monitor their children: three of the mothers—Families 3, 9 and 10—listen from afar when their daughters are watching videos on YouTube, and if they don’t like what they listen they intervene; other parents—Families 6 and 10 mothers, Family 7 father—occasionally check their children’s tablets and if they find apps installed that they find inappropriate, they ask the children to delete them and explain why.

Table 2.2 sums up the relationship between the parental mediation style in each family, classified according to Valcke et al. (2010), and the level of technology use from parents and children, classified according to Livingstone et al. (2015).

The parental mediation model proposed by Valcke et al. (2010) is a matrix formed by two axes carrying between high and low: orientation and involvement. Thus, the *laissez-faire* parental mediation style is characterised by lack of rules and autonomous use by the child, the permissive is based on few restrictions but monitoring and co-using, the authoritarian is very restrictive but parents don’t participate in their children’s digital practices and the authoritative includes rules but also monitoring and co-use.

In our sample, we found a similar number of families using each of the styles, but this classification was difficult, as parents adopt and combine different mediation actions and their style is dynamic, constantly adapting to the practices and reactions of children, and often to family conversations and negotiations. We questioned parents about their most common mediation actions and classified their style according to their predominant actions and perceptions about their own mediation style.

On Table 2.2, we can observe some correspondence between the frequency of usage of digital media of parents and children, corroborating the ‘mirror effect’ described by Lauricella et al. (2015). Also, the parents who demonstrate low involvement have the least frequent digital media users. Their lack of involvement is, therefore, related to a lack of digital and media skills, with the *laissez-faire* group stating that they "*don't know how to teach or help when it comes to the tablet*" (Family 4, mother), and the authoritarian group saying "*I don't really understand what they do with the tablet, so I try to reduce screen-time. I do that for protection*" (Family 11, mother). Permissive and authoritative parents are involved and try to engage in co-use as often as they can, but they differ on their opinions about rules. For authoritative parents, "*rules are negotiated, they are necessary, and I have the final word*" (Family 15, father), while for permissive parents "*we all know that children break rules, so I think they are useless. I monitor, and then we talk, I teach and try to set a good example. I don't think forbidding works*" (Family 7, father). The permissive and laissez-faire styles are based on low orientations, and share overall positive perceptions about digital media, believing that "developing digital skills is fundamental for the future" (Family 3, mother). The authoritarian and authoritative styles share the believe that rules and guidance are necessary and justify it with their negative perceptions about online risks and negative consequences of excessive use.

Conclusions

This study aimed to explore how families with young children are coping with digital practices. We explored how the perceptions of parents about digital media and their own media usage may influence their parental mediation style.

We believe that one important contribution of our findings is highlighting discrepancies and contradictions—between the parents and the children's perceptions and between what parents and children say and what they actually do—that need to be further explored in future research.

These contradictions play a very important role in parental mediation, as parents often decide, and act based on misperceptions:

1. Parents believe their children's digital practices are limited and consequently they are not exposed to online dangers, but young children reveal considerably advanced digital skills for their age, and are exposed on age-inappropriate content and intrusive advertising;
2. Parents perceive excessive use as the main cause of problems, and consequently restrict time of use, but don't monitor or regulate content or activities;
3. Parents value educational apps but recognise that children dislike them and some claim not having time to motivate their use; some parents resort to the tablet as 'babysitter' and children are often left alone to explore their 'best friend'.

Our study also emphasised the influence of the parents' digital practices and perceptions about digital media on their parental mediation—although acknowledging that these are only two factors among many others: individual, relational and contextual. Digital practices, on their turn connected with digital skills and media literacy, bear weigh on the parents' ability to engage in co-use and participate in their children's digital practices. Perceptions are related to orientation, as parents with overall positive perceptions about digital media feel less need of establishing rules and restricting than the parents with overall negative perceptions.

We hope it helps parents find a balance between the rights of protection, participation and provision.

References

Bittman, M., Rutherford, L., Brown, J., & Unsworth, L. (2011). Digital natives? New and old media and children's outcomes. *Australian Journal of Education, 55*(2), 161–175.

Boyatzis, R. E. (1998). *Transforming qualitative information: Thematic analysis and code development*. SAGE.

Braun, V., & Clarke, V. (2006). Using thematic analysis in psychology. *Qualitative Research in Psychology, 3*(2), 77–101.

Chaudron, S., Beutel, M. E., Černikova, M., Donoso Navarette, V., Dreier, M., Fletcher-Watson, B., ... Wölfling, K. (2015). *Young children (0–8) and digital technology: A qualitative exploratory study across seven countries*. JRC 93239/EUR 27052.

Chaudron, S., Di Gioia, R., & Gemmo, M. (2018). Young children (0-8) and digital technology. A qualitative study across Europe. *Pedagogika, 3*, 329–335.

Childwise. (2016). *The monitor report. Children's media use and purchasing*. [online] http://goo.gl/Vdb3b9.

CommonSense Media. (2013). *Zero to eight: Children's media use in America 2013*. [online] http://goo.gl/ddbjnm.

Connell, S. L., Lauricella, A. R., & Wartella, E. (2015). Parental co-use of media technology with their young children in the USA. *Journal of Children and Media, 9*(1), 5–21.

Dias, P., & Brito, R. (2018). *hAPPy kids: Aplicações seguras e benéficas para crianças felizes. Perspectivas das famílias*. Lisboa: Centro de Estudos em Comunicação e Cultura, Universidade Católica Portuguesa. Retrieved from https://goo.gl/9SoMD6

Dias, P., Brito, R., Ribbens, W., Daniela, L., Rubene, Z., Dreier, M., & Chaudron, S. (2016). The role of parents as 'Gatekeepers' of digital technologies: exploring the tension between the rights of access and protection. In V. Coppock, & Gillett-Swan, J. (Eds.), *Children's rights in a 21st century digital world: Exploring opportunities and tensions. Global studies of childhood special issue*, 4(6), pp. 414–427. https://doi.org/10.1177/2043610616676024

Genc, Z. (2014). Parents' perceptions about the mobile technology use of preschool aged children. *Procedia – Social and Behavioral Sciences, 146*, 55–60.

Goh, W., Bay, S., & Chen, V. (2015). Young school children's use of digital devices and parental rules. *Telematics & Informatics, 32*, 787–795.

Kabali, H., Irigoyen, M., Nunez-Davis, R., Budacki, J., Mohanty, S., Leister, K., & Bonner, R., Jr. (2015). Exposure and use of mobile media devices by young children. *Pediatrics, 136*(6), 1044–1050.

Kucirkova, N. (2011). Digitalised early years – Where next? *New Voices, 24*(12), 938–940.

Kucirnova, N., & Sakr, M. (2015). Child-father creative text-making at home with crayons, iPad collage and PC. *Thinking Skills and Creativity, 17*, 59–63.

Lauricella, A., Wartella, E., & Rideout, V. (2015). Young children's screen time: The complex role of parent and child factors. *Journal of Applied Developmental Psychology, 36*, 11–17.

Lindlof, T., & Taylor, B. (2002). *Qualitative communication research methods*. Sage.

Livingstone, S., Mascheroni, G., & Staksrud, E. (2015). Developing a framework for researching children's online risks and opportunities in Europe. [online] http://goo.gl/c3xMPm.

Livingstone, S., Ólafsson, K., Helsper, E. J., Lupiáñez-Villanueva, F., Veltri, G. A., & Folkvord, F. (2017). Maximizing opportunities and minimizing risks for children online: The role of digital skills in emerging strategies of parental mediation. *Journal of Communication, 67*(1), 82–105. https://doi.org/10.1111/jcom.12277.

Livingstone, S., & Third, A. (2016). Children and young people's rights in the digital age: An emerging agenda. *New Media & Society, 19*(5), 657–670.

Luscombe, B. (2015). YouTube's view master. *Time, 186*(9/10), 7–14.

McKenney, S., & Voogt, J. (2010). Technology and young children: How 4-7 year olds perceive their own use of computers. *Computers in Human Behavior, 26*, 656–664.

Morentin, J., Cortés, A., Medrano, C., & Apodaca, P. (2014). Internet use and parental mediation: A cross-cultural study. *Computers & Education, 70*, 212–221.

Nikken, P., & de Haan, J. (2015). Guiding young children's internet use at home: Problems that parents experience in their parental mediation and the need for parenting support. *Cyberpsychology: Journal of Psychosocial Research on Cyberspace, 9*(1).

Nikken, P., & Jansz, J. (2006). Parental mediation of children's videogame playing: A comparison of the reports by parents and children. *Learning, Media and Technology, 31*(2), 181–202.

Nikken, P., & Janz, J. (2014). Developing scales to measure parental mediation of young children's internet use. *Learning, Media & Technology, 39*(2), 250–266.

OfCcom. (2015). *Children and parents: Media use and attitudes report.* [online] http://goo.gl/FQuYAY.

Plowman, L. (2015). Researching young children's everyday uses of technology in the family home. *Interacting with Computers, 27*(1), 36–46.

Plowman, L., McPake, J., & Stephen, C. (2008). Just picking it up? Young children learning with technology at home. *Cambridge Journal of Education, 38*, 303–319.

Ponte, C., Simões, J. A., Batista, S. & Jorge, A. (2017). *Crescendo entre Ecrãs: Uso de meios eletrónicos por crianças (3–8 anos).* Lisboa: ERC – Entidade Reguladora para a Comunicação Social. [Online] Retrieved December 13, from http://bit.do/fkEMX.

Pordata. (2018a). *Composition of domestic households.* [Online] Retrieved December 13, from http://bit.do/fkEBd.

Pordata. (2018b). *Education of the population.* [Online] Retrieved December 13, from http://bit.do/fkEBu.

Ragin, C. C., & Rihoux, B. (2004). Qualitative comparative analysis (QCA): State of the art and prospects. Qualitative methods. *American Political Science Association, 2*(2), 3–13.

Valcke, M., Bonte, S., De Wever, B., & Rots, I. (2010). Internet parenting styles and the impact on internet use in primary school children. *Computers & Education, 55*, 454–464.

Patricia Dias is Assistant Professor of the Faculty of Human Sciences (FCH), Catholic University of Portugal (UCP). She is a researcher at the Research Centre for Communication and Culture (CECC) and at the Católica Research Centre on Psychological, Family and Social Wellbeing (CRC-W). She holds a PhD in Communication Sciences.

Rita Brito holds a PhD in Educational Technology and researches on the use of technologies by families and children up to 6 years old. She is an Adjunct Invited Professor at the School of Education, Instituto Superior de Educação e Ciências. She is a researcher at CRC-W, Universidade Católica Portuguesa.

Chapter 3
Family Mediating Practices and Ideologies: Spanish and Portuguese Parents of Children Under Three and Digital Media in Homes

Mitsuko Matsumoto, David Poveda, Ana Jorge, Raquel Pacheco, Vitor Tomé, Cristina Aliagas, and Marta Morgade Salgado

Introduction

Contact with digital devices has become ubiquitous, and children are going online at younger ages and accessing a wide range of digital devices even from the first months in life (Holloway et al., 2013; Marsh et al., 2015). The technical possibilities of current digital devices (e.g. smartphones, tablets) allow young children to interact with them around any space of their houses or even in other spaces such as cars or restaurants. Discussions about the use of digital media by very young children are caught between discourses that emphasise their potential for current or future development and learning (Marsh et al., 2017; Livingstone et al., 2015;

M. Matsumoto (✉)
Universidad Internacional de la Rioja, la Rioja, Spain
e-mail: mitsuko.matsumoto@unir.net

D. Poveda · M. M. Salgado
Universidad Autónoma de Madrid, Madrid, Spain
e-mail: david.poveda@uam.es; marta.morgade@uam.es

A. Jorge
Universidade Lusófona, Lisbon, Portugal
e-mail: ana.jorge@ulusofona.pt

R. Pacheco
University of Algarve, Faro, Portugal
e-mail: rpcunha@ciac.uab.pt

V. Tomé
Universidade Autónoma de Lisboa, Lisbon, Portugal
e-mail: vitor@rvj.pt

C. Aliagas
Universitat Auta de Barcelona, Barcelona, Spain
e-mail: Cristina.Aliagas@uab.cat

D. Holloway et al. (eds.), *Young Children's Rights in a Digital World*, Children's Well-Being: Indicators and Research 23,
https://doi.org/10.1007/978-3-030-65916-5_3

Sefton-Green et al., 2016), and the risks at various social and psychological levels (Livingstone, 2009). In the context of the home, families are working through this reality across cultural, social and economic backgrounds, adjusting family practices and environments to hyper-connected society. Parents are mediating their young children's engagement with digital technologies often with uncertainty, because these have little precedent in their own experiences and, importantly, because they lack clear guidelines (Chaudron et al., 2018; Livingstone & Blum-Ross, 2018). Very young children, particularly those under the age of three, have been identified as a priority group to be studied further (Gillen et al., 2018; Holloway et al., 2013) as there are currently very few studies (Mascheroni et al., 2018; Ólafsson et al., 2013; Poveda & Matsumoto, 2018).

This chapter thus seeks to advance our understanding of how parents mediate the inclusion of their very young children into this digital society. We examine parental mediating practices and ideologies around digital devices as well as the media ecology of home (cf. Hepp, 2014) with regards to the digital activities of children aged under three, drawing on the cases of five children in Spain and Portugal from the *A Day in the Digital Lives of 0-3 Year-Olds* project developed under COST DigiLitEY Action (Gillen et al., 2019). Data collection was carried out during 2017 and the research protocol consisted of a 6-hour observation/video-recording of the focal child's activity at home, as well as a preliminary and a follow-up semi-structured interview with the parents on attitudes towards technology in relation to their overall child-rearing perspectives, and on basic information regarding the family (see the full methodology in Gillen et al., 2019).

We explore differences in how access to digital devices is granted to children and how digital technologies are taken up by young children. There are significant convergences in the digital technologies available in the homes of the families analysed here, yet also visible differences in them with regards to how children and adults engage with these technologies. From our perspective, this process can be understood in terms of parental mediation practices and we propose an analysis drawing from a specific understanding of parental mediation that will be used to examine the case families.

Parental Mediation and Children as Participants in Family Practice

Work on parental mediation rests on different conceptualisations of mediation with specific methodological and analytical implications. A first approach focuses on *parental mediation styles*. This perspective tends to conceptualise mediation style as a stable, internal trait of the parent (similar to parenting style, Darling & Steinberg, 1993), which can be associated to individual characteristics such as other psychological components or socio-demographic variables such as socioeconomic status, gender, cultural origin, geographical location, educational level, etc. of the parents.

This approach has also been associated with parents' own digital practices; which, in turn, are also conceptualised as relatively stable and measurable (e.g. Nikken & Jansz, 2014; Valcke et al., 2010).

A second perspective focuses on *mediation strategies.* Within this approach, the focus is on how digital technologies are incorporated into broader family routines and the arrangement of the daily lives of children at home. This shift moves attention to child characteristics and mediating variables, such as children's age (e.g. Galera et al., 2016; Livingstone, 2007; Zaman et al., 2016) or family structures (Nikken, 2018).

Finally, more recently, mediation has been understood as an *emergent process*. This approach aims for a much more interactional perspective, as a social and material ensemble involving participants (parents and children), digital/media devices and other artifacts (Jewitt, 2013). In particular, for young children, aged below three, focus has been placed on the interactional ecologies (Erickson, 1996) that are created in homes through the organisation of domestic space, the availability of different digital devices and the organisation of activity, and how these aspects are shaped by parental beliefs.

These approaches to mediation build on different metaphors (and ontologies) around digital media and devices, as well as childhood. From a *mediation style* perspective, digital technologies and media is something to be 'contained', because without supervision from adults, digital devices and screens would overflow and consume most of contemporary children's time and activity. From a *mediation strategy* perspective, digital devices are 'administered': parents and adults provide or subtract devices to arrange daily life and regulate children's behaviour (i.e. punish/reinforce). Lastly, from an *emergent process* perspective, devices become 'objects' within interactional systems alongside parents and children.

In this chapter, we aim to *move towards* a framework closer to the third perspective, as it will allow comparison across cases and examination of how families create distinct arrangements around digital media and digital technologies. We do this through an analytical procedure in which each family/case is examined through the following steps: (a) describe the digital media environment/ecology in each home (Hepp, 2014; Plowman, 2015); (b) examine how digital media practices unfold, and how they are arranged and mediated by parents during young children's on-going activities, focusing on when/how/with whom the case study children use or have access to digital technologies; (c) explore the parental media ideologies and beliefs around digital technologies behind the practices (Gershon, 2010); and (d) analyse how home arrangements, digital media practices and media ideologies intersect within each family and across families.

The Five Case Study Families in Spain and Portugal

Both Spain and Portugal are quickly developing into advanced digital societies. In 2018, Spain ranked 10th among the 29 European countries, while on the whole Portugal was ranked 17th (European Commission, 2018). Yet, Portugal has more than 90% coverage of ultrafast broadband at homes, which makes it third among the EU countries.

Two families (Gloria and Roser[1]) from Spain and three families from Portugal (Tomás, Matias and Vicente) participated in our study. The focal children were two girls and three boys, aged between 19 and 34 months at the moment of data collection. Participating families were considered middle-class; two families (Gloria and Matias) are of mixed nationality and use more than one language in their homes, and one lives in a bilingual region (Roser in Catalonia). All children lived in two-parent heterosexual homes, except Vicente whose parents are separated and lives with his mother. The families resided in the metropolitan areas of Madrid, Barcelona and Lisbon.

Young Children's Media Ecologies, Practices and Processes

Children in this sample possess a similar range of digital technological/media devices (Table 3.1). All the families have at least a standard TV set, a smartphone, and a laptop or PC; and most of the families also have a DVD player, electronic toys, and tablets. The number of devices possessed by families is also similar. The Device/Person Ratio (DPR) ranges from 1.75 to 3.25; for those devices that specifically share mobile and tactile features (underlined), the ratio ranges from 0.66 to 1.5.

A case by case analysis—which we compare at the end of this section—shows how different mediation practices and processes emerge across families within these relatively similar living conditions.

Gloria

Gloria engages very little with digital technologies throughout her day. Her play activities and the objects she uses are primarily non-digital. She only has access to two digital devices, which she can only use with parental assistance: the laptop computer—to watch media/videos from YouTube and the TV/DVD player—to watch cartoons and materials (from home and from the local library). Neither the

[1]We use pseudonyms for the focal child in each family. See Gillen et al. (2019) for a full description of each case.

Table 3.1 (Mobile) Digital technologies possession by the case families

	Gloria (SP)	Roser (SP)	Tomás (PT)	Matias (PT)	Vicente (PT)
Devices available in the home	1 standard TV (not connected to a cable)	1 standard TV	2 standard TVs	1 smart TV	1 standard TV
	1 smartphone	2 Smartphones	1 smart TV	2 smartphones	1 smartphone
	1 laptop computer	1 tablet	3 smartphones	2 tablets	1 tablet
	1 desktop computer	1 laptop computer	1 IPad	1 tablet computer for children	1 PC
	1 E-Reader	1 DVD recorder Electronic toy(s)	1 tablet computer for children	1 PC	1 digital video recorder
	1 digital video player/recorder Electronic toy(s)		1 laptop	1 digital video recorder	1 digital radio
			1 PC Electronic toy(s)	1 digital radio	1 PlayStation Electronic toy(s)
				1 PlayStation	
				1 portable media player Electronic toy(s)	
Total	7	7	11	12	8
DPR[a]	2.33	17.5	2.2	3	2.66
MDPR[b]	0.66	0.75	1.2	1.5	0.66

[a]Device/Person Ratio (DPR); [b]Mobile Device/Person Ratio (MDPR)

family nor Gloria watch broadcast TV programmes as the TV Set is not even connected to a cable or antenna.

Her main type of digital media consumption happens during daily routines. For instance, during breakfast she sits on a high-chair while she watches videos on the laptop. She occasionally watches a video, alongside her mother or father, mostly upon her request. Yet, Gloria frequently observes her parents using digital technologies. The laptop computer emerges as a central device in the family home. It is located in the living room table; it is always on (even when other activities are taking place) and parents use it often (to use the internet, make arrangements and run errands, etc.), but it is strictly for parental 'work'. It is also present—as is her

Fig. 3.1 Father preparing a video on the laptop while Gloria has a snack; watching a DVD with her mother; parents engaged in the laptop while Gloria plays; Gloria and her mother reading a picture book—the laptop is always on, on the table

mother's smartphone—even when the mother engages in a book-reading activity with Gloria (these moments are captured in Fig. 3.1).[2]

Roser

As for Roser, digital devices and media are centrally present in her daily routines (see Fig. 3.2). She has all her meals at home watching cartoons on a tablet device. She watches it alone, or with her father, who helps her eat. Roser also has occasional access to a smartphone. She participates in calls (talking with her mother on the phone) or watches videos of her friends sent by their parents. She also likes to see photos of herself and take selfies.

Roser's daily life is also surrounded by her parents' use of digital devices. The household has the TV on most of the time. This is not meant for her to watch directly, but is rather part of the media 'background' of the home, while she plays with other toys or parents do house chores. Parents also watch a series while feeding Roser. Roser also observes frequent use of the smartphone by her father and predicts some of its uses. At the same time, she is encouraged to and does play with other non-digital toys. For example, during the recorded day her father took her to a park in front of their flat and they also spent the afternoon playing with traditional toys.

[2]In Figs. 3.1—3.5, we have intentionally blurred images so that we protect the anonymity of the participants.

Fig. 3.2 Roser having breakfast while watching a cartoon; getting ready for lunch watching a cartoon; talking to her mother on the phone; taking a selfie on sofa with her father and brother; TV on the background when Roser has yogurt for her afternoon snack; handing over the smartphone to father when it made a sound

Matias

Matias' home has the largest number of digital devices among the five cases but he himself does not use or have access to most of these. Throughout the day he interacted and played with his sister mostly on non-digital artefacts and toys. The only digital technological device they have regular access to is the television—a Smart TV—in the living room. Matias watched broadcast child TV channels, at the request of his older sister. Other devices such as tablets are used very occasionally when the children are travelling (both parents are from outside Portugal and the family travels to their parents' home countries).

TV/media consumption time emerges when the children want to rest from other forms of play or, as described by the parents when they need to be entertained while the parents or the mother have to complete other chores or activities, which is described by the mother as a 'babysitter' role. Daily routines and meals for the children do not involve the use of digital or any other media devices. Nonetheless, Matias is also a witness to the intense use of digital technologies by his parents. In particular, he views his mother who is a photographer working from home on her laptop computer placed in the main living room. This work takes place while Matias

Fig. 3.3 Siblings watching television while the mother prepares lunch in the kitchen; the two siblings having lunch in the kitchen without digital devices; the mother working on her laptop in the living room while attending to Matias

is also in the living room engaging in other forms of play so that he can be monitored by his mother (Fig. 3.3 illustrates these moments).

Tomás

The main devices Tomás has access to are television and smartphones. The digital activities are integrated as part of his and his older brother's play activities, usually on or around the sofa in the living room. Tomás likes to watch movies on the television, either on his own or with his family. However, he needs to negotiate the use of the television with his older brother as they want to watch different content. The television set in the living room is usually on throughout the day even when children engage in other activities such as reading a book with other family members. The boys do not usually use the TV in their bedroom.

Tomás also engages with his mother's smartphone. He uses game apps installed on the phone and asks for his mother's help when pop-ups and ads emerge. As reported by the family, they had bought a tablet for the children in order to keep the 'adult iPad safer'. However, rather than using this child-oriented tablet, the boys prefer to use the parents' smartphones.

Nevertheless, Tomás spends more time throughout the day engaging with other forms of non-digital play and literacy activities. He loves to play with his own dolls, 'reading' books and riding on bicycles. Also, there is a strong connection between his digital and non-digital activities: for example, his passion for animals is

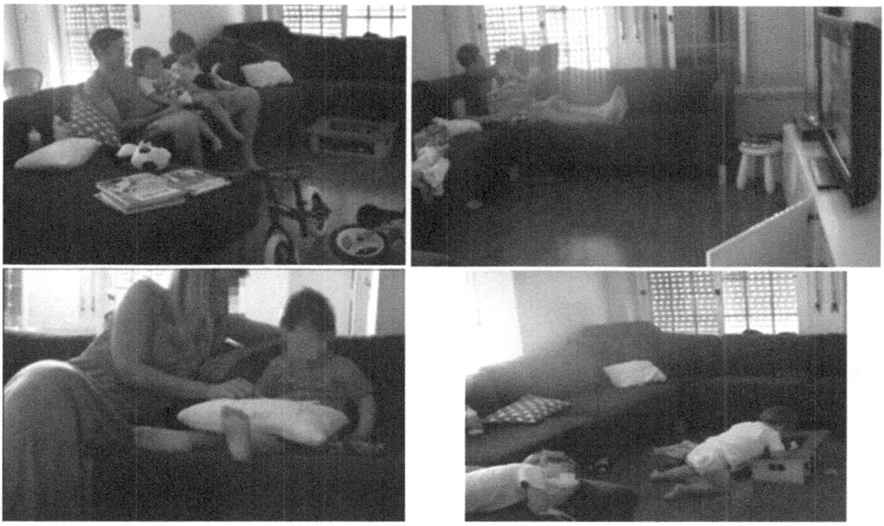

Fig. 3.4 Tomás watching a movie with his family; Tomás is reading a book with his father while the television is on for his sibling; Tomás' mother is helping him with the smartphone when an ad pops up; Tomás playing with his farm toys

demonstrated in his digital activities above with animal-feeding game and love for Lion King, as well as playing with a farm box made of wood and stuffed animals (Fig. 3.4).

Vicente

Vicente has access mainly to a tablet and a television and they are used for a 'babysitting' role throughout the day. After waking up, Vicente and his brother watch TV together in the living room (Fig. 3.5). While he watches TV, he may be engaged in other activities simultaneously, such as drawing on the floor. He also watches TV again in the evening (after he has taken a nap and has been to a park). The mother puts a show for him on TV normally, as he cannot operate the remote control or the TV box on his own.

Vicente also engages with tablet devices in the afternoon, to watch the same cartoons and music videos he watches on broadcast television. However, this use needs to be negotiated with his brother, as his sibling also likes to use the family tablet device but to watch different programs/videos. Therefore, Vicente will use the tablet when he sees it as 'free' or will ask his mother to negotiate use with his sibling. His engagement with the tablet also includes dinner time in the kitchen, watching English music videos, as well as after dinner in the living room, while the mother cleans up the kitchen.

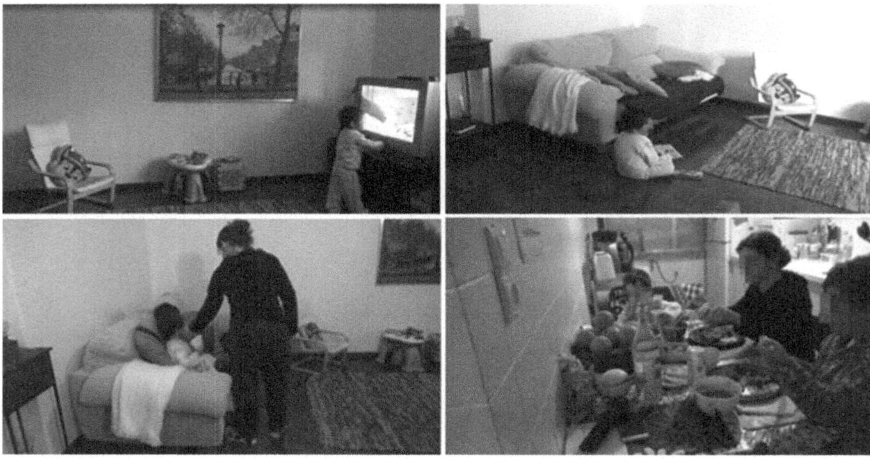

Fig. 3.5 Vicente watching and touching TV after waking up; watching TV while drawing on the floor; Vicente playing with a tablet while brother watches TV; engaging with the tablet while having dinner

Vicente can operate the list of suggestions on the tablet and can choose his next videos within the streaming application. Yet, he needs help to solve other problems and asks his mother for assistance. His older sibling also plays a supportive role with digital technologies despite the fact that this role is less acknowledged by Vicente or his older sibling.

A Summary of Mediation Practices and Processes Across Cases

Table 3.2 summarises the mediation practices in the five case families. By looking at mediation practices as processes, we can see that access given by parents does not necessarily lead to its uptake by the children, such as the case of Tomás with a tablet especially bought for him, or the case of Roser with broadcast television. We can also see that, in some cases (Roser and Vicente), digital activities are clearly integrated into routines (i.e. meals or snack), while in others (especially Matias) they are only part of play activities. Even in families that promote little child digital engagement (Gloria and Matias), the children are surrounded by and observe frequent and intense digital engagement by parents. Siblings, in particular older siblings, also play distinct but important roles in the mediation practices of the family (Matias, Vicente and Tomás).

Table 3.2 Summary comparisons of mediation processes

	Gloria (S)	Roser (S)	Matias (P)	Tomás (P)	Vicente (P)
General description	Limited access to digital technologies but part of family routines. Surrounded by parents' use of technology.	Integrated into meal routines. Surrounded by parents' use of technology.	Limited access to digital technologies Surrounded by parents' use of technology.	Integrated use of TV and mobile phone in play activities. Connection between digital and non-digital activities.	'Babysitter' role. Long, free access, helped by Mother.
Main devices used	Laptop	Tablet	TV	TV (Smart TV) Smartphone	TV (standard) Tablet
When	Breakfast/ snack/play	Meals	Play	Play	Meals/play
How	Father or child choose a program (i.e. YouTube videos).	Father chooses YouTube videos.	When requested by his sister, the mother selects programs for both siblings.	He requests videos to play, navigates YouTube with supervision of his mother. Negotiation with other family members.	He selects and navigates YouTube videos on his own, with help of his mother. Negotiations with brother regarding which device to use.
With whom	Alone or with family.	With father or alone.	With sister.	Alone or with brother and father.	Alone or with brother.
Additional activities	Surrounded by and observes parents' use of laptop and mobile. Co-watch DVD video requested by her on TV.	Surrounded by and observes parents' use of mobile and TV (which is on all the time), but not interested in using for herself.	Surrounded by many digital devices at home that he does not have access to. Observes mother working on computer all 'day'. Father lets children use his mobile occasionally.	TV (children's channel) is on all the time even when he engages with other activities. No interest in tablet bought for them.	

Digital Media Ideologies

The interviews with parents allow us to tap into different dimensions of engagement with digital technologies: (a) rhetoric behind the mediation processes discussed above; (b) the perceived 'effects' of digital technologies on children; (c) attributed learning value; (d) the role of technologies in future uses and learning; and (e) 'media literacy' concerns and aims. Parental beliefs across these themes are threaded into distinct family digital media ideologies (Gershon, 2010), briefly described here for each case.

Gloria's parents actively promote alternatives to digital practices. They see digital technologies as a necessary and valuable learning tool in the future but, they also believe it is easy to learn how to use digital technologies. Therefore, they consider that delaying her contact is not problematic as Gloria will quickly learn how to use them. Their main media literacy concern is helping Gloria have a critical perspective on technologies and technology use.

Roser's parents also want to promote alternatives to digital activity in their child but, at the same time, accept that technology is a part of their family life—although here parental stances vary and Roser's mother sees digital technologies as a much more 'dehumanising' force. They also consider that learning how to use digital technologies is not difficult and can be delayed but situate digital skills alongside other basic skills (such as reading and writing). When turning to media literacy concerns their focus is on learning how to handle privacy and security issues in digital technology.

Tomás' parents have moved a step forward in accepting digital technologies as one more aspect of daily life and turn their concerns and strategies to their child's current engagement with digital technologies. They see digital media as something that may 'consume' children's attention and time, so they actively monitor use. They also see some present learning potential for digital technologies (e.g. English as L2) and do not express future media literacy concerns beyond that later learning of digital technologies will be easy and 'natural'.

Matias' parents are also primarily oriented to their child's present engagement with digital technologies but, in contrast, have a much more negative view of digital technologies. They see digital media as something that can be 'all-consuming' of the child's interests and attention and as having 'zero' learning value. Consequently, they restrict as much as possible Matias' current access to digital technologies and believe that delaying this contact will not have consequences as it will be easy to catch up.

Finally, Vicente's mother articulates her discourses on digital technologies around practical concerns. As discussed above, digital devices are used as a 'baby-sitter' resource and, in this respect, her only current concern is that Vicente becomes 'dependent' on technologies to organise his leisure and activity. Given this use of digital technologies in the organization of daily life, little is expressed regarding the educational value or learning demands of digital technologies. Turning to the future, her main worry regarding digital media is how Vicente will learn the handling of ‘peer pressure’ with and through digital technologies.

In short, the analysis of the different dimensions reveals diversity in media ideologies across families. For instance, although Gloria’s and Roser’s overall family approaches or rhetoric are similar, in the sense that they promote alternatives to digital technologies, their views regarding the effects of technology on children or their primary media literacy concerns are slightly different. Also, across the cases the learning value attributed to technology ranges from explicitly mentioning a ‘zero-value’ to recognising its relevance to different present and future skills.

Conclusions

We bring together analyses on media ecologies, practices and ideologies as a first attempt at understanding parental mediation as emergent processes. While there is not a substantial difference in the presence of digital devices available across the cases (between 6 and 12 devices), there is considerable variation in how access is granted to young children, how use is structured by the parents, and how those devices are (or not) taken up by children. More digital technology available in the home does not necessarily lead to more use, as either access is restricted by parents (Matias) or devices simply are not taken up by the child (Tomás). On the contrary, even though not many devices were available, the use of the available devices may be intense depending on parental ideologies, circumstances, and perhaps the child’s characteristics. Similar patterns of technology use by young children may also, actually, connect to different parental media/digital ideologies, family biographies and parental experiences with digital technologies (Gloria, Roser).

Understanding these mediation practices as an *emergent process* allows us to acknowledge children as actors in the social and material interactions with their parents and digital technologies—and therefore leads us to recognise children’s (digital) rights to access, understand and participate in digital media (Livingstone, 2016; Staksrud & Milojevic, 2017). This chapter made evident the complexities in which these processes occur among children of young age and their cultural situatedness. It highlighted the multilayered relationship of children with digital technologies, as increasingly integrated into everyday life, interacting with physical spaces, objects and experiences. It is thus crucial to interrogate how children’s digital rights are promoted in ways that foster their general healthy development and well-being.

References

Chaudron, S., Di Gioia, R., & Gemo, M. (2018). Young Children (0-8) and Digital Technology. A qualitative study across Europe. EUR 29070. Publication Office of the European Union. *Pedagogika, 68*(3). https://doi.org/10.2760/294383.

Darling, N., & Steinberg, L. (1993). Parenting style as context: An integrative model. *Psychological Bulletin, 113*(3), 487–496.

Erickson, F. (1996). Going for the zone: The social and cognitive ecology of teacher-student interaction in classroom conversations. In D. Hicks (Ed.), *Discourse, learning, and schooling* (pp. 29–62). Cambridge University Press.

European Commission. (2018). *Digital Economy and Society Index (DESI)*. ec.europa.eu/digital-single-market/en/desi.

Galera, N., Matsumoto, M., & Poveda, D. (2016). The place of digital devices in the home and family routines of young children (3-7) in Madrid. *Media Education: Studi, Ricerche, Buone Pratiche, 7*(2), 303–319.

Gershon, I. (2010). Media ideologies: An introduction. *Journal of Linguistic Anthropology, 20*(2), 283–293.

Gillen, J., Marsh, J., Bus, A., Castro, T., Dardanou, M., Duncan, P., ... & Jernes, M. (2018). Digital literacy and young children: Towards better understandings of the benefits and challenges of digital technologies in homes and early years settings. *Policy briefing of DigiLitEY COST Action IS1410 and the Digital Childhoods SIG of the European Early Childhood Research Association*. http://digilitey.eu/wp-content/uploads/2018/08/DigiLitEY-and-EECERA-Digital-Childhoods-Policy-Brief.pdf.

Gillen, J., Matsumoto, M., Aliagas, C., Bar Lev, Y., Clark, A., Flewitt, R., ... & Pacheco, R. (2019). *A day in the digital lives of children aged 0-3*. Full report: DigiLitEY ISCH COST Action 1410 Working Group 1: Digital Literacy in Homes and Communities. http://www.research.lancs.ac.uk/portal/en/publications/-(19b42af9-7828-4950-afca-69fdce62702e).html.

Hepp, A. (2014). Mediatization: A panorama of media and communication research. In J. Androutsopoulos (Ed.), *Mediatization and sociolinguistic change* (pp. 49–66). DeGruyter.

Holloway, D., Green, L., & Livingstone, S. (2013). *Zero to eight. Young children and their internet use*. EU Kids Online, LSE.

Jewitt, C. (2013). Multimodal methods for researching digital technologies. In S. Price, C. Jewitt, & B. Brown (Eds.), *The SAGE handbook of digital technology research* (pp. 250–265). Sage.

Livingstone, S. (2007). Strategies of parental regulation in the media-rich home. *Computers in Human Behavior, 23*(2), 920–941.

Livingstone, S. (2009). *Children and the Internet: Great expectations, challenging realities*. Polity Press.

Livingstone, S. (2016). Reframing media effects in terms of children's rights in the digital age. *Journal of Children and Media, 10*(1), 4–12.

Livingstone, S., & Blum-Ross, A. (2018). Imagining the future through the lens of the digital: Parents' narratives of generational change. In Z. Papacharissi (Ed.), *A networked self: birth, life, death* (pp. 50–68). Routledge.

Livingstone, S., Mascheroni, G., Dreier, M., Chaudron, S., & Lagae, K. (2015). *How parents of young children manage digital devices at home: The role of income, education and parental style*. EU Kids Online, LSE.

Marsh, J., Mascheroni, G., Carrington, V., Árnadóttir, H., Brito, R., Dias, P., ... & Trueltzsch-Wijnen, C. (2017). *The online and offline digital literacy practices of young children*. http://digilitey.eu/wp-content/uploads/2017/01/WG4-LR-jan-2017.pdf.

Marsh, J., Plowman, L., Yamada-Rice, D., Bishop, J., Lahmar, J., Scott, F., ... & Thornhill, S. (2015). *Exploring play and creativity in pre-schooler's use of apps: Final project report*. www.techandplay.org.

Mascheroni, G., Ponte, C., & Jorge, A. (2018). *Digital parenting: The challenges for families in the digital age*. Nordicom.

Nikken, P. (2018). Parenting and young children's media use: A Dutch approach. In C. Ponte (Ed.), *Boom digital? Crianças (3-8 anos) e ecrãs* (pp. 105–112). ERC.
Nikken, P., & Jansz, J. (2014). Developing scales to measure parental mediation of young children's internet use. *Learning, Media and Technology, 39*(2), 250–266.
Ólafsson, K., Livingstone, S., & Haddon, L. (2013). *Children's use of online technologies in Europe. A review of the European evidence base.* EU Kids Online, LSE.
Plowman, L. (2015). Researching young children's everyday uses of technology in the family home. *Interacting With Computers, 27*(1), 36–46.
Poveda, D., & Matsumoto, M. (2018, October 25). *Methodological approaches to research young children (0-8)'s digital literacies and practices: Comparing trends in the DigiLitEY Database and relevant literature reviews.* [Blog Post] https://digiliteymethodscorner.wordpress.com/2018/10/25/methodological-approaches-to-research-young-children-0-8s-digital-literacies-and-practices-comparing-trends-in-the-digilitey-database-and-relevant-literature-reviews/.
Sefton-Green, J., Marsh, J., Erstad, O., & Flewitt, R. (2016). *Establishing a research agenda for the digital literacy practices of young children.* A White Paper for COST Action IS1410. http://digilitey.eu.
Staksrud, E., & Milosevic, T. (2017). Adolescents and children in global media landscape: From risks to rights. *Annals of the International Communication Association*, 1–7. https://doi.org/10.1080/23808985.2017.1387503.
Valcke, M., Bonte, S., De Wever, B., & Rots, I. (2010). Internet parenting styles and the impact on Internet use of primary school children. *Computers and Education, 55*(2), 454–464.
Zaman, B., Nouwen, M., Vanattenhoven, J., de Ferrerre, E., & Looy, J. (2016). A qualitative inquiry into the contextualized parental mediation practices of young children's digital media use at home. *Journal of Broadcasting and Electronic Media, 60*(1), 1–22.

Mitsuko Matsumoto is Assistant Professor at Universidad International de la Rioja (UNIR) and also a researcher and a member of the Contemporary Childhood Research Group of Universidad Autónoma de Madrid, Spain. She holds DPhil in Education from University of Oxford, UK.

David Poveda is a Professor at the School of Psychology of the Universidad Autónoma de Madrid. He uses qualitative and ethnographic research methods to study child and adolescent semiotic practices in a variety of socio-educational contexts. He is the co-coordinator of the research group "Infancia Contemporánea" (www.infanciacontemporanea.com).

Ana Jorge is Research Coordinator at CICANT. She holds a PhD in Communication Sciences from University NOVA of Lisbon. Her research interests cover children/youth and media, audiences digital culture, celebrity. She was a member of COST Actions DigiLitEY and European Literacy Network.

Raquel Pacheco is a Postdoctoral researcher at the Center for Research in Arts and Communication (University of Algarve) and at Foundation for Science and Technology in Lisbon. She completed her PhD in Communication Sciences at University NOVA of Lisbon. She is also the Project Coordinator of Primeiro Plano https://primeiroplano.ciac.pt/

Vitor Tomé is Assistant Professor at Autónoma University of Lisbon, a professional journalist, a researcher at Centre for Research and Studies in Sociology of the University Institute of Lisbon, and an international expert on Digital Citizenship Education. He was a member of the COST Actions: European Literacy Network and DigitLitEY.

Cristina Aliagas is Tenure Track Lecturer (Serra Húnter programme) at Universitat Autònama de Barcelona. She received her PhD in 2011 from the Pompeu Fabra University. Her research interest focuses on the role of literacy in the everyday life of Catalan and Spanish children and teenagers, particularly those that 'resist' schooling.

Marta Morgade Salgado is Associate Professor in Educational Psychology at the Universidad Autónoma de Madrid. In recent years, her work has focused on the development of collaborative methodologies and sensory ethnographies with young people, professional artists, collectives at risk of social exclusion, and adopted children and their families.

Chapter 4
Digital Parenting and Transnational Migration: Cultural and Emotional Drives for Digital Media Use

Teresa Sofia Castro and Cristina Ponte

Introduction

Objects such as smartphones or tablets are still recent in our lives, when compared to traditional media, meanwhile readapted to the digital shift. Research is paying increasingly attention to how these devices are conquering space and time in young children's lives, in families' interactions and parenting daily practices (e.g. Castro & Ponte, 2019; Elias & Sulkin, 2017; Mascheroni et al., 2018; Ponte et al., 2017). However, still little research privileges a longitudinal approach to capture changes and relate causal effects with consequences in the family's environment and processes (Lev et al., 2018).

This article is part of a longitudinal study with 18 Portuguese families and pays particular attention to a group of five families for their transnational and migrant characteristics, as their perspectives are missing in national research on digital parenting (granting and mediating access and use of digital media).

These are families crossed by the intricacies of migration, the rapidly changing media ecology, the effects of globalisation, the diverse configuration of family structure, and the patterns of digital media use. Our aim is to contribute with a socio-constructivist standpoint based on qualitative data collected with these families, over the past 3 years, and answering to two questions: How are migration and spatial mobility affecting digital media appropriation and parental mediation in families with young children? In which ways is parental mediation intersecting with the cultural and emotional significance of digital media in the management of these families' lives, whether for helping cultural adaption/preservation or strengthen emotional bonds?

T. S. Castro (✉) · C. Ponte
Universidade NOVA de Lisboa – ICNOVA, Lisbon, Portugal
e-mail: cristina.ponte@fcsh.unl.pt

D. Holloway et al. (eds.), *Young Children's Rights in a Digital World*, Children's Well-Being: Indicators and Research 23,
https://doi.org/10.1007/978-3-030-65916-5_4

Parenting the Digital in Migrant and Transnational Families

Literature supports that parental mediation and families' decisions regarding digital media can be tense and emotional. At stake is the calculation between what represents good parenting and what are the best choices for the child's development (Zaman et al., 2016; Clark, 2011). To meet these concerns, ultimately, parents reveal three different approaches regarding digital media: (1) a positive and enabling attitude towards technological possibilities, which is embraced in family's daily lives; (2) weighing the pros and cons of digital media with the aim of balancing family's on/offline time; (3) resisting and postponing digital media, preferring to have the family members engaged in non-digital activities and interactions.

Nevertheless, typifying parental mediation is not always this straightforward, as we have to consider the development of the ecological environment of family's life in which interconnections between settings and events happen (Bronfenbrenner, 1979).

For Clark (2014), social class, ethical principles, and concerns shape parents' actions towards the management of digital media in the family context. Parents tend to fall under one of two contrasting principles concerning the management of access and use of digital media: *ethic of expressive empowerment*, in which families embrace the empowering potential of technologies to cultivate children's education and self-development—a response to an increasingly competitive society and more common in upper and middle class families with high education background; and *ethic of respectful connectedness*—attitudes that guide the use of digital media as a way for spending time together 'as a family' (more common among lower income families). Paus-Hasebrink et al. (2019) enhance that in the latter, children's media usage can be a way of compensating everyday life shortcomings, acting as providers of good examples, information and entertainment.

Meanwhile, parents who are avoiders and more restrictive can just control children's exposure to digital media to a certain degree since older siblings, the extended family, and school also play a very influent role in this process of media adoption.

Each family and child are different, as well as their motivations and attitudes towards digital media. In times of deep mediatisation (Hepp & Hasebrink, 2017), new cohorts are embracing parenthood, and new frames guide their parenting principles and behaviours. The child displays a recognised agency and a voice that influences family's decisions. Also, the relations and interactions between family members became increasingly mediated and moulded by screens (Oswell, 2013) for the past 20 years (see Ponte et al., 2019). Digital media are part of families' everyday practices, (inter)actions (e.g. family conversations occurring via multiple platforms or video calling with relatives through WhatsApp), and characterise the modern way of 'doing family' (Morgan, 2011). Family is understood beyond the structure to which individuals belong. It includes the relational practices, in which family members enact, in the fluidity of their everyday life (Assmann, 2018) (e.g. waking up, daily meals, shared activities, playing and leisure).

In the case of families touched by migration flows, digital media's affordances offer and expand possibilities to: (i) communicate and participate in family's daily routines (Ponte et al., 2017); (ii) foster connectedness and bridging physical distance, as proved by Gonzalez and Katz (2016); (iii) assimilate cultural and linguistic aspects of the 'outward' society, while also supporting family's unity and preserving the cultural heritage of the 'inward' society, as Elias and Lemish (2008) substantiate.

A holistic picture of how internal and external factors influence family's lives over time in their relation to digital media (Zaman et al., 2016) (e.g. acquisition of new devices, changes in family composition, changes in family routines) is only achievable using a longitudinal design where changes and continuities can manifest.

Structural, social, technological and economic transformations, in which technology adoption, employment, and migration are dimensions of that change, do have a great influence in the lives of modern families (Sefton-Green et al., 2016). We focus our gaze in the families that represent the "intersection of the global movements of people and cultures" (Livingstone, 2018, p. x), that is the group of migrant and transnational families.

By migrant and transnational families, we consider the families that face similar challenges, namely to adapt to a new country, culture, language, educational and social system of a host country (Shih, 2015). What distinguishes the transnational families is the fact that family members (children, mum or dad) may be physically separated in different nation states (Schmalzbauer, 2004).

The Context: Portugal at the Intersection of Global Movement of People

From the mid-1980s onwards with the entrance in the European Economic Community (EEC), Portugal saw a diversification of its (in and out) migratory flows. Marked by the colonial past and its historical, cultural and economic connections (Baganha et al., 2004), the main immigration came from Lusophone Africa (47%), followed by the European Union (30%), Asia (4%) and Brazil (11%), according to Góis and Marques (2018). Thorough the 1990s is remarkable the growth of immigrants from China and Pakistan, countries with which Portugal had no historical or economic ties. In the 2000s the Ukrainian, Romanian, Moldovan or Russian migrants were at the top of immigration statistics in Portugal—as legal barriers to mobility were progressively dismantled in Eastern Europe (Baganha et al., 2009).

Historically, Portugal is a country of emigration, having an expressive diaspora particularly in the US, Western Europe and Canada. In the recent decades, Portuguese emigration developed into a more diversified geography to include other European and non-European destinations (Marques, 2010), such as Asian countries and former colonies namely, Angola, Brazil and Mozambique (Santos, 2013). From the beginning of the twenty-first century, Portugal became one of the European countries with the highest rate of an exodus of highly qualified people leaving the country as a result of the economic crisis, the lack of employment opportunities, low wages, and as a result of increased investments in educational qualification (Gomes, 2015). When Portugal began its economic recovery, in 2016, migratory flows to Portugal increased again, in particular coming from Brazil, following the Brazilian political and economic crisis (França & Padilla, 2018).

The Longitudinal Study 'iTec Families': Aims, Methods and Ethics

iTec Families is an ongoing study, initiated in June 2017 with 18 Portuguese families with children aged from zero to eight (age at the beginning of the study) with the aim of gaining a comprehensive understanding of how digital incorporation and mediation is (re)configured in the domestic setting over time.

To address this, a qualitative approach, using ethnographic and participatory strategies is being privileged: in-depth interviews with parents to gain a grounded angle of family media processes and to unveil (in)consistencies and changes (domestic scenario, perceptions, mediation, and practices); observant participation, creative (drawings, play), ludic (games), and visual strategies (images) to engage and reach children's voices, motivations and practices, prioritising their protection and participation rights in research (e.g. anonymisation of data, use of pseudonyms, collection of pictures and videos without showing the child's face, and the respect for the right of the child to not participate in research tasks).

An informed consent was signed before the first interview and is negotiated before each visit to the family and with children before research tasks that involve their participation.

The recruitment process aimed to reach as much diversity in families as possible in terms of family's composition, SES or ethnicity, and children's gender and age.

Fieldwork has been a reflexive and evolving exercise in which the parents proved to be important allies, opening their home and intimacy, giving their time, helping building rapport with the children, and acting as field collaborators, in particular with pre-verbal toddlers.

Five families, crossed by a migrant and transnational experience, are analysed here (Table 4.1): one is a single child family, three have two children, and one has four children. A total of seven children are aged between zero and eight. In this group of families, we could not achieve a gender balance among children. Thus, we have one girl and six boys—one child was born during the study. Two families live in Portugal, one lives apart (mother and children live in Portugal and the father works and lives in Angola), and two live in the UK. Two fathers are from Cape-Vert and other two from Brazil. In two families, parents are aged 35 and over and in three families, parents are aged below 35. Family names were replaced by pseudonyms:

Table 4.1 Transnational families

Family name	Household composition	Education (parents)	Country of birth	Country where they live
Freitas	Mum, Dad 2 boys, born in 2014 and 2018	Mum: Higher Education (HE) Dad: HE	Dad: Brazil Mum: Portugal Children: Portugal	Portugal
Guimarães	Mum, Stepfather* 1 boy, born in 2016	Mum: HE Dad: Secondary school	Dad: Cape Vert Stepfather: Brazil Mum: Portugal Child: Portugal	Portugal *Stepfathers' daughter (born in 2016) lives in Brazil with the mother
Henriques	Mum, Dad 1 boy, born in 2004; 1 girl, born in 2010	Mum: HE Dad: HE	Portugal	Portugal and Angola
Sousa	Mum, Dad 2 boys, born in 2010 and 1997	Mum: HE Dad: Secondary school	Mum: Mozambique Dad: Angola Older child: Germany; Younger child: Portugal	UK
Teixeira	Mum, Dad 2 boys, born in 2013 and 2015 2 girls, born in 2007 and 2009	Mum: HE Dad: Secondary school	Mum: Portugal Dad: dual nationality: Cape Verdean and Portuguese Boys: United Kingdom (UK) Girl: UK	UK

Source: iTec Families Study

The five families are different in terms of their structure (un/married, living together apart, nuclear, blended), composition (number of family members), age of the children, nationality and the country they live in.

Families' socio-economic status also differs. One family experiences more precarious and temporary/part-time jobs and one mother is unemployed. Mothers have higher academic qualifications than fathers, who have more unqualified jobs.

So far, between three and five visits have been made, depending on the availability of the family, and in between visits we keep in touch via mobile phone.

Data analysis is a dynamic process as part of becoming more familiar with the data as new information is added, and new insights emerge from the dialogue between different theories and concepts. To organise the information and identifying themes and relationships among the data, we have been privileging thematic analysis. To minimise the influence of the researcher, the findings have been put under scrutiny in conferences, other academic and community events, and to blind review.

Some of the provisional findings are presented and discussed in the following section.

Digital Parenting and Digital Media Management at the Intersection of Transnational and Migration Flows

Families' accounts offer perspectives on the management of digital media in their daily lives, crossed by the challenges of geographical movements and its social, cultural and emotional peculiarities. To answer the research questions this analysis and discussion is organised under two topics: (i) The appropriation and mediation of digital media considering characteristics of migration and mobility of the families; (ii) the intersection of parental mediation with the emotional and cultural meaning of digital media to adapt to a new culture or to preserve a sense of identity with the country left behind.

Digital Media Appropriation and Mediation

The five families use different approaches to digital media in terms of appropriation and mediation.

Family Henriques (nuclear family that lives apart) reveals motivations strongly attached to their migratory condition in line with what is sustained by Gonzalez and Katz (2016), that is to maintain emotional ties and participate of family life: "I'm always with them on Skype for dinnertime. And I help my daughter with her homework" (father). Besides the migration factor, the family stimulates children to use technology to develop skills like solving technical problems or learning English. These two abilities are highly valued by the parents, as key competencies to succeed professionally (*ethics of expressive empowerment*).

On the opposite side, family Freitas is very influenced by a digital-free nostalgic feeling. Parents postpone and avoid as much as possible their son's interactions with digital media—"he doesn't watch TV every day at home"—they privilege instead, "books, spending time outdoors together, and listening to music on the radio during dinnertime". Nevertheless, this avoiding attitude is challenged by external factors: at school, where the son watches cartoons on television; at grandparent's house (mother's family), where with grandfather's help, the boy gave his first steps watching videos on YouTube; and with his aunt, the two bond and have fun taking selfies with Snapchat.

For both families living in the UK, although not all members have a personal smartphone (just parents and older siblings), the feeling of being far from the extended family exacerbates a sense of isolation and insecurity that is fulfilled by affordances enabled by communication-based devices:

"Here (in the UK), it's just the four of us. I'm always concerned with our safety. The older one knows it's very important for me to receive a call or a text, just assuring me that he's ok" (Mum, Family Sousa).

Family Guimarães also proves that their son's screen time is an emotional and tense challenge as the child spends time with cousins, at grandparents', at school or when he goes to his dad house (they are separated):

"I don't want him to be left behind other children his age, but I uphold that balance is the best path. However, I know that his dad uses game consoles and TV to babysit him, which I don't approve, and I think we have to come to an agreement about rules". (Mum, family Guimarães)

There are differences in terms of mediation and in the relation the family develops with digital media. For family Teixeira, in particular the television is a way to spend time together (*ethic of respectful connectedness*):

"We have four children and we don't have much money. So, the kid who gets more stars (ranking system in a board) for getting household tasks done during the week can choose the movies we'll watch on Sunday. We spend Sunday in our pyjamas around the television". (Mum, family Teixeira)

In terms of brokering digital risks, the Family Sousa counts with the technological expertise and the critical perspective of Sérgio, the older son, to mediate his younger brother, Simão:

"He [Simão] watches YouTube on the Smart TV. That way I can easily keep an eye on what is watching. I don't like some YouTubers and I explain to him that some behaviours and language are not appropriate. He respects what we tell him and his big brother opinion". (Mum, family Sousa)

Besides helping his parents in mediation decisions and mentoring his younger brother, Sérgio shares two concerns: "the type of people he may meet online" and "game addiction". To other parents he leaves some advice: "restricting just creates more problems", and "just consulting PEGI[1] is not enough. Parents need to research about the content and the game's community."

Digital media appropriation and mediation is a process that changes over time, as the child grows older or as new events create different dynamics inside the family. For instance, despite the avoiding attitude of family Freitas, in the second visit the mother recognised that Francisco (the 5 years old son) is becoming digitally more autonomous: "he already knows how to open Spotify on the computer".

Family Henriques is starting to deal with their daughter's (now, aged nine) claims for more digital autonomy and participation. The girl wants to have a profile in the video-sharing social networking Tik Tok, because her friends are also using it and she is feeling left out. Parents are giving it a though "to weigh the pros and cons and set some rules" before they agree on that. In the case of family Guimarães, the mother observes that Guilherme's (with 3 years and a half) interest for cartoons and animated movies is recent and very much related with the moment his stepfather moved in with them, in the past year: "I once showed him Snow White on TV, via YouTube, and now he enjoys watching cartoons on TV." Guilherme used to watch

[1]Pan European Game Information or PEGI is a European content rating system for electronic games, and other entertainment programs for computers or other platforms.

role-playing videos starred by children or music video clips on the mother's tablet or smartphone.

Except for families Freitas and Guimarães, whose children do not possess their own personal device, the five other children have their own tablet (these ownerships were prior to the study). Each family has rules that they adapt accordingly to situations or child's behaviour (e.g. to avoid boredom, used to reward or punish a behaviour). Families Henriques and Teixeira use technical strategies to block unapproved content: "I get a notification on my e-mail with everything they do on the tablet" (Mum, family Teixeira). Screen time is a big concern for Freitas and Guimarães families. Both families use communication-based Apps to talk with family living in Brazil, but the parents are the ones who do the call. In the case of family Henriques, whose dad is living in Angola, Helena "knows how to call me on Skype since she was very young [before age six, when the family was recruited], I taught her."

Digital Media as Drives for Cultural and Emotional Management

While parents are in between managing their adjustment to a new culture (outward) and keeping in touch with the inward culture, for their children—who were born or are growing in a different country—keeping in touch with parent's culture and being in touch with an outward culture can be a very complex exercise.

Depending on the migration characteristics of the family/family member, the five families offer different scenarios that are worth of looking into its idiosyncrasies.

The balance between the adaption to a new culture and the preservation of the culture of birth is less noticeable among the families that live in Portugal, in which just one of the parents is a migrant.

The examples of family Freitas and family Guimarães provide some clues. In both families, video calls are regular practices for the fathers to keep in touch with the family left behind. They use it to strengthen ties and have a sense of physical proximity. However, these affordances do not seem to have the same importance for the younger ones. Families report that they "feel curiosity, but do no interact much" (Mum, Family Freitas). Francisco (at the time aged 4) simply stared at the camera. These reactions may have to do with the age of the child: "I believe he doesn't understand much what is going on" (Family Freitas). In family Guimarães, communication difficulties may have to do with language differences, as the mum explains: "he doesn't understand his (stepfather's) daughter talking to him in Portuguese Brazilian. She speaks very quickly".

The fathers emigrated from Cape Vert and Brazil stimulate their children's contact with their left behind culture by co-viewing music, cartoons and contents from their inward culture, usually on YouTube. The children do not resist to such acculturation, they incorporate the language (mixing words from two countries) and the contents in their internet searches and digital activities.

To shrink the distance, family Freitas revealed that grandparents, in Brazil, purchased better smartphones, so they can make more often video calls with their grandchildren, creating bonds and simulating emotional closeness with them.

For family Henriques, communication and internet-based technologies have a major role for the family, also narrowing the distance and facilitating the father's participation in everyday-shared practices as referred previously, during meals or helping studying. They value the easiness it enables to reach each other whenever needed. Digital presence "gives the sense that dad is here", adds Helena. The use of digital media in this family is very much related with emotional motivations, and not so much for cultural adaptation/preservation.

Among the families living in the UK, there are different levels of cultural adaption/preservation: a) the parents and the older children, who emigrated during childhood/adolescence; and b) the children that emigrated while babies or were already born in parents' outward country.

Parents and older children feel in-between cultures. The children left behind friends with whom most of the bonds fade over time. In their daily lives they use digital media to create a balance between the inward and outward cultures: "At dinnertime we watch the UK news, and afterwards we watch the Portuguese news" (Family Sousa). To keep in touch with family and friends they use Facebook Messenger, WhatsApp, and regular mobile phone calls.

In the case of younger children who were born or grew up most of their life in the outward society it is more difficult for parents to stimulate their connection with the Portuguese language and culture. And it gets even more challenging when they enter in the educational system, are included in the peer group, and build friendships. The children from the families we interviewed in the UK spoke in English most of the time and Portuguese is their second language. This situation may impact in keeping in touch with the left-behind family and maintaining these long-distance ties, driving to a disconnection with left-behind family and cultural heritage. Nonetheless, young generations can be very helpful assisting parents in their adaptation process to a new language and culture. In these two families, the contents children watch are part of the UK culture, namely YouTubers, soap operas, idols, or music. However, as happens with children living in Portugal or elsewhere, they enjoy watching cartoons that are also a world phenomenon, strongly connected to YouTube's increasing power as a digital entertainment provider for children (e.g. PJ Masks, Paw Patrol, Pegga Pig).

Summary and Final Thoughts

The voices of families marked by migrant and transnational flows are missing in national research on digital parenting—granting access and mediating use of digital media—and on what digital media affords in terms of emotional and cultural connectedness between the inward and outward culture.

The five households presented here are part of an ongoing longitudinal study and represent the realities lived by families crossed by the experience of geographical distance and mobility, and changes crossed by time.

The data captured in different moments of their lives provide information on (i) how in these families experience digital media appropriation and mediation at home; and (ii) how digital media represents an emotional and cultural drive to preserve cultural heritage, foster connectedness with the relatives left behind, and supporting adaptation to the host country.

Although these are preliminary findings, we would like to highlight that depending on parenting values, digital media imprints different experiences for the families, in which children have different levels of digital autonomy and digital motivations: sense of closeness, leisure, security, empowerment, emotional support or family's together time.

In the cases where only one parent is a migrant, the cultural baggage models children's digital experiences: long-distance video calls, and influencing the contents watched. Nevertheless, the level of adaptation and preservation of new or former cultures is closely related with being aware of to whom this matters and why, when we consider two generations experiencing different spectrums of migration. As parents' roots are grounded in another country, digital media are important means of communication and information that help them assimilate the new culture, without losing their emotional support and cultural connectedness with the inward country.

In the case of the children, the ones who were born or arrived into a host country at a very young age, the dialogue between the inward and outward society is more complicated, as the parents' outward country is theirs inward country. They have, however, an important role as facilitators of parents' adaptation to a new language and culture.

Cross analysing these five families, the two research questions are very entwined. Nevertheless, all families recognise the added value that digital media brought to foster communication and connection in and over borders in a cheap and interactive manner (when compared to previous generations of e/immigrants). It also revealed how always-connected devices brought new possibilities for children to access multiculturalism just a click away, facilitated by video calls and YouTube.

Acknowledgments 'iTec Families' is a postdoctoral study funded by Portuguese Foundation for Science and Technology (FCT)—European Social Fund and Portuguese national funding from MCTES (SFRH/BPD/116279/2016).

References

Assmann, C. (2018). *Doing family in second-generation British migration literature* (Vol. 25). Walter de Gruyter GmbH & Co KG.

Baganha, M. I., Marques, J. C., & Góis, P. (2004). The unforeseen wave: Migration from Eastern Europe to Portugal. In M. I. Baganha & M. L. Fonseca (Eds.), *New waves: Migration from Eastern to Southern Europe* (pp. 23–39). Luso-Americana Foundation.

Baganha, M. I., Marques, J. C., & Góis, P. (2009). Imigrantes em Portugal: Uma síntese histórica. *Ler história, 56*, 123–133.

Bronfenbrenner, U. (1979). *The ecology of human development: Experiments by nature and design*. Harvard University Press.

Castro, T. S., & Ponte, C. (2019). Não há parentalidades perfeitas: Caminhos e desafios do digital no dia a dia das famílias modernas. *Crianças, famílias e tecnologias. Que desafios? Que caminhos?* 171–183.

Clark, L. S. (2011). Parental mediation theory for the digital age. *Communication Theory, 21*(4), 323–343. https://doi.org/10.1111/j.1468-2885.2011.01391.x.

Clark, L. S. (2014). Mobile media in the emotional and moral economies of the household. In G. Goggin & L. Hjorth (Eds.), *The Routledge companion to mobile media* (pp. 320–332). Routledge.

Elias, N., & Lemish, D. (2008). Media uses in immigrant families: Torn between 'inward' and 'outward' paths of integration. *International Communication Gazette, 70*(1), 21–40.

Elias, N., & Sulkin, I. (2017). YouTube viewers in diapers: An exploration of factors associated with amount of toddlers' online viewing. *Cyberpsychology: Journal of Psychosocial Research on Cyberspace, 11*(3). https://doi.org/10.5817/cp2017-3-2.

França, T., & Padilla, B. (2018). Imigração brasileira para Portugal: Entre o surgimento e a construção mediática de uma nova vaga. *Imigração brasileira para Portugal: entre o surgimento e a construção midiática de uma nova vaga, 2*, 207–237.

Góis, P., & Marques, J. C. (2018). Retrato de um Portugal migrante: A evolução da emigração, da imigração e do seu estudo nos últimos 40 anos. *E-cadernos CES, 29*.

Gomes, R. M. (Ed.). (2015). *Fuga de cérebros. Retratos da emigração portuguesa qualificada*. Bertrand Editora.

Gonzales, C., & Katz, V. (2016). Transnational family communication as a driver of technology adoption. *International Journal of Communication, 10*, 2683–2703.

Hepp, A., & Hasebrink, U. (2017). Researching transforming communications in times of deep mediatization: A figurational approach. In A. Hepp, A. Breiter, & U. Hasebrink (Eds.), *Communicative figurations: Transforming communications in times of deep mediatization* (pp. 15–50). Palgrave MacMillan.

Lev, Y. B., Elias, N., & Levy, S. T. (2018). Development of infants' media habits in the age of digital parenting: A longitudinal study of Jonathan, from the age of 6 to 27 months. In G. Mascheroni, C. Ponte, & A. Jorge (Eds.), *Digital parenting. The challenges for families in the digital age* (pp. 103–112). Nordicom.

Livingstone, S. (2018). Foreword. In E. Gee, L. M. Takeuchi, & E. Wartella (Eds.), *Children and families in the digital age: Learning together in a media saturated culture* (pp. x–xi). Routledge.

Marques, J. C. (2010). *A emigração portuguesa em tempos de imigração* (Vol, 20, pp. 115–129). POLÍGONOS: Revista de Geografía.

Mascheroni, G., Ponte, C., & Jorge, A. (2018). Introduction. In G. Mascheroni, C. Ponte, & A. Jorge (Eds.), *Digital parenting. The challenges for families in the digital age* (pp. 9–16). Nordicom.

Morgan, D. (2011). Locating 'family practices'. *Sociological Research Online, 16*(4), 14. https://doi.org/10.5153/sro.2535.

Oswell, D. (2013). *The agency of childhood: From family to global human rights*. Cambridge University Press.

Paus-Hasebrink, I., Kulterer, J., & Sinner, P. (2019). *Social inequality, childhood and the media: A longitudinal study of the mediatization of socialisation. Transforming communications – studies in cross-media research*. Palgrave MacMillan.
Ponte, C., Simões, J. A., Batista, S., Castro, T. S., & Jorge, A. (2017). *Crescendo entre ecrãs: uso de meios eletrónicos por crianças (3–8 anos) (Públicos e Consumos de Media)*. Entidade Reguladora para a Comunicação Social.
Ponte, C., Pereira, S., & Castro T. S. (2019). Parenting young children in changing media environments with twenty years apart. In G. Bolin & F. Colombo (Eds.), *Generations, time and media, Comunicazione Sociali, journal of media, performing arts and cultural studies* (n. 2, pp. 276–288). https://doi.org/10.26350/001200_000064
Santos, I. D. (2013). L'émigration au Portugal, avatar d'un pays "semi-périphérique", métropole postcoloniale. *Hommes & migrations. Revue française de référence sur les dynamiques migratoires, 1302*, 157–161.
Schmalzbauer, L. (2004). Searching for wages and mothering from afar: The case of Honduran transnational families. *Journal of Marriage and Family, 66*(5), 1317–1331.
Sefton-Green, J., Marsh, J., Erstad, O., & Flewitt, R. (2016). *Establishing a research agenda for the digital literacy practices of young children*. A White Paper for COST Action IS1410.
Shih, K. Y. (2015). Transnational families. In C. L. Shehan (Ed.), *The Wiley Blackwell encyclopedia of family studies* (1st ed., pp. 1–7). Wiley. https://doi.org/10.1002/9781119085621.wbefs177.
Zaman, B., Nouwen, M., Vanattenhoven, J., de Ferrerre, E., & Looy, J. V. (2016). A qualitative inquiry into the contextualized parental mediation practices of young children's digital media use at home. *Journal of Broadcasting & Electronic Media, 60*(1), 1–22.

Teresa Sofia Castro is developing her Post-Doctorate research "iTec Families" at Universidade NOVA de Lisboa – ICNOVA, with the supervision of Cristina Ponte. Teresa holds a European Doctorate in Educational Technology and is a member of EU Kids Online and the ySkills Project. Her main research interests are children's digital lives, digital parenting, and qualitative research.

Cristina Ponte, Professor of Media and Journalism Studies at Universidade NOVA (NOVA FCSH), has published extensively on children and media. Co-editor of the book Digital Parenting. The Challenges for Families in the Digital Age (Nordicom, 2018), currently she is a MC member of the ySkills Project and of the EU Kids Online network.

Chapter 5
Children's Rights to 'Good' and 'Bad' Screen Time: Parental Narratives of How Children Do Family Online

Helga Sadowski and Lina Eklund

Family Life and the Right to *Screen Time*

For many contemporary families, internet and communication technologies (ICTs) have become an indispensable part of their intimate family life, both within and across households. Smartphones and other digital communication technologies help children, parents, and other relatives to connect during school/work hours or to organise family gatherings through chat-groups created specifically for that purpose. Even communicating with grandparents overseas has become much more vivid with video-telephone applications. Having ICTs integrated into families' everyday lives means new and exciting ways and opportunities to connect and interact in order to 'do family' (Schier & Jurczyk, 2008), but also comes with a bundle of new micro type tasks, responsibilities, and potential troubles. Indeed, each family member has to take part in these micro-tasks—such as texting family members after one arrives at a travel destination or sending grandparents a video clip of a grandchild's first steps—and do their share in what we have named *digital intimate family work* (Sadowski & Eklund, 2020). Intimate family work consists of finely tuned social interactions, practical and symbolical everyday actions which create and maintain family. Today, these also happening in digital and social media, and include children getting accustomed to new media technologies as well.

The omnipresence of communication technologies encourage scenarios in which each family member is expected to take part in communicative 'kinwork' (Di Leonardo, 1987). In other words, while it traditionally used to be the wife/mother's responsibility to make sure family members stay in touch, for example by sending out holiday cards (Di Leonardo, 1987), today each family member has their own

H. Sadowski (✉) · L. Eklund
Uppsala Universitet, Uppsala, Sweden
e-mail: helga.sadowski@im.uu.se; Lina.eklund@im.uu.se

D. Holloway et al. (eds.), *Young Children's Rights in a Digital World*, Children's Well-Being: Indicators and Research 23,
https://doi.org/10.1007/978-3-030-65916-5_5

network that they create and maintain. As we show, this also applies to the youngest family members who use, their own or others, digital devices to be part of and maintain their own social networks.

The tools used in intimate family work today, such as laptops, tablets, and mobile phones, are also the bearer of many negative associations, especially when used by children. They contradict the romantic ideals of 'natural childhoods' free from technology (Buckingham, 2000). Involuntary, children thus find themselves in a paradoxical situation. On the one hand, they are supposed to use digital devices in order to be available and approachable for family communication (Sadowski & Eklund, 2020). On the other hand, parents often define the spatial time windows in which the children have access and are allowed to use the devices. These negotiations oftentimes happen along with the notion of 'screen time' (Blum-Ross & Livingstone, 2018), and it is exactly this context in which questions of children's rights as digital citizens come to the fore. Young children should have the right to access digital devices and contents in order to socialise. Yet, their access is restricted and we know little about the social implications of 'screen time' and the underlying approaches of parents to balance between their children's digital empowerment and their protection from what they consider unhealthy usage patterns.

Drawing on interview data with six extended Swedish families spread over 18 households (29 adults, 8 young children, and 3 pre-teen siblings), this chapter investigates how parents and grandparents manage, relate to, and assist in children's intimate family work. It delivers insights into the question: How are children's ICT uses, as part of intimate family work, controlled, encouraged, enabled, or disabled? For this purpose, the present chapter is focusing on how parents reflect on their children's ICT uses and how they understand and utilise 'screen time' in their digital parenting practices in contemporary Sweden. Our results show that not all types of screen time are alike, the same screen use by children can be seen as 'good' or 'bad' in different contexts. In our data and in relation to communication, 'good' meant screen time that brought children together with other family members. 'Bad', on the other hand, were individual uses that were considered divisive and isolated children from other family members.

Children's Screen Time and Digital Preferences

Sweden is a highly connected and tech-savvy country. This also applies to its children, 48% of 2-year-olds, 85% of 6-year-olds, and 98% of 8-year-olds use the internet frequently. Probably because of its portability and supposed ease of use, 72% of the very youngest (0–8 year-olds) do so via a touchpad device. However, considering these high usage numbers and that this is one of the first generations growing up in a digitalised world, many worry about its impact on childhood. Buckingham argues that the relationship between childhood and electronic media often has been construed in essentialist terms. He argues that children 'are typically seen to possess inherent qualities that are somehow uniquely related to the inherent characteristics of particular communications media (Buckingham, 2000, p. 41). On the one hand, he argues, this is seen as something negative. Electronic media are understood to 'exploit children's vulnerability, to undermine their individuality and to destroy their innocence' (Buckingham, 2000, p. 41). On the other hand, more recent discourses arose in which children are seen as digital natives who possess 'a powerful form of 'media literacy', a spontaneous natural wisdom that is somehow denied to adults' (Buckingham, 2000, p. 41) and which provides them with previously unknown opportunities for creativity, for community and self-fulfilment.

Related to this latter perception, Livingstone and Blum-Ross divide between two optimistic parental narratives of children's digital media use: The romantic narrative and the instrumental one. The romantic narrative underlines the affordances digitalisation offers children to express themselves, 'to act, create and self-actualise' (2018, p. 41). Instrumental narratives, on the other hand, focus on the usefulness of digital knowledge and the ability for successful futures: digital skills and knowledge are seen as a guarantee and necessities for future careers.

However, when it comes to parent-child communication, digital media scholar Turkle speaks of a vicious circle (2016):

> Parents give their children phones. Children can't get their parents' attention away from their phones, so children take refuge in their own devices. Then, parents use their children's absorption with phones as permission to have their own phones out as much as they wish. (p.76)

As a consequence, she argues, family communication and intimacy suffer. In fear of this development, many parents lay down rules for their children and increasingly, also for themselves. This is often done by implementing 'screen time'. Screen time is a term used to describe activities done in front of a screen, such as watching TV, working on a computer, playing video games, surfing the internet, or using digital devices for communication. But importantly, the term represents parental values and inner-family politics. Here, it becomes a controversial topic, how much screen time is 'too much' for children? What about children's right to access, when parents are the 'gatekeepers' (Hamilton et al., 2016)?

As Little (2018) has pointed out, screen time is a more complex issue than one might assume initially. For example, it has to be taken into account that screen time in front of the computer is not the same as passively watching TV, nor is scrolling through a newsfeed while sitting in the back of the car or playing *Pokemon Go* while walking through a city. The interactivity, omnipresence, multimediality and ordinariness of contemporary social media call for different criteria. Likewise, the term 'screen time' is used inconsistently in public discourse and tied to opposing parental advice. This means that digital parenting, as the 'concept that refers both to how parents are increasingly engaged in regulating their children's relationships with digital media [...], and how parents themselves incorporate digital media in their daily activities and parenting practices' (Mascheroni et al., 2018, p. 9), becomes a confusing mission.

Our Study (Methods)

In order to learn about ICT use and family life, we conducted interviews with six extended Swedish families, spread across 18 households with 40 family members aged 1–86. We used stratified quota sampling (Kothari, 2004) with three predefined sampling locations: the Swedish countryside, one of Sweden's larger cities, and cities along the US west coast with emigrated Swedish families. The goal was not to compare but to gain contested views and analytical depth by widening the range of informants. Most participants were born in Sweden, a few in other Scandinavian countries, and some of the children in the United States, but everyone spoke Swedish at home. We reached interview informants via community representatives and social media channels. In our call for participation, we asked for families of all types and sizes and explained that the interviews would be about family communication. As we consider family as a set of flexible social practices instead of static entities (Morgan, 1996), we chose not to define what family is but asked our informants to define what family means to them. What they mentioned were the relationships to parents, children, grandparents, grandchildren, aunts, and uncles, but also parents-in-law, siblings-in-law, and sometimes friends. Our only requirement for participation was that more than one generation would be interested in taking part in each family. The interviews took place individually or in groups, depending on informants' preferences. Most were set in informant's homes, but a few also at workplaces during lunch and via Skype.

All interviews were held as in-depth, open structure interviews (Hayes, 2000) focussing on five themes concretised as five open questions followed by questions attuned to each interview. The themes were: the family and the meaning of family for the interviewee in general, the 'why's and how's' of contact in the family, technology use in the family and communicative practices within the family, particularly around holidays such as Christmas and birthdays. We interviewed grandparents, parents, children, and siblings depending on the structure of the family in question, and depending on who agreed to take part in the study. Twenty-nine adults were officially contacted and agreed to be interviewed, but in addition, 11 children were present during the interviews, offering opinions, talking to their parents or the interviewer. However, to protect our young informants we do not present any names in the results or use any quotes from children as, in this study, it was the parents who agreed to the interviews and who agreed to the informed consent. As mentioned above, our study is consequently focussing on parental narratives of children's uses and the use of 'screen time' in the family.

Of our adult informants, around 60% had a university degree or equivalent. Most were working or retired, two women and one man were on parental leave, and one woman was a homemaker. Their professions were as diverse as McDonald's employee, lumber size controller, architect, and biomedicine analyst. All adults and children over the age of 12 had a mobile phone; only six of these were not smartphones. Everyone had access to the internet at home, although one retired couple was unaware of this fact. They owned a tablet that their children had connected to the internet.

Of the present children, five were girls and six boys. Eight were under the age of eight (five under five). All children used digital technology to communicate with extended family in one way or the other, many video-calling or phoning. Mostly they talked to grandparents, great grandparents, cousins and aunts/uncles who lived some distance away, but also as close as a 1-hour-drive away.

For our analysis, we transcribed the interviews, then compared them for a first understanding of the sample. Initially, the theory was 'put in brackets' to focus on the informant's everyday understanding (see Aspers, 2009). In other words, in a first analytical step, we studied the life-world of the informants as they described it, using their own words. The data set was then extensively and manually coded with an inductive approach using a qualitative data analysis software (NVivo 11). Coding focused on a wide variety of categories, initially defined in discussions but added to in the course of the coding phase: e.g. types of communication technology (e.g. SMS), relationships (e.g. siblings), family rituals (e.g. holidays), relationship to ICTs (e.g. frustration) et cetera with 59 coding categories and almost 2000 coded sections. Coding categories included for example everyday uses and frequencies of family members; children's uses in particular; technological scepticism and frustrations with ICTs; and specifications of which devices were used (TV, smartphones, tablets, etc.). In this analysis, we mainly draw on the following coding categories and their subcategories: (1) the everyday understanding of the data from the perspective of the informants, (2) the inductive coding and (3) the theoretical concept of doing intimate family work (Sadowski & Eklund, 2020). Based on these we refined and abstracted three overarching themes that capture the practices, attitudes, and strategies of these families in relation to ICT use for family contact. These were (1) practices of screen time for family bonding, (2) managing ideals and norm surrounding screens and childhood, (3) strategies for managing screen time. The result is narratives of digital parenting: The data gives insights into how parents interpret their family's everyday ICT practices, and the role 'screen time' plays therein.

Screen Time as a Bonding or Separative Family Practice

Screen Time for Family Bonding

The children in our sample, were, as public discourse has it, growing up with digital devices and often learnt from an early age how to 'do family' with technological means. As this father reflects on, talking about his 4-year-old daughter's use of Skype:

> Yes, I think it works really well. And they [the children, H.S.] understand, it's intuitive in a way, they see someone talking and so they talk. Our oldest is beginning to understand the phone, so you can just give her the [phone] like that. (...), just some months ago, I thought it

was too difficult, she heard the sound, but she just wanted to look at the screen. Funny, really.

Father, 35 years old, white-collar worker, university-educated.

This man is fascinated with his daughters' adoption process: To him, it seems as that she could not find a use for a smartphone one day and suddenly she can use it just 'like that'. Throughout the interview material, the idea that the adjustment to digital and social media for children comes 'intuitively' and that they automatically find it enjoyable or pleasurable was very widespread. This is something that digital childhood scholar Sonia Livingstone (2018, n.p.) has argued to be a 'digital myth'. Children are often constructed as digital natives who 'know it all'. In tandem, their parents are constructed as 'digital immigrants', who have major struggles adjusting to fast-paced digital developments. These assumptions do not hold true: Children might be more confident and experimental, but do not necessarily know more about the underlying mechanisms (Helsper & Eynon, 2010). Research also has shown that children often turn to their parents for help and guidance, which they try to provide as much as possible (Livingstone & Blum-Ross, 2018).

However, all parents to young children argued that it is easy and natural for their children to communicate with relatives via ICTs. Indeed, they underlined how they are counting on the children doing their part of intimate family work, for example by spending a lot of 'screen time' with grandparents:

I think it is nice to see them [her parents], and I believe that they think that is very good to be able to see the grandchildren. So, it's a lot of reading facial expressions and it almost feels like you can see each other and they can be quite long, the sessions, so sometimes when you have time to goof around, I mean, it can take an hour, an hour and a half.

Mother, 36 years old, white-collar worker, university-educated.

This statement underlines the perceived 'naturalness' of digital forms of communication. Maintaining binary thinking of on/offline, this mother underlines how it 'almost feels like you can see each other'. Indeed, the sessions become such mundane events that the participants let down their guard and start to goof around, meaning that an informal, intimate atmosphere can be created in such contexts.

Elsewhere, we have argued that ICTs play a particular role in overcoming intergenerational distances (Sadowski & Eklund, 2020). That is, we have shown how grandparents often put an extra effort to get accustomed to modern ICTs in order to create closeness to their grandchildren. Scheduled Skype sessions with the grandparents, or other relatives, became even more important with increased distances between households:

Interviewer: So, you started to use Skype more when your son was born?

Father: Yeees

Mother: Yes we did

Father: So that they [the grandparents] can see him all the time, it is important (...) with grandchildren which they don't see more than twice a year and they develop all the time

> Mother: Mm, and then he [her son] can show like, here is my new truck. And this is how the new curtains that grandmother made looks like hanging up, and these sort of small things which are important and matters for this contact that you have as a family.
>
> Father: 55 years old, white-collar worker, university-educated.
>
> Mother, 43 years old, homemaker, university-educated.

The examples above point out that screen time that is social, communicative family work is encouraged, and in this context, a negative impact of technologisation and digitalisation is not feared. This becomes particularly evident when it comes to relationships to grandparents which are maintained through different digital communication channels. Here, the affordance of visibility that digital media such as Skype enables is particularly important, since it allows for the remediation of everyday life situations, such as presenting one's curtains, easily.

Screen Time, 'Natural Childhoods', and Parenting Guilt

The notion of these 'good' uses of digital technology contrast sharply with 'bad' uses. Bad use was foremost related to digital technology as entertainment devices and perceived as separative and harmful for family life. In the quote below the mother of a 1-year-old reflects on how she perceives the negative aspects of omnipresent digitalisation:

> But actually, I wish that it would be like it was for us when we were small... you run around and play with your friends, [...] I mean, when a three-year-old already knows everything from the computer to the iPad, I think that is a bit too early, [...] they should not sit inside and look at a screen and get square eyes.
>
> Mother, 25 years old, blue-collar worker, practical education.

In this quote, norms on what a 'proper' childhood should contain are used to shape the ways in which young children are allowed to engage with digital devices. Running around and playing with friends is seen as favourable, and incompatible with ICT use. For this and similar reasons, parents, particularly middle-class ones, laid down ground rules for how much and how long their young children were allowed to use screens. In this context, children are perceived as 'victims' of omnipresent ICTs and digitalisation of everyday life, who need help, guidelines and rules of ICT usage.

Technologically deterministic discussions like this, mirror fears from earlier decades. In the 1950s, for example, TV was often understood as affecting children and family life in negative ways. TV consumption was thought to make children 'aggressive and irritable as a result of over-stimulating experiences, which leads to sleepless nights and tired days' (Uhls, 2015, n.p.). A similar moral panic evolved around video gaming in the 1980s (Karlsen, 2015). However, it seems that also video-gaming screen time falls into a binary scheme: Though not a focal point in our interviews, a family mentioned playing a dancing game on the *Nintendo Wii*, a rather interactive video console, and describe it as an explicitly social activity. In another family, cousins living in Sweden and the US respectively played video games together, while chatting on Skype on a second computer. This social gaming was approved and supported by the parents and stands in contrast to negative views on gaming as isolating (Eklund & Roman, 2019).

In this vein, 'good' use was social use and 'bad' use was engaging with the technology alone. Sometimes, parents want their children to use digital technology to watch videos or play games. The entertainment value and distractive nature of these technologies in these contexts were used as an active parenting technique, for moments when parents needed a break:

> Interviewer: You have to bribe a little with the tablet?
>
> Father: No, just distract, it is not a bribe because we do not say 'do this, then you get to do that' like it should not be a reward. It will be later [meaning when the child is older], but ... right now it's not a reward but now it's just ...
>
> Mother: ... now it's a babysitter. *all laugh* We are horrible parents *said as half-joke*. [...] We try to limit it as much as we can but sometimes we need to have it.
>
> Father: 55 years old, white-collar worker, university-educated.
>
> Mother, 43 years old, homemaker, university-educated.

This is a good example of how parents engage in the regulation of children's uses. Here, they practice both use and non-use when utilising tablets and phones as 'babysitters' for their children. When placing the child in front of an entertainment technology like this, to calm them down or to get some free time to cook dinner, parents in our data feel guilty, as they are contributing to what they perceive as bad screen time. This, they feel, translates to bad parenting. They obey to actual or imagined norms around which use patterns are appropriate for children. Nevertheless, in the management of everyday life with small children, using screens to occupy children was common, and balancing between valuable and unfavourable use practices a major part of digital parenting.

Actively Balancing Screen Time

Screen time in our data was often contrasted with family time, and many families had 'screen-free weeks' during particular periods, such as the summer holidays. Here, parents and children alike had to put their digital devices away in favour of other activities, as a means to 'reset' the family and experience shared time together. Another strategy to counteract the potentially family-eroding, 'bad' screen time entailed one family trying to turn screen time into a family event by turning to more traditional technology, the TV. The mother of three elaborates:

> Earlier we had, like, that everyone was doing their thing, meaning that the one sat down with one tablet and the other one with the other tablet, and someone was using the phone and another one was sitting in front of the computer so that it [the family] was much more spread

> out. With the TV it's a bit different, we gather a bit and watch together. [...] we call it 'social screen time'.
>
> Mother, 41 years old, white-collar worker, university-educated.

She goes on to explain that this 'social screen time' was allowed even when children had 'used up' their own screen time for the day. This confirms a double standard when it comes to distinguishing between different kinds of uses. Screen time does not have a negative connotation when it means that families spend time together. On the contrary, the family in the quote above recently bought a TV to invite another opportunity for spending time together into their homes. Thus, when it comes to screen time, binary thinking is prevalent: the (usually smaller) screen of the children's devices encourage individualised use, which in the context of intimate family work/family coherence becomes 'bad' use. While social use on (usually) bigger screens becomes favourable because it becomes part of intimate family work.

Concluding, we can say that not all screen use is 'bad' use. Indeed, it is 'good' when it is part of what we call doing intimate family work. This applies not only to family spending time together in one household but also in cross-household contexts when the children communicate, for example, with grandparents or cousins. In that case, it is highly encouraged to spend much time with the device, and children are praised for their ability to connect 'intuitively'. That is, that they grasp new technologies quickly and are not shy to use them. Family members engage in active and conscious interplays of use and non-use. They switch back and forth between those two modes, influenced by what they associate with the different technologies. In other words, it is not in the technology but the use that these normative values come to the fore. The same tablet used for playing games becomes a different technology when it is used for video-calling grandparents. Moreover, it mostly lies within parents' hands to decide which uses their children have access to.

Forced Disconnections

In this chapter, we have shown the contradictory nature of many parents' approaches to children's ICT uses. These uses are framed in terms of 'good' and 'bad', meaning that in different contexts the same device and its affordances can have opposing connotations. Generally, screen time is valued when it is understood to contribute to the family as a collective, and communicative sessions such as video chats are part of intimate family work. 'Bad' uses, on the other hand, are when children use devices alone for what is perceived as solely entertaining aspects, and when they do so without parental control.

But the intimate family work done here is not only toddlers spending digital time with relatives. Parents, as the guardians of screen time, is involved in active management of the children's—and oftentimes other family members'—use. This is not an easy process, for many parents, it is unclear 'how much is too much', and social norms or constructs of 'ideal childhoods' cause some of them to worry and feel guilty.

Considering the context of networked individualism and thus the individualised responsibility for one's own (digital) networks, attempts to control family members' ICT uses through screen time or other means, can be read as attempts to prevent erosions of family cohesion. In other words, while families in the 'old days' would often communicate as a unit, for example by having one landline or sending out a Christmas card in the name of the whole family, today family communication more often consists of individual messages, photos, video chats and the like, and attempts to control individual screen times might be seen as attempts to curb these individualising tendencies.

This happens through conscious negotiations of use and non-use. But importantly, non-use does not mean being passive, it is actively doing something in the sense that it is intimate family work to constantly negotiate and re-negotiate boundaries, to change habits, get accustomed to technologies, and the like.

The most critical aspect is the question of children's rights when screen times are negotiated above their heads. Disconnection becomes a forced disconnection when decisions of both use and non-use are made for children, without their input. In other words, it could be argued that 'screen time' regulations sometimes mean an unjust limitation to children's rights to maintain, or disengage from, their own social relationships online.

References

Aspers, P. (2006). *Markets in fashion: A phenomenological approach*. Routledge.

Blum-Ross, A., & Livingstone, S. (2018). The trouble with "Screen Time" rules. In G. Mascheroni, C. Ponte, & A. Jorge (Eds.), *Digital parenting: The challenges for families in the digital age* (pp. 179–187). Göteborg: Nordicom.

Buckingham, D. (2000). *After the death of childhood: Growing up in the age of electronic media*. Polity Press.

Di Leonardo, M. (1987). The female world of cards and holidays: Women, families, and the work of kinship. *Signs, 12*(3), 440–453.

Eklund, L., & Roman, S. (2019). Digital gaming and young people's friendships: A mixed methods study of time sse and gaming in school. *Young: Nordic Journal of Youth Research, 27*(1), 1–16.

Hamilton, K., Spinks, T., White, K. M., Kavanagh, D. J., & Walsh, A. M. (2016). A psychosocial analysis of parents' decisions for limiting their young child's screen time: An examination of attitudes, social norms and roles, and control perceptions. *British Journal of Health Psychology, 21*(2), 285–301. https://doi.org/10.1111/bjhp.12168.

Hayes, N. (2000). *Doing psychological research: Gathering and analysing data*. Open University Press.

Helsper, E. J., & Eynon, R. (2010). Digital natives: Where is the evidence? *British Educational Research Journal, 36*(3), 503–520.

Karlsen, F. (2015). Analyzing game controversies: A historical approach to moral panics and digital games. In *The dark side of game play: Controversial issues in playful environments* (pp. 15–32). Routledge.

Kothari, C. R. (2004). *Research methodology: Methods and techniques*. New Age International.

Little, S. (2018). Drawn in all directions: Heritage language families' use of technology. In G. Mascheroni, C. Ponte, & A. Jorge (Eds.), *Digital parenting: The challenges for families in the digital age*. Nordicom.

Livingstone, S. (2018, October 10). *Six myths about children in the digital age*. [Blog post]. https://blogs.lse.ac.uk/parenting4digitalfuture/2018/10/10/six-myths-about-children-in-the-digital-age/.

Livingstone, S., & Blum-Ross, A. (2018). Imagining the future through the lens of the digital. In *A networked self, and birth, life, death* (pp. 50–68). Routledge.

Mascheroni, G., Ponte, C., & Jorge, A. (Eds.). (2018). *Digital parenting: The challenges for families in the digital age*.

Morgan, D. H. J. (1996). *Family connections: An introduction to family studies*. Polity Press: Blackwell Publishers.

Sadowski, H., & Eklund, L. (2020). Creative forms of family intimacy: managing family bonds across distances. In J. Mikats, S. Kink-Hampersberger, & L. Oates-Indruchová (Eds.), *Creative families: Gender and technologies of everyday life*. forthcoming.

Schier, M., & Jurczyk, K. (2008). "Familie als Herstellungsleistung" in Zeiten der Entgrenzung. *Sozialwissenschaftlicher Fachinformationsdienst soFid*, (Familienforschung 2008/1), 9–18.

Turkle, S. (2016). *Reclaiming conversation: The power of talk in a digital age*. Penguin.

Uhls, Y. (2015). Parents, chill. Technology isn't destroying teens' brains. *UCLA Newsroom, 2*. http://newsroom.ucla.edu/stories/parents-chill-technology-isnt-destroying-teens-brains.

Helga Sadowski is a Postdoctoral researcher in Sociology, Örebro University, Sweden. She is currently working on how digital technologies are transforming family communication and their impact on sustainable work-life balances.

Lina Eklund is a Lecturer at the Department of Informatics and Media, Uppsala University, Sweden. Her work focuses on uses and practices of digital technologies in managing families, the impact of anonymity on digital sociality as well as the role of digital games in museums.

Chapter 6
Taking over the Home: Children's Mobile Media Play in Domestic Space

Will Balmford, Larissa Hjorth, and Ingrid Richardson

Introduction

This chapter explores scenarios of use concerning the play practices of children on mobile devices, and the subsequent domestic tensions that arise out of such practices. The home has a long history of being a key site for the study of new media technologies (Horst 2012; Morley, 2000, 2006; Silverstone & Haddon, 1996). In this dynamic media landscape, mobile media has increasingly featured in the contested and contradictory practices of the home (Morley, 2000). Mobile media constantly blur distinctions between public and private, work and leisure—and it is in the home that the tensions around this 'presence bleed' become most palpable (Gregg, 2011).

For many families, the renegotiation of the 'digital hearth' (Flynn, 2003, 2004) often creates tensions between generational expectations—such as unwanted noise, imbalanced usage, issues over sharing and damage to expensive technological equipment. As Froes and Tosca note in their discussion of children's use of mobile media, often a key part of media literacy involves what adults deem as playful subversion (2017). In this chapter we will draw on fieldwork from the Games of Being Mobile (GoBM) Australian Research Council (ARC) project—a 3-year, nationwide study of Australian mobile gaming practices—to unpack how these tensions manifest within the domestic environment.

In order to best examine this tension, the chapter will initially review key literature on the relation between domestic space and media practices. Such a review contextualises the history of media technology in the home and considers how households have been reshaped by, and with, media. A brief section will follow

W. Balmford · L. Hjorth · I. Richardson (✉)
RMIT University, Melbourne, Australia
e-mail: william.balmford@rmit.edu.au; larissa.hjorth@rmit.edu.au; ingrid.richardson@rmit.edu.au

D. Holloway et al. (eds.), *Young Children's Rights in a Digital World*, Children's Well-Being: Indicators and Research 23,
https://doi.org/10.1007/978-3-030-65916-5_6

this review, covering the key techniques of the GoBM project. The second part of the chapter, through two key case studies, details how domestic usage of mobile devices by younger children are challenging earlier models of media use in the home.

For this chapter we have chosen to define 'younger children' as those aged 8 and under, in line with the age categories explored elsewhere in this volume. In addition, the chosen fieldwork examples involve two scenarios of use—one child who owned their device, the other that didn't. In our research, in cases where children did not own devices, they instead borrowed parents' mobile phones or had access to family tablets and iPads. The negotiation of mobile media ownership reveals the uneven dynamics and variable agencies of adults' and children's media practices in the constitution the home. The final section will provide a summary of key insights from the two case studies, contextualise them in the wider GoBM findings and offer further provocations for research concerning the impact of children's mobile media usage within the domestic environment. The phenomena we explore in this chapter highlight the shifting needs and agencies of children in a digital world, a key focus of this volume. Furthermore, the various familial tensions are indicative of the difficult balancing act parents face between ensuring safety, and encouraging digital literacy, for their children.

Domestic Media Usage

In her 2002 book ***At Home With Computers***, Elaine Lally analyses the domestication of computers in Australia, discussing how from around the mid-1990s home computers became a common household item. However, she notes that among the Australian populace there was significant rejection of these items as leisure objects—instead they were largely seen as a technology used "only for work" (Lally, 2002, p. 61). Jock Given and Gerard Goggin expand on this evolution, arguing that the new millennium brought a proliferation of new ways to use and engage with the internet to include "blogs and wikis, broadband internet, social networking systems, social media [and] the rise of mobile media such as smartphones" (2012, p. 59). As mobile media have become ubiquitous, they are now a key part of the complex and contested terrain of domestic space. Through mobile media we can learn much about how these conflicts are informed by gendered and generational understandings of both media practice and the habitudes of home life.

Like the introduction of the radio and television into the home (Flynn, 2003; Silverstone & Hirsch, 1992), mobile media have expanded upon tensions first elucidated by Raymond William's notion of mobile privatisation (1974). Cultural studies scholars such as David Morley have identified that media, including mobile devices, have not only seen a further 'domestication of technology' but also a dislocation of domesticity (2006). Within the Australian context, media studies has explored the ways in which new media technologies in Australian homes can reshape behaviours, spaces and locations (Lally, 2002; Hollows, 2008).

Early media literature frequently highlighted users' concerns around how new media technologies (and greater access to the internet) might disrupt or change their everyday arrangements (Goggin, 2004). A key text in this area is *Consuming Technologies: Media and Information in Domestic Spaces* (Silverstone & Hirsch, 1992). Central to this book is an emphasis upon the nuanced relationships that take place in—and help to define—domesticity in the modern world (Silverstone & Hirsch, 1992). Barry Wellman and Caroline Haythornthwaite's *The Internet in Everyday Life* (2002) recounts several scenarios of home internet use, largely through desktop or laptop computers, with wired internet access demarcating particular sites within the home and defining them accordingly through terms such as 'home office' (2002, p. 484).

However, over the last decade, this designation of specific 'places' for internet use is increasingly negotiated, and 'dislocated' through mobile devices. For parents of younger children, how they display, store and share their mobile phone works to define their domestic spaces around media access. These boundary-making practices have implications for how the home is understood and the role that mobile media use plays in its reshaping.

A key concept within research examining domestic integration and engagement of media technologies is 'double articulation' (Silverstone & Hirsch, 1992). Double articulation refers to the ability of media technology to link the public and private spheres of life within the domestic environment—and the meanings ascribed onto the technology, and specific domains and temporal routines within the household—in the process (such as the home office, the 'play' computer or the shared family iPad) (Silverstone & Hirsch, 1992). The dominant discourse among media studies scholars engaging with domestication theory and double articulation argues that many media technologies exhibit some form of 'double articulation' and consequent generation of meaning in the home (Miller & Slater, 2000; Horst, 2012; Horst & Miller, 2012).

The first articulation focuses on the device in its material contexts and its functional significance—in the case of this chapter, such an articulation might be the type of mobile device and its instrumental use within the home. The second articulation concerns the symbolic significance of the object (Berker et al., 2005), and the role and rhythms of media technology in everyday life. In our analysis, this second articulation includes patterns and routines of use such as who has access to the device (and when), and how a device works as an affective conduit to the public sphere through online games, social media services and the like. Also important in the second articulation are the various communicative practices and pathways that are enabled by mobile devices. (Silverstone & Hirsch, 1992; Park et al., 2011).

As Sonia Livingstone has argued, children's media habits are becoming increasingly difficult to monitor as their main points of access become mobile devices that are capable of being used in unsupervised spaces, "particularly the bedroom but also the study, playroom, and kitchen, thereby spreading both spatially and temporally—from defined and prioritized spaces and times to casual use throughout the home and throughout the day" (2009, p. 156). In the case of mobile device usage by younger children, such interpretations (from both parent and child) manifest as a tension between conflicting modes of usage and parental desires to manage their child's media engagement. Sun Sun Lim has also done significant work into parental management of children's device usage (2016). Lim argues that the 'always on' nature of mobile media is having a significant impact upon communication between parents and children (2016). Lim sees this impact through the rise of 'transcendent parenting'—a move beyond traditional physical concepts to "incorporate virtual and online parenting and how these all intersect" (2016, p. 21). The notion of 'transcendent parenting' aligns well with research into children's online play and the complicated notions of 'where', 'who' and 'how' it is taking place (Aarsand, 2013).

A key issue noted by Aarsand and other scholars is seen in the 'who' of play and concerns the gender of the player (regardless of their age) (Aarsand, 2013; Kafai et al., 2008). Mazzarella notes that game playing on digital devices such as mobiles has been considered one of the 'primary differences in girls' and boy's use' (of said technology) (2013). This conception has meant that as mobile media technology became appropriated into household throughout the 2010s, the nuances of such appropriation has become a topic of frequent research (Berker et al., 2005). A useful example is Heather Horst's *New Media Technologies in Everyday Life* (2012). The data from this research comes from Horst's ethnographic exploration of the role of computer technologies in Silicon Valley homes in San Francisco. Horst explores how these technologies were brought into the home and adapted in differing ways across households, from families setting up laptops in the kitchen where they could be supervised, to having separate computers for 'work' and 'play' (2012).

However, as more and more technological functions become associated with mobile devices, predominant modes of media engagement require reassessment. For such a task, ethnographic research, similar to Horst's (2012) work, offers the best means of collecting data about lived domestic media practices. Ethnography is a key method in media studies, particularly for those seeking to explore the effects of digital media in everyday life contexts (Pink et al., 2015), and more recently has been an effective way of gathering experiential data on networked spaces while also recognising the entangled presence of offline practices and social contexts. By deploying such techniques within the home, our research sought to unpack how mobile media impacts upon home life; in this chapter, our focus is on the role of mobile media in reshaping the everyday configurations and rhythms of domestic space in Australian families with younger children.

Data Collection

Fieldwork was conducted across five Australian cities (Brisbane, Perth, Melbourne, Adelaide and Sydney), with 12 initial households recruited within each city. Further households were recruited as research trends developed. Research was conducted for a period of 3 years, with each household engaged with on three occasions. The initial engagement consisted of an informal interview to establish the household's history of mobile media usage and mobile gaming. Subsequent meetings were adapted

Table 6.1 GoBM participants discussed in this chapter

Family	Members (gender, age)	City
Silk	Lynn (female, 40) Jayden (male, 45) Arthur (male, 8)	Melbourne
Burton	Emily (female, 28) Howard (male, 31) Anna (female, 6)	Brisbane

around the particularities of a household to include play sessions with participants and observations of daily household life through 'home tours' and other exercises. These 'play sessions' consisted of playing videogames with (and chosen by) participants. The sessions proved to be highly effective in revealing domestic practices of use and how participants configured their homes through and around play. 'Home tours' involved participants walking us through their homes, highlighting media conduits and sites of use, such as where mobile phones were placed to charge, or where iPads might be stored overnight.

The scenarios described in this chapter comprise a small subset of the data produced through the GoBM project, based primarily on the experiences of two families with young children, in order to engage in a deep analysis of the relationship between mobile device usage and domestic configurations in the context of young children's use and parental supervision (see Table 6.1).[1]

Arthur and the Tension of Volume

Arthur, aged 8, lives in a Melbourne apartment with his parents Jayden and Lynn. Arthur does not own his own mobile device but has access to the family's ageing Android tablet. The tablet is housed in a bright blue plastic cover, with a plastic screen protector over the screen. Arthur is restricted both in how and where he uses the device. He is usually only allowed to play in the living room, after school and before dinner (although certain concessions have been made for occasions like long car trips). When playing, he has a preferred position on the corner of the couch, laying the tablet flat along the couch's arm, resting his chin in one hand while tapping and swiping the screen with the other.

In addition to these place-based restrictions, Arthur also finds his device usage negotiated through another parameter—sound. Arthur is a fan of several prominent YouTube personalities such as *SkyDoesMinecraft* and *Captain Sparklez*, whose videos he likes to watch. Arthur also often plays games such as Fruit Ninja and Minecraft on the family tablet. During an interview, Arthur commented that he liked to play with the sound on, but this is often not allowed. When questioned, he elaborated that: "I play with sound sometimes, but usually dad says, 'put the

[1]All names have been changed to ensure anonymity.

sound down'". He went on to detail how he is told the sound is disruptive to other members of the family using the common space and might bother others in the neighbouring apartments. Arthur is also not allowed to use headphones, as Jayden and Lynn found that Arthur tended to be unresponsive to their verbal requests—such as calls for dinner—when using them.

These aural tensions coincide with perceptions of the family home and being a good neighbour. Sophie Kerr and her co-researchers have explored how family noise levels can come into conflict with others in apartment living, arguing that "households with children are on the front line of such tensions" (Kerr et al., 2018, p. 1). They suggest that "the materiality of sound and built form interact... [influencing how] spaces are understood and inhabited" (Kerr et al., 2018, p. 1). Such scholarship aptly describes Arthur, Lynn and Jayden's family apartment living, where noise and its location within the home is carefully negotiated.

The impact of mobile device sounds—from ringtones, music and conversations, to the beeps of texts and notifications, and fluctuating noises of mobile game sound effects and soundtracks—upon mobile media usage and modes of engagement in the urban environment has been discussed at some length (see Bull, 2007). To date, however, there is scant research on the effects of mobile media sound upon household practices and how domestic spaces are consequently negotiated and renegotiated. Several scholars have discussed the physical placement of new media devices more generally within the home, in terms of their impact upon the management and navigation of domestic space (Hollows, 2008; Green, 2010; Horst, 2012). Yet in the case of Arthur's tablet usage, the restricted placement of the device has reciprocal effects on the mode of engagement, and his perceptual or sensory experience of mobile media content and app-based affordances. As a brief example, Arthur was unable to watch a *Captain Sparklez* video one night as his parents were trying to read (and Arthur was unable to read the subtitles attached to the video).

Media interfaces and network devices within the home typically 'reshape' domestic space; homes are now spatially configured by dedicated media spaces such as computer desks, console hubs, and wireless router requirements. These technologies and the spaces they inhabit dynamically shift the organisation and soundscape of the home, as they become embedded in the cadence of everyday family life (Green et al., 2004; Horst, 2012). In their apartment, Lynn and Jayden sought to maintain a quieter atmosphere, both out of respect for their neighbours and their own enjoyment of their domestic space. Jayden in particular found the noises of Arthur's games jarring and random, especially without the visual context to predict and contextualise the audio component. In response, Arthur would at times strategically challenge the sonorous boundaries imposed by his parents, seeking to test, bend or break them.

Other media use also impacted on Arthur's ability to watch videos on the tablet, such as watching television, which requires volume to be engaged with, or listening to vinyl records, one of Jayden's weekend hobbies. On these occasions Arthur would not be allowed to watch videos with the sound. Instead he could watch them without the sound (a less than favourable activity, as discussed above), or play games with the sound off. To this extent Arthur's engagement with mobile devices on the weekend was more limited by sound than time. Lynn explained that on weekends "(he) can pretty much play as much as he likes... Until we get annoyed with it". Jayden quickly added that this play was done with the sound off. In this way mobile games afforded more flexible patterns of use within the home as they were more readily adaptable to the needs and desires of others within the space.

Such interactions are becoming increasingly prevalent in major Australian cities such as Melbourne that are currently experiencing rapid urban densification. With families such as Arthur's now living in closer proximity to others, the travelling noise of devices has become a much larger concern. Mobile media in particular were an issue for several of the GoBM participant families, due to both the wide range of noises, pitches and cadences emitting from devices, and because this noise could be transported by occupants as they moved throughout the home (as opposed to the strategic placement of the television to limit the noise pollution to neighbours and cohabitants). In addition, the private screen of the mobile tablet means there is less opportunity for visual context, an issue Arthur's father found particularly discomfiting.

Jayden and Lynn's solution involved spatially and aurally restricting Arthur's use of the mobile device to provide a more harmonious and peaceful living environment for both themselves and their neighbours. An important affordance here is the ability of mobile games to be played without sound. Here we can see how the multiple affordances of mobile media have varying effects on household practices and spatial organisation—from the offending intrusion of sound and the consequent volume restrictions, to the ease of place-based restrictions and physical removal (positive affordances for Jayden and Lynn), and the ability to engage with media content without sound, such as games, so as to extend usage time (a positive affordance for Arthur).

The next example builds on this discussion concerning the impact of mobile device usage upon domestic space and family life, turning to another significant aspect of these interactions: the negotiation of 'sharing' mobile devices between parents and their children, and the careful temporal and spatial negotiations that ensue.

Anna and the Tension of Sharing

Howard Burton, his wife Emily and their 6-year-old daughter Anna live in Brisbane's outer suburbs. In their three-bedroom house, the family has plenty of space including an office for Howard, who often works from home in order to look after Anna when she is not at kindergarten. In Howard's office is a desktop computer, which Anna is not allowed to use. However, Anna does often borrow her father's iPhone to play games such as Minecraft, Crossy Road and Temple Run. During our initial interview with the family Anna particularly enjoyed playing Minecraft, which her father had downloaded on his phone for him and his daughter to play. Howard occasionally played without his daughter, though admitted he was mainly interested in co-playing with her, as he felt not only that they were able to spend 'quality time together' but that he was able to teach his daughter 'good mobile habits' such as limiting hours of usage per day, posture and online etiquette.

This shared usage has occasionally been problematic for the family. Howard, also a videogame player, enjoys seeing his daughter take an interest in a hobby they can enjoy together. He explained how sometimes the two would play together, with Howard helping Emily build structures in the game of Minecraft. Such co-present and co-located parent/child play is a common practice in domestic mobile game usage (Balmford & Davies, 2019). Anna was also allowed to play on her own, most often over the weekends. However, these boundaries and co-playing rules became problematic when Anna sought to play at inappropriate times. As Howard explained: "We wake up in the morning and she asks if it is time to play [Minecraft on the phone]. I do try to restrict her since that is my device for using Twitter or the internet or whatever".

Howard's sense of frustration—concerning both Anna's constant eagerness to play and the restrictions sharing imposed on his own usage—was expressed by many GoBM participant parents. This frustration is particularly felt during times such as weekend morning sleep-ins, when Howard would much prefer to use his device at his own leisure, to check Twitter or browse the internet before rising. To lessen the issue of sharing, Howard and Emily have established several parameters around when Anna can use the mobile alone (only on weekend afternoons), and where the device is stored (it 'sleeps' on the bedside table in Howard and Emily's bedroom).

Recent media studies researchers such as Jason Farman (2015), Deborah Chambers (2016) and Christian Licoppe and Yoriko Inada (2016), have explored the impact of mobile games on our experience of, and in relation to, place. Such scholarly focus calls attention to how specific locations become associated with or against game usage. Chambers' analysis of how relationships form in media rich homes echo Licoppe and Inada's work into the ability of mobile games to inform "a particular form of urban space through the entanglement of variously mediated experiences of place" (2016, p. 281). Within the Burton household, such entanglements of experience are critical to how Anna engages with games. As mentioned above, Howard has a home office where he works. While he is in there working, the family have established a rule that Anna cannot borrow his phone (under the justification that he needs it to make and receive phone calls). His office is cast within the home as a place of work that the parents have attempted to establish as separate from the otherwise entangled media usage of the home. However, the complete spatial separation of work and leisure is not always successful; Howard admitted that he often checks Twitter on his phone when "he should be working" and on occasion Anna had convinced him to play Minecraft within his office.

Alison Harvey's *Gender, Age, and Digital Games in the Domestic Context* (2015) discusses how mobile games complicate the boundaries between work and leisure—similar to Howard's experiences with his home office becoming a site of play for his young daughter. Harvey argues that regulations of digital play in domestic settings are conceptualised through both common shared spaces such as living or 'rumpus rooms' and more sequestered or private areas, such as Howard's office, becoming private places of gameplay within the domestic sphere (2015). For Emily and Howard, playing games is an important leisure activity for both them and their daughter. However, the pursuit of such leisure time is balanced against a myriad of factors including their own levels of enjoyment, the desires of their daughter and what Harvey identifies as structural factors; the time and location of play within the home. As Chambers argues "the growing mobility of media devices is making it difficult for parents to monitor their children's media uses" (2016, p. 81). Chambers also discusses the issues of sharing space and time in the home in relation to media, arguing that the growing presence of media interfaces (both mobile and stationary) is having a significant impact on how children acquire media literacy (2016).

In the Burton family's case, monitoring Anna's mobile media use became difficult when Howard felt 'compelled' to give his daughter his phone. When we asked Howard how often this occurs, he estimated that he acquiesced to her "around twice a day on average", often to keep her from distracting him while he was working. During these times Howard felt unable to monitor his daughter's growing media literacy, as distinct from the benefit he perceived when they were able to co-presently talk, play and learn together on the phone.

The ways in which Howard and his daughter 'share' Howard's phone provide insight into how mobile devices influence domestic configurations and reveals the double articulation of mobile devices. Mobile phones are technologies used for both work and leisure, and their meaning and functionality shifts according to each user's interpretation and use. The device that Howard uses on the couch to play with his daughter is the same device that must be sequestered and out-of-bounds in his office to avoid her constant requests. Yet the mobile device remains an important tool for Howard in assisting and guiding his daughter's digital literacy, and their shared negotiation of it within the home is a key part of such an education. In this scenario, the parents' management of the device, interwoven with the negotiation of work and play within the home, often revolved around employing spatial and temporal restrictions, and ensuring clearly demarcated boundaries around where and when Anna could play.

Conclusion

This chapter has explored the relation between mobile media practices and domestic configurations among families with younger children. It has drawn attention to how these relationships can cause tension within the domestic environment, and the way the various affordances of mobile media devices are deployed to negotiate these tensions. Using two detailed scenarios of use from GoBM fieldwork, we have highlighted how mobile media practices become sites for intergenerational tensions around engagement, expectation and practices.

The two case studies discussed in this chapter show how mobile gaming presents a convergence of locative, social, and mobile media, an echoing of the overall findings of the GoBM project. The project offers a lens through which we can see the changing nature of mobility, play, and communication as it moves through broader sociocultural, technological, and economic dynamics. Within the various assemblages of mobile gaming, we see new types of intimacy and communicative practice emerging around gameplay and new ways in which co-present forms of play and place-making are enacted. Our findings here focus on domestic configuration and communication, with mobile games affording both ambient and located social connection, and counteracting the emotional and physical isolation experienced by many people—young and old—in playful and innovative ways.

Despite the emphasis placed on the capacity of mobile media to be carried on the body as it moves through urban space, the intimate portability of mobile media results in particular forms of immobility in the home—especially in terms of intergenerational usage and understandings of etiquette. For example, when the noisy iPad becomes a source of frustration and annoyance for parents, young children's usage is consequently restricted to particular times and places, and the kind of content or mobile application (e.g. YouTube video or gameplay) is also delimited. So too, issues around sharing devices highlight different generational understandings of ownership and privacy. Indeed, mobile media become vehicles for understanding the complex ways in which household power relations and dynamics play out.

The scenarios discussed in this chapter highlight the need for ongoing research into the changing domestic rhythms of Australian homes as they are increasingly inundated by mobile media. Emerging technologies such as smart assistants will continue to inform and transform how younger children co-opt media into their everyday domestic routines, play practices and pathways of communication. Longitudinal studies that capture broad shifts in family device ownership over time may also effectively highlight trends in children's individual and shared usage and consequent impact on the organisation of the home. With a particular focus on Australian family life, this chapter contributes to the nexus of mobile communication and childhood studies to better understand how the experience, management and negotiation of early childhood coalesces with mobile media practices in complex and ever-evolving ways.

References

Aarsand, P. (2013). Children's digital gaming cultures. In D. Lemish (Ed.), *The Routledge international handbook of children, adolescents and media*. Routledge. https://doi.org/10.4324/9780203366981.

Balmford, W., & Davies, H. (2019). Mobile minecraft: Negotiated space and perceptions of play in Australian families. *Mobile Media & Communication*. https://doi.org/10.1177/2050157918819614.

Berker, T., Hartmann, M., Punie, Y., & Ward, K. (2005). *Domestication of media and technology*. Open University Press.

Bull, M. (2007). *Sound moves: iPod culture and urban experience*. Routledge.

Chambers, D. (2016). *Changing media, homes and households: Cultures, technologies and meanings*. Routledge.
Farman, J. (Ed.). (2015). *Foundations of mobile media studies: Essential texts on the formation of a field*. Routledge.
Flynn, B. (2003). Geography of the digital hearth. *Information Communication & Society, 6*(4), 551–576.
Flynn, B. (2004). Games as inhabited spaces. *Media International Australia Incorporating Culture and Policy, 110*(1), 52–61.
Froes, I., & Tosca, S. (2017). Playful subversions. *European Journal of Cultural Studies, 21*(1), 39–58.
Given, J., & Goggin, G. (2012). Australian internet histories: It's time [online]. *Media International Australia: Incorporating Culture & Policy, 143*(1), 57–62.
Goggin, G. (2004). *Virtual nation: The internet in Australia*. UNSW Press.
Green, L. (2010). *The Internet: An introduction to new media*. Berg Publishing.
Green, L., Holloway, D., & Quin, R. (2004). @Home: Australian family life and the internet. In G. Goggin (Ed.), *Virtual nation: The internet in Australia* (pp. 272–300). UNSW Press.
Gregg, M. (2011). *Work's intimacy*. Polity.
Harvey, A. (2015). *Gender, age, and digital games in the domestic context*. Routledge.
Hollows, J. (2008). *Domestic cultures*. McGraw-Hill Education.
Horst, H. (2012). New media technologies in everyday life. In H. Horst & D. Miller (Eds.), *Digital anthropology* (pp. 66–82). Berg Publications.
Horst, H., & Miller, D. (2012). *Digital anthropology*. Berg Publications.
Kafai, Y. B., Heeter, C., Denner, J., & Sun, J. Y. (2008). *Beyond Barbie and Mortal Kombat: New perspectives on gender and gaming*. MIT Press.
Kerr, S., Gibson, C., & Klocker, N. (2018). Parenting and neighbouring in the consolidating city: The emotional geographies of sound in apartments. *Emotion, Space and Society, 26*(1), 1–8.
Lally, E. (2002). *At home with computers*. Berg Publishers.
Licoppe, C., & Inada, Y. (2016). Mobility and sociality in proximity-sensitive digital urban ecologies: 'Timid encounters' and 'seam-sensitive walks'. *Mobilities, 11*(2), 264–283.
Lim, S. S. (2016). Through the tablet glass: Transcendent parenting in an era of mobile media and cloud computing. *Journal of Children and Media, 10*(1), 21–29. https://doi.org/10.1080/17482798.2015.1121896.
Livingstone, S. (2009). Half a century of television in the lives of our children. *The ANNALS of the American Academy of Political and Social Science, 625*(1), 151–163.
Mazzarella, S. R. (2013). Media and gender identities: Learning and performing femininity and masculinity. In D. Lemish (Ed.), *The Routledge international handbook of children, adolescents and media*. Routledge. https://doi.org/10.4324/9780203366981.
Miller, D., & Slater, D. (2000). *The internet: An ethnographic approach*. Berg Publications.
Morley, D. (2000). *Home territories: Media, mobility and identity*. Routledge.
Morley, D. (2006). 'What's 'home' got to do with it?': Contradictory dynamics in the domestication of technology and the dislocation of domesticity. In T. Berker et al. (Eds.), *Domestication of media and technology* (pp. 21–39). Open University Press.
Park, D., Jankowski, N., & Jones, S. (2011). *The long history of new media: Technology, historiography, and contextualizing newness*. Digital formations, Peter Lang.
Pink, S., Horst, H., Postill, J., Hjorth, L., Lewis, T., & Tacchi, J. (2015). *Digital ethnography: Principles and practice*. Sage.
Silverstone, R., & Hirsch, E. (1992). *Consuming technologies: Media and information in domestic spaces*. Routledge.
Silverstone, R., & Haddon, L. (1996). Design and the domestication of information and communication technologies: Technical change and everyday life. In R. Mansell & R. Silverstone (Eds.) *Communication by Design: The Politics of Information and Communication Technologies* (pp. 44-74). Oxford University Press.
Wellman, B., & Haythornthwaite, C. (2002). *The internet in everyday life*. Wiley-Blackwell.
Williams, R. (1974). *Television: Technology and cultural form*. Fontana.

Will Balmford lectures within RMIT's Bachelor of Design (Games) program. His current research involves exploring the impacts of domestic videogame usage. His work examines the everyday use of new media through innovative ethnographic methods and an interdisciplinary approach to collaboration.

Larissa Hjorth is an artist and digital ethnographer. She has two decades experience working in cross-cultural, interdisciplinary, collaborative creative practice and socially innovative digital media research to explore intergenerational literacies around play and sociality. She is a Distinguished Professor, and is currently the Design & Creative Practice ECP Platform Director at RMIT University. She is experienced in alternative ethnographic methods and also knowledge exchange.

Ingrid Richardson is Professor of Digital Media at RMIT University. She has a broad interest in the 'human-technology relation', and has published on topics such as scientific technovision, games, mobile media, remix culture and web-based content creation and distribution. Recent books include Ambient Play (MIT Press, 2020) and Mobile Media and the Urban Night (Palgrave, forthcoming).

Chapter 7
Think of the Parents: Parental Controls in Digital TV and Family Implications

Maureen Mauk

Introduction

Media regulations have traditionally been framed as '*for the children*' with the expressed intent to protect society's most vulnerable, yet often fail to consider the needs of the parents playing the role of familial gatekeeper, contending with exigent public scripts on kids' media consumption. As we consider the design and practice of children's participation in the digital space, it is also necessary to take into account the implied roles and responsibilities of families as they navigate their children's interactions with media content.

The surge in digital children's content evidenced by the continued investment in new media platforms and internet distributed television from *Sesame Street* moving to the HBO app, to the 2019 launch of Disney Plus, indicates, from a media research perspective, a dominant objective within the industry towards capitalisation of the child consumer as a commodity (Lobato, 2019, p. 17; Cunningham & Craig, 2019, p. 27; Lotz, 2017, p. 2). With children (and their parents) positioned as subjects within this targeted market, research analysing the discursive practices being produced by the industry through its interface will better help conceptualise how the production of children's entertainment in the digital space is affecting families. Utilising discursive interface analysis (Stanfill, 2015), this case study couples Amy Jordan's (2016) concept of 'psychological space' with the norms and parental expectations of content moderation to interrogate Netflix's platform affordances against its claims of easing parent/subscriber burdens with its 'informed viewing' tools (Netflix Media Center, 2018) to examine the cultural implications of parental controls in this largely unregulated space.

M. Mauk (✉)
University of Wisconsin-Madison, Madison, WI, USA
e-mail: mtmauk@wisc.edu

D. Holloway et al. (eds.), *Young Children's Rights in a Digital World*, Children's Well-Being: Indicators and Research 23,
https://doi.org/10.1007/978-3-030-65916-5_7

This chapter considers the normative principles and associated values of public interest against the backdrop of Netflix's interface to examine the affordances built or denied by Netflix in its presentation of parental controls and maturity ratings. My research works to begin reframing the questions of content creation and moderation within platformised entertainment (Cohn, 2019; Gillespie, 2018) to understand how the digital TV environment is shaping, augmenting or contributing to parental responsibilities. Is the platform utilising algorithmic technology towards helping families? Are digital platforms like Netflix utilising algorithms and affordances to create utility in parental controls and maturity ratings? My findings indicate that the allure of customisation (Andrejevic, 2007) and mythology of the algorithm (Bucher, 2018) generated by contemporary media discourse belies the actual utility of the platform's parental controls. While Livingstone and Blum-Ross (2017) have sensibly encouraged parents to tailor their media environments at home to suit their children's individual needs and support their family values, I argue that some of the onus should be relegated away from parents and towards the media providers to better support family needs and preferences. I conclude by offering a paradigm shift to the way we approach platform governance, thinking about content moderation beyond government standards, recognising that current digital media culture and policies ask already anxious and overwhelmed parents to "transcend" (Lim, 2016, p. 21) beyond the mystique of the algorithm, policing content themselves in order to reconcile the disparities between the gluttony of unchecked entertainment content, faulty algorithmic recommendation systems aimed at our youngest consumers, and a lack of regulatory oversight. In questioning the promise of the algorithm and emphasising the problematic standards and infrastructure of parental controls—parents, scholars and regulatory watchdogs have the ability to push companies like Netflix to offer its subscribers far greater transparency and utility to customise the platter that their kids' media is served upon, creating new allowances in time, mental space and sanity 'for the parents.'

Literature Review

Parents and Protocol

In order to consider the affordances of Netflix's parental controls in the digital space, we first have to understand the foundations of U.S. media regulation and its influence on family culture. American media policy roots are grounded in mores where children are considered "vulnerable subjects" (Sefton-Green, 2006, p. 282). Jordan (2008) discusses the 'unenviable' position government media policy makers walk between "the best interests of a capitalist, speech-protected society and the best interests of the vulnerable, developing child" (p. 236). U.S. Federal Communications Commission (FCC) policy changes often arise following new executive directives after election years, coupled with pressure from parent advocacy groups, as was the case in the 1990s when the FCC moved to establish the TV parental ratings guidelines for all television broadcasters to rate individual show content based not only on age or maturity levels but also with descriptors to point out innuendo, sexual situations, mature language and violence (Heins, 2007). The U.S. Telecommunications Act of 1996 established a TV ratings system for all sets manufactured after 2000 to contain the V-chip, giving parents the physical means to filter problematic content according to their personal family standards using passcode protections. The implementation of labeling content with V-chip ratings codes (i.e. TV-PG) was, and is, however, considered voluntary, allowing TV networks to loosely work together to design and self-regulate its means of labeling shows for age and content. Decades later, with television carrying over to a streaming environment, digital platforms distributing content carry no legal obligation to utilise V-chip labels as their content does not fall under the purview of the FCC when it comes to content regulation (FCC Act 551 (e)(1), 1996). To meet the historically-based cultural expectations of their subscribers, and in some instances, sponsors, platforms such as Netflix, Hulu, and Disney+ *do* offer some levels of parental filtering tools while also assigning maturity ratings to content in the same vein as V-chip ratings, presumably to offer guideposts and consistency to viewers.

Given the power generated through algorithmic classification (Cheney-Lippold, 2017), protocols (Lotz, 2017), and curation (van Dijck, et al., 2018) within the platformised entertainment ecosystem, small markers such as maturity ratings have the potential to be useful guideposts for parents working to organise their children's media offerings. In an algorithmic environment, "those without access to the database are left with the 'poor person's' strategies for cutting through the clutter: gut instinct, affective response, and 'thin slicing' (making a snap decision based on a tiny fraction of the evidence)" (Andrejevic, 2013, p. 17). The paratextual meanings (Gray, 2010) created by maturity ratings allow parents to manage information overload using these "visceral shortcuts" (Andrejevic, 2013, p. 5). V-chip or maturity ratings have become an aid to navigation utilised as a beacon for parents steering through a sea of television content.

There exists an opportunity to better connect research on the platformised entertainment environment to family and children's media scholarship. Past work on family television viewing can be utilised to begin to thread the conversations regarding algorithmic affordances and platformisation with family media culture to conceptualise the emotional and technological negotiation and guilt that parents contend with in their role as watchdogs for their children's media consumption. Hoover, Clark, and Alters explain how parents' sense of identity as a family is often organised around the ways they negotiate media through a 'subtle script' wherein "how we deal with media defines how we are as parents and family" (2012, p. 5). Public scripts on media effects have led parents to experience "a heavy sense of accountability" and a feeling that it was "ultimately their isolated duty to restrict their kids' media" (Hoover et al., 2012, p. 76). These tensions have only intensified with the abundance of digital content and 'infoglut' of the digital marketplace (Andrejevic, 2013). This recent scholarship has begun to bridge a better understanding children's online practices, parental mediation, and screen time rules.

In negotiating the role of content regulation, both historically and presently, parents balancing these duties have cited feelings of guilt. 'Good parenting' does not include time in front of a TV or computer in the public scripts of parenting. Blum-Ross and Livingstone (2018) note, "time and again we heard parents of young children struggle to balance the convenience of screen time with their worries about being a 'good' parent. Sentiments of guilt about feeling 'lazy' in 'letting' children consume television content were a frequent narrative in their research" (p. 183). Public scripts of 'responsibilisation' have emerged within the neoliberal consumer practices encouraged by the media that forgive the minimal industry oversight by coupling children's media oversight to the duties of "good parents" (Willett, 2015, p. 51).

Black Box Mythology and Algorithmic Culture

Social discourse has placed exceptional emphasis on technology to help navigate more technology, with its power often mythologised by the mystery and legend of the algorithm. In a "risk-filled and time-poor society" prevalent parenting technology has been introduced and adopted to counter problems of information overload and media saturation (Clark, 2013, p. 210). But as families download and adopt shared calendar apps, child tracking devices, and Netflix's PIN codes on individualised programming, they are also acquiescing to the schemes of capitalism supporting an increasingly commercialised consumption culture which creates a paradigm where parents are again working to find the balance between children's empowerment and protection—buying into the promises of smart tech in order to avoid perceived pitfalls of media technology. 'Algorithmic culture' (Striphas, 2012) which has evolved to include societal reverence at the altar of this 'black box' lends much of its mythology to the 'datafication' of viewing practices (Van Dijck et al., 2018, p. 4). Value is extracted by media corporations via data surrounding the viewing practices of adults and children alike, with children implicated as they are measured not as a generation of consumers or learners but as 'digital data assemblages' (Lupton &Williamson, 2017, p. 4). In our highly datafied, media-rich environment, the personalisation and parental controls provided by opaque algorithms within the black box have become uncontested conditions of family entertainment viewing which "alleges to be a mirror of our interests and desires" (Finn, 2017, p. 96). My research refracts the promise of the algorithm as it relates to Netflix's parental controls to consider its efficacy in serving as a reflection of parents' needs.

Netflix and Families

While substantial scholarly discourse surrounding Netflix and Hulu and the societal ripples of their primetime serials is available, researchers are only beginning to tackle the influx and development of high-quality kids' television content and its cultural implications. Netflix's Ted Sarandos announced in 2016 that they were "doubling down on kids and families" (Flint, 2016). An expanding corporation, Netflix has not shied away from feeding into the demand for children's programming. "We have a ton of data on what they watch, so we know what properties resonate and when they don't—when (viewers) have had enough of a franchise and when they can't get enough of it," said Andy Yeatman, Netflix's Director of Global Kids Content. "We always try to feed that demand so they always want more" (Harwell, 2016). Netflix claims its algorithmic technology helps its members be "better informed, and more in control, of what they and their families choose to watch and enjoy on Netflix" (Hastings, 2018). As users "watch and enjoy" Netflix, the company also enjoys the mining and watching of its users' viewing habits and preferences, all being tracked within Netflix's 'digital enclosure' (Hastings, 2018; Andrejevic, 2002). In addition to AI data mining, human content moderators also work to screen, tabulate, and organise Netflix's Kids content. Netflix Kids Content Taggers are employed to tag and classify content to account and measure for various themes and cultural sensitivities which helps the company better categorise its own content and make recommendations based on its various themes, tone, storyline and characters to harmonise with viewer preferences (Lead Kids Content Tagger, Netflix, n.d.; Grothaus, 2018). Ideally, one might imagine that the themes tagged and accounted for in the children's content would give parents more guidance and options when setting up parental controls—presenting a digitally enhanced and human-moderated interface to allow for easy curation and mediation of children's content, however, my analysis shows that is not the case.

There are few safeguards in terms of government regulations or platform governance creating any sort of standard for the ways in which parental controls should work in the digital sphere. Netflix is a U.S. based company focused on international growth and while its content is not subject to broadcast regulations, recent inquiries have been made by advocacy groups to discern the role of digital platforms in media regulation, including V-chip rating accuracy and the efficacy of the entire system. Heavy lobbying by the Parents Television Council (PTC) resulted in the 2019 directive by Congress to the FCC to review the U.S. television ratings system on the "extent to which the rating system matches the video content" and the ability "to address public concerns" (FCC Docket 19-41). Amidst confusion on how to govern social media for children and concerns on digital privacy for kids as a whole, organised oversight is beginning to percolate to hold the digital industry more accountable. As Schofield Clark notes, "If media industries were regulated and held accountable for their role in *fostering the public good* rather than regulated and held accountable for their ability to maximise profits for corporate shareholders, we might have a different media landscape" (2013, p. 225). Global platforms need not wait for stakeholders or regulators to fill the void and inflict regulation upon digital content to provide parents with some relief.

Case Study: Netflix's Parental Controls Interface as Discourse

This case study aims to identify the way that family audiences perceive and navigate through Netflix's parental controls and algorithmic affordances. The broad inferences made towards parental norms of use and familial applications of digital media are by no means exhaustive but situated within family and media researchers' theoretical framework to further the exploration and scholarly probes into how parents may labour through Netflix's operating model and options. Netflix's present parental controls allow for individualised PIN code settings as well as more prominent displays of their version of V-chip ratings—'maturity ratings'—over their content (Hastings, 2018). The maturity ratings are used as textual frames that a parent can take into consideration as they approach mediation of their children's content. First, I broadly consider the interface norms one might expect to be offered by a media behemoth such as Netflix with a vast expanse of family viewership and then analyse what actually is offered via classification and affordances. In an era where Netflix and its algorithms not only tell us, but also our children, what to watch, we have to consider how it classifies its material and how we access it. "Roughly half of our 65 million members around the world regularly watch kids content and with the addition of these seven diverse original series we are continuing our strong investment in kids programming," notes Netflix Director of Global Kids Content, Andy Yeatman (Spangler, 2015). With the tens of millions of members, presumably a fair portion of them children, tuning into kids' content, platform tools which offer the ability to customise and curate the SVOD would seem like an acceptable offering for subscribers interested in mediating their child's viewing. On affordances, Guins writes, "Not long ago it would have been an absolutely absurd action to purchase a television or acquire computer software to intentionally disable its capabilities, whereas today's media technology is marketed for what it *does not contain* and *will not deliver.* Control is the ability to disable" (2009, p. 171). My methodology analyses the tools of control offered by Netflix utilising Mel Stanfill's analytic method for examining a website using discursive interface analysis considers the assumptions built into an interface and allows us to view it "as both reflecting social logics and non-deterministically reinforcing them" (2015, p. 1060). The productive power approach allows us to review Netflix's interface as it relates to its latest parental controls to "illuminate the norms of use" (p. 1061). Employing discursive interface analysis, I approach Netflix's functional affordances, its allowances in terms of legibility and placement, the assumptions on the users, and its classifications and explanations to better situate the ideologies undergirding the platform's construction while also research how those affordances reinforce specific actions and beliefs through the interface functionality (2015, 1071).

When a Netflix subscriber in the U.S. seeks to utilise parental control tools, they will likely find such features fairly buried. The tools are certainly not obvious upon log in or set up but searching the 'Help Center' leads users to the 'Profile-level (soft)' and 'Account-level (hard)' parental control options. After searching for 'parental controls' in the Help Center pull-down menu, which is only accessible via logging in to Netflix on a home computer or laptop, and not accessible via a search via smart TV, a Help Center article explains its levels of parental control. On the 'soft' profile level, a parent can create a user profile for a child and set a maturity level, choosing between four options ranging from 'for Little Kids only' to 'All Maturity levels.' There are no gateways, however, that prevent a child from simply selecting their parent's profile if they wanted to view content outside of the maturity level assigned to their profile. The instructions explicitly state, "You, the parent or guardian, ensure your child uses only the correct profile" (How do I set parental controls, 2019). From the perspective of children's rights as agential citizens within a digital world and parents' empowerment and protection strategies, Netflix's entire strategy relies on parental supervision. This is where the algorithm fails. We know they have content tags from Kids Content Taggers but are not harnessing this data to put parents in the drivers' seat.

Netflix also provides 'hard' controls, where parents can set a four digit PIN code to not only prevent the playback of specific categories of content based on their maturity ratings, but also individual movies and series to allow parents "more specific control over what children can watch on the service" (Hastings, 2018). In order to target shows or films one might want to block from their children's screens, a parent would have to already be aware of the existence of said show and they can only specify titles to block beyond maturity rating categorisation by typing in the specific title and setting a PIN code (Netflix Help Center, 2019).

In addition, when we analyse what affordances Netflix allows parents to block, we must also consider that which they cannot control. The parental controls do not prevent the profile-user, children included, from viewing Netflix's full catalogue of titles. In other words, even if a child may not be able to hit 'Play' on a particular text due to a parent's PIN code protection, they can still see the title, read the show summary, view the title's main page and skim through the main screen of individual episodes. For instance, as of 2019, with the youngest 'Little Kids' PIN parental control set, a user in the Kids profile will still have access to see the main page offerings and imagery of titles such as *Fuller House, Glee, Coraline,* and *Unbreakable Kimmy Schmidt.* Further research is needed to better gauge the negotiations within Netflix subscriber homes around the world as a parent might defend their choice of blocking a program to their children (or give in) to apply a PIN code to a particular series. Netflix's vivid child interface has heroically positioned its world as a child's oyster... it's the parents shucking away the pearls of mature content who become the antiheroes.

As the algorithms offer young viewers programs not suitable for their maturity level or not congruent with their family's philosophy, a parent's solution lies within Netflix's buried controls, requiring a time-intensive process of labelling individual properties, which a child would still see as an option within their profiles, but be unable to view. In this case, while the rhetoric is that algorithmic technology makes life easier, in fact, it creates more work and negotiations within families.

While Netflix's PIN code offerings were posed as significant at the time of their rollout, Netflix additionally announced an 'enhancement' of more prominent displays of its V-chips, which it calls maturity level ratings. Not only will the rating now be displayed when a member hits ‘play’ on a title, they are also “available in other parts of the experience,” as the company wants “to ensure members are fully aware of the maturity level as they begin watching” (Hastings, 2018). Netflix seemingly appears to be hedging its bets through the provision of its maturity ratings throughout subscribers’ paratextual experience as a way to assuage recent complaints from parent advocacy groups, and potentially, other subscriber complaints (Eggerton, 2015). Netflix offers that, through the provision of its ratings before and at the start of viewing, it can alter a user’s textual experience by making the user ‘fully aware of the maturity level.’ In order to grasp how this label, which can channel a user’s “speculative consumption,” will affect their experience, we need to first understand how the maturity ratings are determined (Gray, 2010, p. 24). Netflix arrives at its maturity ratings on all content through a combination of internal and external assessments. Netflix’s site notes that “Maturity ratings are either determined by Netflix or by a local standards organisation. Netflix determines maturity ratings by the frequency and impact of mature content in a TV show or movie” (Netflix Help Center, 2018). With no direct pipeline or rubric for organising maturity ratings within its content, Netflix provides little actual *standards* to the traditional TV framework of standards and practices.

Attending to Stanfill’s discursive interface analysis, I note the importance of also specifying what *isn’t* possible on a site (2015, p. 1061). In the case of Netflix, parents or other concerned viewers lack the opportunity to flag inappropriate content. Unlike digital platforms such as YouTube or Facebook, users cannot point out texts that appear to be rated inaccurately or even post or file complaints or concerns to the company. The method of viewing largely impacts the interface and a subscriber’s ability to interject, flag or interact with the company. The Netflix parental controls interface configures users to acquiesce to ‘enhancements’ such as the ability to individually block Netflix’s expansive library from viewing, while still presenting its full, unfiltered—though sometimes PIN protected—catalogue to all users.

Discussion

Netflix’s maturity ratings and parental control interface, while meant to assuage parental concerns, carries the potential to actually increase the pressures users experience with regard to mediating children’s content. While the power to block a specific text from full viewing ensures direct sentry to a text, parents still face many negotiations when interacting with the interface. Parents often must defend their choices to the very people, the children, they are trying to protect in relation to the public scripts of other people’s opinions about a specific show, film, genre or maturity rating. Evidence of the interface’s deficiencies lies in the way it configures users into the detailed personalisation of individual assignment of PIN codes throughout the Netflix library, while still making every title and front screen visually available for consideration to all, fails to engage the customer and protect their well-being in the digital space.

Netflix's clean interface offers a clutter-free facade, an intimate construction based on user data for users to interact with their curated content. The expected norms of Netflix's platform, however, relies not on the technology that *should* algorithmically 'know' us after all its careful monitoring, but on dutiful, guilt-laden parents to invest great time and human labour in order for children's eyes and ears to be protected. When it comes to children's digital rights and protections, once we strip away the subscriber's prescribed and necessary human labour, 'thin-slicing' and PIN coding, Netflix's algorithm is deficient in understanding what parents want for their children. Either the algorithm does not understand families' needs, or it forces content to its child viewers regardless of what it knows about the parents' preferences. Without the human side involving the vigilant, zealous parents whose work involves the comparison of Netflix's impotent maturity ratings to the actual content combined with the placement of PIN codes and deployment of hard and soft controls, the bungled algorithm and parental control design is a flop.

In the scrutinisation of Netflix's rhetoric, coupled with society's expectations for parental control in the digital space, a clearer picture of the significant nuances behind a technology's affordances and gaps reveals meaning surrounding the power structures involved, revealing the platform as another site of "cultural struggle" (Ang, 1995, p. 43). The cultural effects of Netflix's faulty algorithms and parental control offerings not only carry over to children watching content on the platform but to the parents carrying the load of content review in the digital space. Bearing the burden of their child's everyday development, safety, education, and media curation, a Netflix parent *may* benefit from the provision of maturity ratings. These ratings, however, need to be not only visible, but also accurate and exclusionary.

Self-Preservation Through Self-Regulation

Media companies have asked parents to rise above the scope of time and space to manage a deluge of content avaricious for our children's eyes. From an academic perspective, in addition to asking parents to reconsider screen time guidelines to better suit their families, we may also want to consider asking questions of the digital streaming industry. What if Netflix and its counterparts were asked to transcend their traditional models to preserve their integrity and shareholder value while servicing the needs of the public? Could Netflix provide more transparency to parents through its content tagging tools and more useful parental controls to uncloud the opacity of its black box technology? Further considerations by the digital industry for better platform affordances and parental utility has the potential to be productive for families while engendering goodwill from subscribers. The current digital media culture and policies ask already anxious and overwhelmed parents to transcend beyond the mystique of the algorithm, to police content themselves in order to reconcile the disparities between the gluttony of unchecked entertainment content, faulty algorithmic recommendation systems aimed at our youngest consumers, and a governmental lack of regulatory understanding and oversight. Lawmakers and digital content providers need reminded that they should also consider protecting another vulnerable and burdened population: the parents.

References

Andrejevic, M. (2002). The work of being watched: Interactive media and the exploitation of self-disclosure. *Critical Studies in Media Communication, 19*(2), 230–248.

Andrejevic, M. (2013). *Infoglut: How too much information is changing the way we think and know* (1st ed.). Routledge.

Andrejevic, M. (2007). *iSpy: Surveillance and power in the interactive area*. Lawrence, Kansas: University of Kansas Press.

Ang, I. (1995). *Living room wars: Rethinking media audiences*. Routledge.

Blum-Ross, A., & Livingstone, S. (2018). The trouble with "screen time" rules. In G. Mascheroni, C. Ponte, & A. Jorge (Eds.), *Digital parenting: The challenges for families in the digital age* (The international clearinghouse on children, youth and media) (pp. 179–188). Nordicom.

Bucher, T. (2018). *If... then: Algorithmic powers and politics*. Oxford University Press.

Cheney-Lippold, J. (2017). *We are data: Algorithms and the making of our digital selves*. New York: New York University Press.

Clark, L. S. (2013). *The Parent App: Understanding families in the digital age*. Oxford University Press.

Cohn, J. (2019). *The burden of choice – recommendations, subversion and algorithmic culture*. Rutgers University Press.

Cunningham, S., & Craig, D. (2019). *Social media entertainment: The new intersection of Hollywood and Silicon Valley*. New York University Press.

Eggerton, J. (2015, February 12). PTC seeks complaints against 'Family Guy.' *Broadcasting & Cable*. https://www.broadcastingcable.com/news/ptc-seeks-complaints-against-family-guy-137981.

FCC Docket 19-41. (2019). *Media Bureau seeks comment on TV ratings*. https://www.fcc.gov/document/media-bureau-seeks-comment-tv-ratings.

Federal Communications Commission Telecommunications Act of 1996. Act, § 551(e)(1).

Finn, E. (2017). *What algorithms want: Imagination in the age of computing*. The MIT Press.

Flint, J. (2016, January 17). Netflix to ramp up originals targeting kids. *The Wall Street Journal*. https://www.wsj.com/articles/netflix-to-ramp-up-originals-targeting-kids-1453058812.

Gillespie, T. (2018). *Custodians of the Internet: Platforms, content moderation, and the hidden decisions that shape social media*. Yale University Press.

Gray, J. (2010). *Show sold separately: promos, spoilers, and other media paratexts*. New York University Press.

Grothaus, M. (2018, March 28). How I got my dream job of getting paid to watch Netflix. *Fast Company*. https://www.fastcompany.com/40547557/how-i-got-my-dream-job-of-getting-paid-to-watch-netflix.

Guins, R. (2009). *Edited clean version: Technology and the culture of control*. University of Minnesota Press.

Harwell, D. (2016, March 28). Netflix is coming for your kids. *The Washington Post*. https://www.washingtonpost.com/news/the-switch/wp/2016/03/28/netflix-is-coming-for-your-kids/

Hastings, M. (2018, March 5). Introducing PIN protection and other enhancements for informed viewing. *Netflix Media Center*. https://media.netflix.com/en/company-blog/introducing-pin-protection-and-other-enhancements-for-informed-viewing.

Heins, M. (2007). *Not in front of the children: Indecency, censorship, and the innocence of youth* (2nd ed.). Rutgers University Press.

Hoover, S. M., Clark, L. S., & Alters, D. M. (2012). *Media, home and family*. Routledge.

Jordan, A. B. (2008). Children's media policy. *The Future of Children* 18(1), 235–253. Princeton University. Project MUSE database https://www.jstor.org/stable/20053126.

Jordan, A. B. (2016, Dec 1). Presidential address: Digital media use and the experience(s) of childhood: Reflections across the generations. *Journal of Communication, 66*(6), 879–887. https://doi.org/10.1111/jcom.12265.

Lead Kids Content Tagger, Netflix. (n.d.). Retrieved June 3, 2019, from https://www.linkedin.com/jobs/view/lead-kids-content-tagger-at-netflix-113056443/.

Lim, S. S. (2016). Through the tablet glass: Transcendent parenting in an era of mobile media and cloud computing. *Journal of Children and Media, 10*(1), 21–29.

Livingstone, S., & Blum-Ross, A. (2017). Researching children and childhood in the digital age. In P. Christensen & A. James (Eds.), *Research with children: Perspectives and practices* (pp. 54–70). Routledge.

Lobato, R. (2019). *Netflix nations: The geography of digital distribution*. New York University Press.

Lotz, A. (2017). *Portals*. Maize Books.

Lupton, D., & Williamson, B. (2017). The datafied child: The dataveillance of children and implications for their rights. *New Media & Society*. https://doi.org/10.1177/1461444816686328.

Netflix Help Center. (2018). *How does Netflix decide the maturity rating on TV shows and movies?* Retrieved March 22, 2018, from https://help.netflix.com/en/node/2064.

Netflix Help Center. (2019). *How do I set parental controls on my Netflix account?* Retrieved February 22, 2019, from https://help.netflix.com/en/node/264?ui_action=kb-article-popular-categories.

Netflix Media Center. (2018). *Netflix adds seven new original series for older kids bringing its global original line up to more than 35 series for kids of all ages*. Retrieved March 14, 2018, from https://media.netflix.com/en/press-releases/netflix-adds-seven-new-original-series-for-older-kids-bringing-its-global-original-line-up-to-more-than-35-series-for-kids-of-all-ages-migration-1.

Sefton-Green, J. (2006). Youth, technology, and media cultures. *Review of Research in Education, 30*, 279–306.

Spangler, T. (2015, October 1). Netflix orders 7 original kids' series, including Lego's 'Bionicle' and DreamWorks' 'Croods.' *Variety*. Retrieved February 6, 2019, from https://variety.com/2015/digital/news/netflix-orders-7-original-kids-series-including-legos-bionicle-and-dreamworks-croods-1201607216/.

Stanfill, M. (2015). The interface as discourse: The production of norms through web design. *New Media & Society, 17*(7), 1059–1074. https://doi.org/10.1177/1461444814520873.

Striphas, T. (2012). What is an algorithm? *Culture Digitally*. http://culturedigitally.org/2012/02/what-is-an-algorithm/.

Van Dijck, J., Poell, T., & de Waal, M. (2018). *The platform society: Public values in a connective world*. Oxford University Press.

Willett, R. (2015). The discursive construction of "good parenting" in the case of children's virtual world games. *Media, Culture & Society, 37*(7), 1060–1075. https://doi.org/10.1177/0163443715591666.

Maureen Mauk, a Doctoral Candidate studying Communication, Media and Cultural Studies at the University of Wisconsin-Madison, researches parents, policy, and industry as it relates to television history and the current platformized media landscape. She previously spent a decade in Los Angeles serving as a Television Standards & Practices executive.

Chapter 8
Screening Language Acquisition Skills in a Mediated Childhood

Kylie J. Stevenson, Lelia Green, Donell Holloway, and Kelly Jaunzems

Overview of ARC Project

This paper arises from the research project funded by the Australian Research Council titled *Toddlers and Tablets: exploring the risks and benefits 0–5s face online*. Children aged between 0 and 5 are experiencing an extraordinary shift in media consumption practices. Approximately one third of children between two and five living in affluent countries have their own touchscreen device (Marsh et al., 2015), while almost all such children now have intermittent access to a parent or caregiver's device. These dynamics have prompted a shift from researching older children's technology use to research, and the subsequent development of a policy-ready evidence base, in relation to 0–5s and touchscreens. The *Toddlers and Tablets* project investigates family practices and attitudes regarding very young children's touchscreen technology use in Australia and the United Kingdom, with an aim to contribute to public debate and practice in Australia, the United Kingdom and Ireland.

Methodology

This project investigates the contribution of intergenerational scaffolding of very young children's digital literacies through provision of resources and sharing of values and attitudes. In examining the role that touchscreen technology plays in the

K. J. Stevenson (✉) · L. Green · D. Holloway · K. Jaunzems
Edith Cowan University, Perth, WA, Australia
e-mail: k.stevenson@ecu.edu.au; l.green@ecu.edu.au; donell.holloway@ecu.edu.au; k.jaunzems@ecu.edu.au

D. Holloway et al. (eds.), *Young Children's Rights in a Digital World*, Children's Well-Being: Indicators and Research 23,
https://doi.org/10.1007/978-3-030-65916-5_8

everyday life of very young children and their families, the project adopts a social constructionist methodological approach (Burr, 2003). It draws on the perspectives, insights and practices of children, parents, educators and other caregivers. These are positioned as the current practising "experts in this field in terms of the everyday uses of these technologies by very young children" (Holloway et al., 2015, para. 8). The project used a conceptual framework that combines social constructionism with a co-creation of knowledge (Fong, 2005) and social learning (Wenger, 2000) framework, which links through to the 'communities of practice' literature (Wenger, 1998). This co-construction of knowledge approach recognises that parents and other caregivers are formulating understandings and attitudes through their own everyday experience of touchscreens, and through sharing these experiences with others, particularly their children. Sometimes, however, parents are not best placed to appreciate the extra-ordinariness of their child's digital skills and competencies. This paper describes one such example in a case study of Lavinia, a young *Peppa Pig* fan, collected as part of the 18 family-based case studies (Yin, 2014) within the *Toddlers and Tablets* research project.

Lavinia is an ardent fan of *Peppa Pig* and loves everything that brings *Peppa Pig* into her life. What astonished the researchers, however, was Lavinia's skill in playing *Peppa Pig* in Mandarin on an iPad in parallel with sourcing the same episode in English in streaming video mode on the family's television set. Whilst talking to her mother, the researchers watched Lavinia set the entire system of media retrieval and replay into action without help from, or reference to, an adult. The mother, in her interview, expressed surprised that this was not a competency generally shared by other 2-year-olds.

Interview Strategies

In the *Toddlers and Tablets* project, two researchers visit the family home where semi-structured interviews are conducted with one or both parents, and the child is engaged in participant observation play with a range of toys which may include digital activities. Parents and everyday routines dictate the choice of possibilities open to the child during the interview—some children, for example, do not have a personal touchscreen but may be allowed to play with a parent's phone when the parent is busy with something else, such as talking to a researcher. In this case, the primary role of one of the two researchers is to observe and note the child's activities, to express interest, and to engage with the child in play if the child invites them to do so. With an early verbal child such as Lavinia, aged just over 2, activities such as colouring-in, asking her to demonstrate her favourite apps, and other child-led activities, allow researchers to engage the child in a two-handed parent–child interview process. This approach allows the researchers to capture an ethnographic picture that echoes the research participants' everyday life through a fluid "free narrative" (Cameron, 2005, p. 601) interview conversation with the child and parent.

This interviewing strategy is informed by Vygotsky's work, both in his conceptualisation of play (2004) as an interpretive activity by the child to explore and make sense of her cultural setting; and in his concept of the zone of proximal development (Vygotsky, 1978) whereby the child's abilities around touchscreen use progress incrementally. One aim of the two-handed interview process with parent and child is to capture the child's digital multiliteracies (Cope & Kalantzis, 2000; Sefton-Green et al., 2016) around touchscreen devices as they happen, positioning the development of these skills in the context of the child's everyday life. This chapter now turns to details of the focus family (de-identified for ethics and privacy reasons).

A Brief Ethnography of the Zhang/Chen Family

The Zhang/Chen family live in a three-bedroom home in an established suburb east of Perth. Stanley Zhang (36) works for a mining company and his wife Rita Chen (33) works in information technology management for one of the nation's largest banks. Stanley and Rita both hold tertiary degrees. While Stanley and Rita are highly proficient in English, they mainly speak Mandarin at home with Lavinia, and with Rita's parents, whom they dine with every weeknight. Both Stanley and Rita were born in China but moved to Australia separately before meeting there. They have one child, Lavinia, who is 28 months old. Even though both parents are skilled IT users, there is no real computer presence in Lavinia's home, apart from a laptop used by Stanley for web-browsing after work.

Rita's parents, Sun and Lily Chen, are highly involved grandparents. Both in their late 50s, they moved to Australia from China after the birth of Lavinia and are Australian permanent residents. They live near to Stanley and Rita's home, also in a three-bedroomed house, and are Lavinia's primary caregivers during the working week. Additionally, they provide a home for two high-school age foster children who are mother tongue English speakers. Sun and Lily Chen mainly converse with their family members in Mandarin and have limited English language skills themselves.

Family Life

Lavinia spends her weekdays with her grandparents at their home and is taken by Grandma Lily on regular excursions to library story–times, swimming and play-group. Grandma Lily keeps a regular schedule with Lavinia, making sure to be home for an afternoon nap at 2 pm. Rita and Stanley eat at Rita's parents' home every night after work. The arrangement suits the family, both because Rita and Stanley have busy working lives and also because it allows Lavinia access to a multi-generational family structure. Lavinia has a dedicated play area at her grandparents' home and is very comfortable there. Grandma Lily does not encourage touchscreen use and generally hides the iPad. However, throughout the day, Grandma Lily will take photos and upload them to *WeChat* (a Chinese version of *WhatsApp*) to keep Rita, Stanley and Sun involved in Lavinia's activities during the day. On weekends, Rita and Stanley organise social activities that involve friends with children and child-centred outings. While Stanley's family still live in China, he calls them via video chat every Saturday. Lavinia communicates with her paternal grandparents in this way. Rita acknowledges a need for Lavinia's English to improve but is also concerned that her Mandarin may suffer as a result of her developing bilingual competency and starting kindy, and then moving through to school. Rita and Stanley are determined that Lavinia will be fully bilingual after starting school, at least verbally, acknowledging that her written Mandarin may not keep up with her spoken language skills once she becomes fully immersed in an English-language school system.

A Fluid Multimodal Digital Engagement, Scaffolded by Parents

Lavinia (28 months) was introduced to the iPad and a LeapPad-style learning pad, both bought specifically for Lavinia's use, at about 26 months. She knows to ask her mother for help to access apps. She is aware of the password protocol but does not know how to unlock the device herself. Once in the device, she knows how to open apps, search for her favourite videos and use the camera on an iPhone.

> We have an iPhone, iPad and I won a little learning pad for her, for Lavinia herself [like a Leap Pad. . .] the password we control, she doesn't know how to enter the password. But she knows the apps she's looking for. [Rita (mum), personal interview, 2017]

Lavinia was able to open her chosen app herself, but sought her mother's assistance to unlock the iPad. Lavinia demonstrated that she could swipe to select different apps, and she knew how to close one app and navigate to another page. As the interview progressed, Lavinia entertained herself, moving through various platforms and selecting her own media according to her preferences. Lavinia began her digital engagement with an English language *Peppa Pig* video streamed from the *ABC Kids* app to the family television set via her mother's phone. During the interview, Lavinia opened the same *Peppa Pig* video on her iPad, but this time in Mandarin. Rita explained that Lavinia self-selects whether she wants to watch in Mandarin or English, flowing in and out of both.

> Yeah, she knows which one is in Mandarin, which one is in English but [*ABC Kids*] has got all English [language videos] so I searched and there's some cartoons in Mandarin that I can only search for on YouTube. [Rita (mum), personal interview, 2017]

The functions on Livinia's 'little learning pad' seemed to be intuitively learned as neither her mother nor (reportedly) her grandparents recall teaching her. Indeed, they say she knows the device functionality better than they do. This concurs with findings from a study by Marsh et al. (2017) in which:

> Parents reported that their children moved fluently across media in their meaning-making practices and, in some cases, were more confident than parents in this regard. (p. 53)

In this way, Lavinia is operating as a multimodal learner. Sefton-Green, Marsh, Erstad and Flewitt "propose that multimodality offers a rigorous analytic framework for investigating how children learn to read multimodal texts" (2016, p. 20). They suggest that:

> The three underlying theoretical premises for multimodality are: 1. Representation, communication and interaction draw on multiple modes, all of which contribute to meaning; 2. Sets of semiotic resources (modes) are socially shaped over time to articulate individual, affective and social meanings; [and] 3. People intentionally choose and configure modes to orchestrate meaning through multimodal design. (2016, p. 20)

Lavinia is able to draw on multiple modes of representation and interacts with multiple devices in multiple ways. The resources available to her include those shaped by her parents, grandparents, and her local library, and by the producers of texts she encounters in her everyday life such as *ABC Kids*, and a bilingual card game of Mandarin/English flashcards sourced by her parents. Lavinia is an active agent in her multimodality. Rita said that Lavinia knows that one device, the LeapPad, is more for learning, whereas the iPad is more for entertainment, though this semi-segregation of usage is something that Rita seems to have scaffolded herself:

> she doesn't play games but she does some drawings—there is learning on that iPad, a drawing thing, but to learn numbers and letters it's on that learning [LeapPad] one. [Rita (mum), personal interview, 2017]

During the interview, Lavinia engaged in virtual and embodied play where she made her physical dolls dance to a song that she played on her iPad. She had previously lined up the dolls on the couch to 'observe' the interview, and then engaged her mother in a dance to a song that she played on her LeapPad. Rita reported that Lavinia sometimes acts out events and images she sees online, e.g., pretending to be a train after watching *Thomas the Tank Engine*. Lavinia also played with her blocks on the floor with the second interviewer and drew with crayons supplied in the interview.

Throughout the researchers' visit, Lavinia constructed her play through engagement with real and virtual modes, through both English and Mandarin, and on digitally–produced texts and real world toys in ways that reflect "'multimodal' . . . combinations of words, still and moving images, icons, screen layout and so on" (Sefton-Green et al., 2016, p. 13). The way Lavinia seamlessly navigated between the different modes she used accords with Sefton Green at al.: "from a multimodal perspective, all modes (including language) are conceptualised as fluid" (2016, p. 21).

The ways in which Lavinia transitioned between screens, physical toys, singing, activity and other modes reflected Carrington's (2013) suggestion that technology use is eclectic and fast-paced, and should be thought of as an assemblage of practice rather than a system of practice. Sefton-Green et al., citing Carrington, propose:

> the use of the term 'assemblages' (following Deleuze and Guattari, 1987, p. 4) to account for the poly-centricity and multi-layeredness of media supersystems: [. . .] assemblages dismantle and reassemble in different combinations as context and requirements shift. (Carrington, 2013, p. 209, cited by Sefton-Green et al., 2016, p. 19).

Sefton-Green et al. further refine Carrington's concept, noting: "If the concept of assemblage is applied to children's digital meaning-making, then we should recognise the messiness and complexity of literacy in the digital world" (2016, p. 19). Although it might have looked messy, chaotic and complex to an observer, Lavinia's fast-paced, free-flowing engagement with the resources at her disposal represented a repertoire of activities which integrated her own play practices with the experience of her family's participation in a research project about very young children's digital competencies and play practices.

Intergenerational Scaffolding

Lavinia's activity-based assemblage, however, was not the eclectic outcome of her interests alone. Her mother Rita demonstrated thoughtful deliberation behind her decisions to provide Lavinia with access to a range of touchscreen devices such as her own LeapPad, with an overt focus on educational benefit. Marsh et al. (2017) argue that parents are:

> central to the development of the children's growing understanding about multimodal texts and practices, whether they felt they were being overt about this or not. As with traditional, print-based literacy practices, parental engagement in digital literacy was crucial to the children's experiences and could provide a solid foundation for future learning in schools. (p. 55)

Rita considered herself a moderately strict parent when it came to device use. She set limits on the use of technology, but also acknowledged that the use of iPads in schools is prevalent and wanted Lavinia to be able to use the technology when she starts her educational career. Rita and Stanley have no tolerance for gaming, however, and see it as an addictive activity with little merit. It appeared that they considered digital devices to be 'necessary evils', doing their best to promote digital technology use for educational purposes only, choosing apps and cartoons with learning content:

> It's just addiction, those ones that they can just open and become so addicted to, it's so, even with the games. I don't put any games on my apps for her. I don't know if cartoons are an addiction or not but at least you know, the *Peppa Pig* program. I just find that the cartoon is teaching a lot of things, not just you know purely cartoon and the stupid images or this, they teach you what Christmas is about, they teach you to know how to do a teapot and then because they got teapot and she knows got that's teapot, how to make a tea party so it's education, more about education. [Rita (mum), personal interview, 2017]

Rita and Stanley were demonstrating a parenting style that highly values education. Knowing that Lavinia was only a year away from a kindergarten-based education where iPads are used, Rita expressed awareness of the need for Lavinia to be exposed to certain types of technology. She made choices about what Lavinia could watch, taking into account the educational value of the app or video. Lavinia was allowed to watch *Peppa Pig*, because it was available in Mandarin and English, and also because Rita perceived educational elements within the cartoon. Rita acknowledged a personal desire that Lavinia should have appropriate Australian cultural references in order to socialise with other children at playgroup, future daycare and kindergarten. Rita preferred Lavinia to be culturally knowledgeable since she acknowledged that Lavinia may be in the minority in her forthcoming chosen kindergarten.

The main benefit Rita saw in using digital devices was their ability to reinforce Lavinia's bilingualism. As Lavinia is mainly in a Mandarin-speaking environment at her parents' and grandparents' homes, Rita saw benefit in Lavinia being able to access content in both languages, with her favourite videos and learning games being in Mandarin and/or in English. The main downsides to touchscreen technology use, according to Rita, include exposure to advertising (on YouTube, not television, as the family does not watch commercial TV) and possible damage to Lavinia's eyesight, even though Rita takes steps to reduce the brightness of the screen for Lavinia.

Cultural systems of belief around the benefits and downsides of technology can be seen as "parental ethnotheories" (Kenner et al., 2008, cited by Marsh et al., 2017, p. 58), and these influence the multimodal texts made available to children in a family. Marsh et al. call attention to the "significance of these intergenerational digital literacy practices in terms of scaffolding children's understanding about literacy, or in this case multimodal communication, as a social practice" (p. 58). Rita, Stanley, and Lavinia's Grandma Lily all contribute to an intergenerational ethnotheory of multimodal digital practice, and each play a role in scaffolding Lavinia's learning. Like the parents in Marsh et al.'s (2017) study of young children's induction into digital practices, Lavinia's family "did this overtly at times, in that [the] parents utilised didactic pedagogies to teach specific skills [...] At other times, scaffolding was such an integral part of everyday life that [the] parents found it difficult to point out when such teaching occurred" (p. 54).

Lavinia's Digital Multiliteracies

Conducting the interview with Lavinia's family in a manner that readily accessed Lavinia's everyday use of touchscreen technology as part of family life took into account a concept of the family's digital multiliteracy; which is to say that digital literacy is embedded within the family's lived social and cultural practices. The term 'digital multiliteracies' (Cope & Kalantzis, 2000; Sefton-Green et al., 2016) acknowledges the inherently multimodal nature of the family practices encountered around the use of touchscreens in family environments such as the Zhang/Chen home. The interview with Rita and Lavinia revealed that the family had set into action an integrated system of media retrieval and play that enabled Lavinia's digital multiliteracy development.

Sefton-Green et al. define digital literacy as "a social practice that involves reading, writing and multimodal meaning–making through the use of a range of digital technologies" (2016, p. 15). They suggest that while digital literacy experiences involve digital technologies, they also involve non-digital practices. Rita was able to identify the growth of Lavinia's digital literacy over three phases, starting with non-digital encounters via the nursery rhymes she encountered in the local library toddler book session attended with Grandma Lily. This led to Lavinia finding *The Wiggles*, then *Peppa Pig*:

> I think that very early on she wasn't interested in *Peppa Pig* 'cause it was on the kids' channel and because she didn't like [that]—she started with *Wiggles*, the sounds, rhymes and songs really. 'Cause she's likes the library's rhyme songs time, she started watching this video about kids' nursery sounds, and then we found *Wiggles* and then this [*Peppa Pig*] is actually the third stage. [Rita (mum), personal interview, 2017]

Sefton-Green et al. identify a significant difference between literacy in relation to traditional books (such as those encountered at the library book sessions) and multimodal texts, such as *The Wiggles* and *Peppa Pig* online videos, which require a different 'reading' approach:

> the act of reading a text or producing a text in a digital medium can be profoundly different to the act of writing on paper or reading from a printed page. [. . .] In multimodal texts the pattern of reading from left to right (or whichever established reading path depending on the linear script system) is no longer strictly adhered to, and images, words and layout interact in complex ways. Reading multimodal texts therefore involves imposing order and relevance on what is presented. (Sefton-Green et al., 2016, pp. 22–23)

Drawing on Green's 3D (1988) model of literacy as a social practice, with the three crucial elements being operational, cultural and critical, Sefton-Green et al. (2016, pp. 15–16), suggest that this is the foundation for the child's reading, alongside "added elements that frame children's engagement in digital literacy" (p. 17). Further, they consider the "contexts of young children's digital literacy practices as [occurring] at a micro, meso and macro level" (p. 17) (Fig. 8.1).

The micro level refers to the child herself, for example, Lavinia's interest in *Peppa Pig* and *The Wiggles*, and her abilities in mastering the touchscreen devices in her family environment. It references Lavinia's identity as a bilingual child, already manifested in her practice of playing *Peppa Pig* in both Mandarin and English, and which also contributes at this micro level.

The meso level in Sefton-Green et al.'s model refers to "the wider influences of home, including parents and siblings, and the community and society the child lives in" (2016, p. 18), Lavinia's parents and grandparents (and the different ways they model literacy practices), the story time sessions at the local library, and the English-speaking foster children in the grandparents' home, all contribute to the meso level of Lavinia's digital multiliteracies.

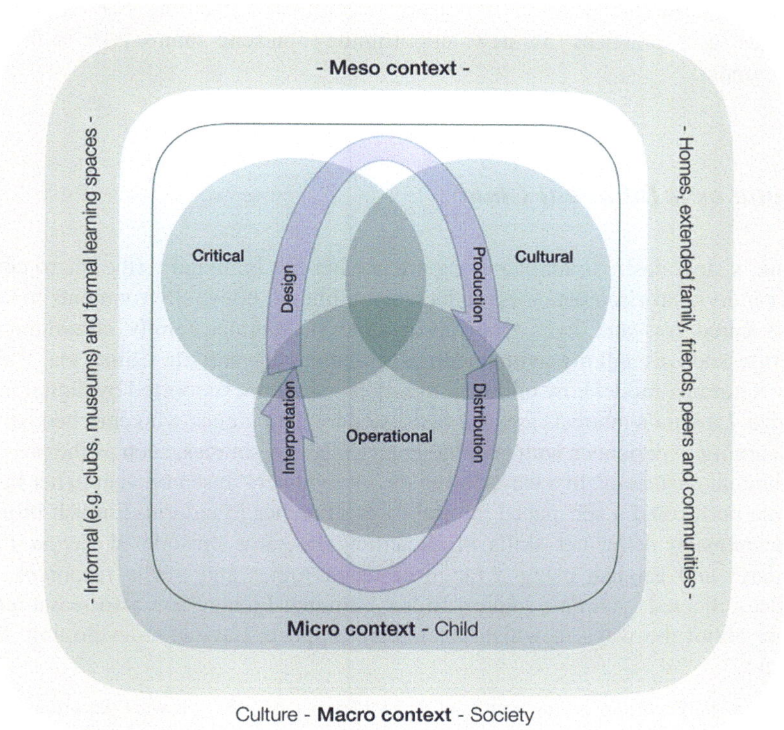

Fig. 8.1 The processes of, and contexts for, children's digital literacy practices (adapted from Colvert, 2015, in Sefton-Green et al., 2016, p. 18)

The macro level of wider social and political influences on Lavinia's digital multiliteracies are less immediately obvious in the interview with her family, but could include specific cultural attitudes to education including the use of iPads in kind; social expectations regarding educational achievement relevant to social-economic status; and national and state educational and literacy policy and practices. Given Rita's identification of Lavinia's forthcoming kindergarten career, and the council-provided library activities, as significant, national educational and literacy policies are clearly impacting upon Lavinia's multiliteracy development and experiences.

The overall picture of Lavinia's developing digital literacies as a bilingual 2-year-old represents:

> 'emergent literacy', where young children's familiarity with literacy merges through observation of and engagement in a range of literacy-related activities in different social domains (home, school, community, work), and in diverse networks of social practices, and where children learn about literacy as part of everyday life, in family and community networks (Kress, 1997; Gregory et al., 2004), and in 'diverse literacy ecosystems' (Kenner, 2005). (Sefton-Green et al., 2016, p. 14).

The child in this position is poised to continue expanding her repertoire of literacies and practices as new opportunities present themselves within her environment.

Lavinia as a Bilingual Child

Lavinia's clear desire to learn and experience her life bilingually (the micro context of her individual circumstances) reflects the bilingual home environments in which she is cared for, and lives (the meso context). Regular family communication patterns, such as talking with paternal grandparents still in China via *WeChat* every Saturday, model how bilingual family practices are supported by digital family practice. Lavinia's interests are also facilitated by her parents, who enriched her play and learning experiences with supplementary media resources, such as the provision of bilingual flashcards. In a way, during the interviewers' research visit to her family, Lavinia performed a self-paced tutorial to practice her Mandarin–English bilingual comprehension using her skills in streaming the same episode of *Peppa Pig* in Mandarin and English using a range of technologies and media resources. This practice reflected Lavinia's interest in her bilingual literacy and also reassured her mother about the ways in which the family supports Lavinia's developing digital literacy:

> I'm not really reluctant giving her all this technology as long as I know what she's doing [...] I think she can switch between languages to known animals and that kind of stuff so I hope it helps. [Rita (mum), personal interview, 2017].

Rita identified Lavinia's digital practices as vital in expanding her exposure to the English language in her home environment:

> I think the most [benefit of technology] for her is languages so we're trying to you know just encourage her to speak both languages but at this stage, because she's mainly cared by my parents who speak Mandarin and my family all speak Mandarin, the only chance for her is go to library and listen to other teachers and other kids. So I think some apps can help her to develop some language so it's got opportunity, this is bilingual [indicating the LeapPad], most of the apps are bilingual. [Rita (mum), personal interview, 2017].

Such technological practices also carried another benefit in Rita's mind, since she hopes that Lavinia's experiences with contemporary children's media will help smooth her entry into shared popular culture:

> Beyond the other [benefits] is just catching up some, you have to be updated with what's happening even with the cartoons so *Peppa Pig* is very popular so when she goes to kindy or pre-kindy and other kids talks about it, I hope she knows about it too. I don't want to know that she feels lonely and that there's no socialising with other kids. [Rita (mum), personal interview, 2017].

The ways in which the Zhang/Chen family inducted Lavinia into a bilingual world corresponds with the study of family digital literacy practices by Marsh et al. (2017) which showed not only that "digital practices were embedded in children's daily lives" but also that such "practices were embedded across the languages children spoke at home, thus promoting the children's bilingual/biliterate skills" (Marsh et al., 2017, p. 52). Sefton-Green et al. suggest such embedded digital practices have implications for the future knowledge society:

> One key effect of these pressures [for a knowledge society] is on the way small children grow up in the 'culturalisation of leisure' (Buckingham & Scanlon, 2002), how forms of entertainment, 'edutainment', play and even supposedly free time for young children have now been scrutinised, packaged and sold (Seiter, 2005) as in some ways helping the child to get ahead and to compete in an uncertain and increasingly scary world. (Sefton-Green et al., 2016, p. 11).

Interestingly, one of the advantages that Rita sees in Lavinia's multimodal literacies is that these offer a potential avenue through which Lavinia will make friends as she ventures into Sefton-Green et al.'s 'uncertain world'. Thus, in Rita's mind, there is more at stake than getting ahead, and competing: there is the creation of future friendships based on shared cultural experiences.

The bilingual and social competencies that Rita hopes are instilled through facilitating Lavinia's consumption of multimodal, multimedia texts accords with the development of digital multiliteracies as advocated by Marsh et al.:

> It is necessary but no longer sufficient for children to develop competence in relation to written texts; they also need to be able to engage successfully with multimodal, multimedia texts if they are to acquire the range of skills, knowledge and understanding necessary to navigate the knowledge economy of the 21st century. (2017, p. 48)

Conclusion

This case study is based on observation and engagement with Lavinia, and with her media and play practices. The media and play-based dataset is complemented by an in-depth interview with Rita, Lavinia's mother. Lavinia demonstrated a sophisticated engagement with a range of technical systems and used these skills to deliver a unique multilingual engagement with Mandarin and English versions of the same *Peppa Pig* episode, demonstrating both her technical fluency and her desire to practice, build, and demonstrate her bilingual competency.

The macro-, meso- and micro-processes of Lavinia's digital literacy practices represent her unique context as a child growing up in Australia, yet deeply embedded within her Chinese cultural heritage. Lavinia has close engagement with one set of grandparents as weekday caregivers, who use *WeChat* as a means of updating Lavinia's parents as to her daily activities while her parents are at work. Regular Saturday video chats keep Lavinia in close contact with her other grandparents in China. At the same time, Lavinia's family is taking active steps to provide pathways through which Lavinia will be integrated within English-speaking Australian society. Lavinia's grandmother also takes her to the local library so that she can participate in toddler story time sessions, exposing her to key features of young children's popular culture. Lavinia's playful engagement with her bilingual digital environment directly reflects her parents' priorities.

Although they are generally dismissive of 'addictive' games, and don't have games on the iPad available for Lavinia's use, Lavinia's parents encourage engagement with *Peppa Pig* and other aspects of contemporary Australian childhood. This supports both Lavinia's English language acquisition and her future integration into children's social circles at pre-kindy, kindy, pre-school and school. In this respect Lavinia's multiliteracies, and her capacity to share informed opinions about *Peppa Pig* and *The Wiggles*, are positioned as a social good as much as an educational one. Reflecting her family's desire to develop multilingual competencies, Lavinia has adapted and harnessed her multiliteracy practices to provide a unique and age-appropriate way of supporting her dual language development. While the interviewers were talking with her mother, and without adult assistance, Lydia set up simultaneous access to both Mandarin and English versions of the same *Peppa Pig* episode. It may be relevant that this activity was in parallel with Lavinia's mother being interviewed at home in the medium of English. It is even possible that Lavinia was using her digital skills to underline the crucial importance of both Mandarin and English in her family context.

In these ways, and through this case study, Lavinia and her family have communicated the importance of multiliteracies to their own lives and have underlined what it is to engage creatively and constructively with technology and culture, even at a very young age. This case indicates the role of the family as a coherent and integrated learning system that inducts young children into domestic practices and attitudes. Reflecting her parents' commitment to their Chinese cultural heritage within the context of an Australian cultural future, Lavinia's self-directed practices enabled her to demonstrate in an everyday way that she was both embracing and living that family-based vision. For parents who construct their children's digital activities as problematic, this case study may highlight the ways in which their own digital engagement has impacted upon their child's relationship with the digital. For families where children's digital engagement has resulted in outcomes aligned with parental priorities, the child's activities may seem so in tune with parental expectations that they only appear remarkable to outsiders.

This study contributes to the field by teasing out the ways in which the cultural priorities and practices of older family members are communicated to members of the newest generation. The invisibility of Lavinia's expertise to her immediate family circle indicates her seamless integration within that shared cultural context. The work suggests that conflicts between parents and their early years' offspring around digital media use might usefully prompt parental reflection upon the behaviors and attitudes that they have communicated to their child, encouraging changes in adult digital engagement as a precursor element for providing the child with an alternative framework for their digital participation. It also suggests that very young children's problematic technology use in early childhood education settings may reflect family-based issues around digital practices and literacies that might helpfully be addressed by whole-of-family discussions rather than focusing solely on behavioral modification with the child.

Acknowledgement This work was supported by the Australian Research Council via Discovery research funding [grant number DP150104734].

References

Buckingham, D., & Scanlon, M. (2002). *Education, entertainment and learning in the home*. Open University Press.

Burr, V. (2003). *Social constructionism* (2nd ed.). Routledge.

Cameron, H. (2005). Asking the touch question: A guide to ethical practices in interviewing very young children [3-5]. *Early Child Development and Care, 175*(6), 597–610. https://doi.org/10.1080/03004430500131387.

Carrington, V. (2013). An argument for assemblage theory: Integrated spaces, mobility and polycentricity. In A. Burke & J. Marsh (Eds.), *Children's virtual play worlds: Culture, learning and participation* (pp. 200–216). Peter Lang.

Colvert, A. (2015). *Ludic authorship: Reframing literacies through peer-to-peer alternate game design in the primary classroom*. Doctoral dissertation. https://roehampton-online.academia.edu/AngelaColvert.

Cope, B., & Kalantzis, M. (2000). *Multiliteracies: Literacy learning and the design of social futures*. Routledge.

Deleuze, G., & Guattari, F. (1987). *A thousand plateaus: Capitalism and schizophrenia*. Continuum.

Fong, P. (2005). Co-creation of knowledge by multidisciplinary project teams. In E. Love, P. Fong, & Z. Irani (Eds.), *Management of knowledge in project environments* (pp. 41–56). Elsevier.

Green, B. (1988). Subject-specific literacy and school learning: A focus on writing. *Australian Journal of Education, 32*(2), 156–179.

Gregory, E., Long, S., & Volk, D. (2004). *Many pathways to literacy: Young children learning with siblings, grandparents, peers and communities*. RoutledgeFalmer.

Holloway, D., & Green, L. (2013). Using ethnography to understand everyday media practices in Australian family life (Chapter 16). In R. Parameswaran (Ed.), *The international encyclopedia of media studies Vol. 3: Audience and interpretation in media studies* (pp. 1–22). Wiley-Blackwell.

Holloway, D., Green, L., & Stevenson, K. (2015). Digitods: Toddlers, touchscreens and Australian family life. *M/C Journal, 18*(4). www.journal.media-culture.org.au.

Kenner, C. (2005). Bilingual families as literacy eco-systems. *Early Years, 25*(3), 283–298.

Kenner, C., Ruby, M., Jessel, J., Gregory, E., & Arju, T. (2008). Intergenerational learning events around the computer: A site for linguistic and cultural exchange. *Language and Education, 22* (4), 298–319. https://doi.org/10.1080/09500780802152572.

Kress, G. (1997). *Before writing: Rethinking the paths to literacy*. Routledge.

Marsh, J., Hannon, P., Lewis, M., & Ritchie, L. (2017). Young children's initiation into family literacy practices in the digital age. *Journal of Early Childhood Research, 15*(1), 47–60. https://doi.org/10.1177/1476718X15582095.

Marsh, J., Plowman, L., Yamada-Rice, D., Bishop, J., Lahmar, J., Scott, F., Davenport, A., Davis, S., French, K., Piras, M., Thornhill, S., Robinson, P. & Winter, P. (2015). *Exploring play and creativity in pre-schooler's use of apps*. Final project Report. Retrieved September 12, 2017, from www.techandplay.org.

Sefton-Green, J., Marsh, J., Erstad, O., & Flewitt, R. (2016). *Establishing a research agenda for the digital literacy practices of young children*. A white paper for COST Action IS1410. Retrieved September 12, 2017, from http://digilitey.eu.

Seiter, E. (2005). *The Internet playground*. Peter Lang.

Vygotsky, L. (1978). *Interactions between learning and development*. Reprinted in Gauvin M. and Cole M. (1997) *Readings on the Development of Children* (pp. 29–36). Freeman & Company.

Vygotsky, L. (2004). Imagination and creativity in childhood. *Journal of Russian & East European Psychology, 42*(1), 7–97. https://doi.org/10.1080/10610405.2004.11059210.

Wenger, E. (1998). *Communities of practice: Learning, meaning, and identity*. Cambridge University Press.

Wenger, E. (2000). Communities of practice and social learning systems. *Organization, 7*(2), 225–246. https://doi.org/10.1177/135050840072002.

Yin, R. K. (2014). *Case study: Design and methods* (5th ed.). Sage.

Kylie J. Stevenson is an interdisciplinary researcher in the Centre for Learning and Teaching at Edith Cowan University in Perth, Western Australia. Kylie's diverse research is primarily concerned with the lived experience and spans children and technology, welfare, creativity, reflective practice, arts education, and higher degree by research pedagogy.

Lelia Green is Professor of Communications in the School of Arts and Humanities at Edith Cowan University, Perth, Australia. With the support of the Australian Research Council, and in collaboration with a range of colleagues from Australia and overseas, she has been researching children's digital lives for over 20 years.

Donell Holloway is Senior Research Fellow at Edith Cowan University, Australia. She is a Chief Investigator on three Australian Research Council grants the ARC Centre of Excellence for the Digital Child; *The Internet of Toys: Benefits and risks of connected toys for children*; and *Toddlers and Tablets: Exploring the risks and benefits 0–5s face online*.

Kelly Jaunzems is a doctoral candidate at Edith Cowan University where she also works as a research associate on Australian Research Council-funded projects, including *Toddlers and Tablets: Exploring the risks and benefits 0-5s face online*. She is co-editor of the book, *Digitising Early Childhood*.

Part II
Learning and Teaching

Chapter 9
Media Arts in Early Childhood: A Framework for Developing Young Children's Creative Participation in the Digital World

Judith Dinham

Introduction

Media studies first appeared in the school curriculum when the evident effectiveness of mass media propaganda during the World Wars, the advent of television, and the growing use of visual media by advertisers, fuelled concerns about the unwitting consumption of media messages. The curriculum focused on immunising children against perceived negative influences (Dinham, 2020). When research in the 1980s highlighted the social, cultural and educative dimensions of children's television consumption (Hodge & Tripp, 1986; Singer et al., 1981; Zuckerman & Zukerman, 1985), analysing and learning *about* media became a focus within the English curriculum.

The birth of the digital age has again changed society's relationship to mass media because the diverse affordances offered by the internet, and the concomitant proliferation of media communication tools, have made the media space available to all. This has radically restructured our personal, local, and global networks; the way we engage socially, culturally and politically; and the way we understand ourselves and our place in the world (Postill & Pink, 2012; Hopkins, 2016).

In this digital world, children not only need to be skilled at analysing and evaluating the way meaning is constructed, and safely navigating the virtual social space, they increasingly require the capability to think *with* media (Ohler, 2008) as they construct meaning through their own creative production and communication.

Policy makers and educators have grappled with the implications of this fast-changing world. There has been focus on the 'digital divide', with initial concerns about equity of children's access to technology being replaced by concerns about the

J. Dinham (✉)
Curtin University, Perth, Australia
e-mail: j.dinham@curtin.edu.au

D. Holloway et al. (eds.), *Young Children's Rights in a Digital World*, Children's Well-Being: Indicators and Research 23,
https://doi.org/10.1007/978-3-030-65916-5_9

equity of their access to bandwidth (Hilbert, 2011). Another focus has been on 'digital citizenship' and developing children's safe and responsible engagement with digital technologies (Ohler, 2010; Ribble, 2015). We have also seen the introduction of 'digital literacy' education to develop children's capacity to navigate digital contexts. Overall though, little attention has been given to developing children's capabilities to be *creative and active participants* who benefit from the opportunities that digital access offers (Jenkins et al., 2006; Ohler, 2010).

In 2014, the United Nations' Committee on the Rights of the Child initiated discussion about digital media in the context of children's rights—including the right to freedom of expression and the right to be heard. Research findings presented to the Committee highlighted the importance of ensuring that children were able to benefit from the opportunities of digital access (Third et al., 2014). The lead researcher, Amanda Third (2014) noted that "... in our efforts to ensure that children's right to be protected is upheld, we may in fact be curbing their right to participate, and subsequently formulate their own opinions and new ways to use technology" (Institute for Culture and Society, UWS, 2014).

With the digital world profoundly transforming the ways we think, communicate, and make meaning—a number of scholars and researchers describe the evolution of a new literacy (epiliteracy). Many of the observed characteristics, such as a participatory and navigational approach to learning and expression, are in accord with the thinking and work practices employed by artists. The first proposition in this chapter is that young children's engagement in arts learning experiences will facilitates their development of the habits of mind and authorial practices of the digital age.

In the Australian Curriculum, Media Arts as one of the five disciplines in the Arts Learning Area. The second proposition in this chapter is that media arts offer valuable and playful ways for children, from a very young age, to participate actively and creatively in the digital space. Media Arts has its roots in the invention of photography in the nineteenth century. It evolved in the twentieth century to include mass media forms such as film and television, and now in the twenty-first century incorporates 'new media art'—a term used for arts practices that explore and employ the affordances of digital and other new technologies: multimodality, interactivity, robotics, computer graphics, and computer animation. Since the Arts curriculum is focused on children telling their own stories and communicating their own ideas through different modes of expression (Dinham & Chalk, 2018), the Media Arts curriculum strenuously supports children's right to freedom of expression and their right to be heard in the digital context.

The third proposition in this chapter is that the principles for integrating arts learning into the teaching programme outlined in the ArtsIN Framework (Hartle et al., 2015) provide a pedagogical model for the early childhood educator to support young children's engagement in the arts in ways that develop their ability to actively and creatively participate in the digital world.

The Digital Ecology

The advent of digital technology is overrunning the familiar text-based world and demanding "new habits of mind, new ways of processing culture and interacting with the world around us" (Jenkins et al., 2006, p. 21). This seismic change in the way we think and the way we know ourselves and engage in our world—our being, belonging and becoming (Peers & Fleer, 2014)—cannot be ignored nor addressed by merely introducing some digital devices into the classroom.

In fact, because digital technologies are uniquely 'malleable' (Hopkins, 2016) 'affordances' (Gibson, 1986), with the potential to be used by different people in different contexts at different times in different ways, they have become deeply integrated and embedded in our world and in our interpersonal, social and cultural practices. This is creating a new ecology: a network of experiences and practices emerging in relational ways across natural, socio-cultural, built, and virtual environments (Stokols, 2018).

Very young children's active participation in this ecology is supported by several factors. Firstly, the core devices are generally affordable, portable and available. In Australia, for example, one third of pre-schoolers have their own tablet or smartphone (Royal Children's Hospital, Melbourne, 2017), and many more have access to them. Secondly, devices are increasingly easy to manage and very young children can readily draw, take photos and play games by using their fingers to touch, swipe and press a touchscreen. Thirdly, the multiple modes (visual, aural, textual) dissolve age, literacy and language barriers. Very young pre-literate children can effectively create and communicate their ideas using the technology around them.

In light of this reality, Barr (2019) suggests we need to move beyond thinking of media in the early childhood context as a 'nuisance variable' and instead accept it as a fundamental part of the context in which children's development occurs. By working *with* media and the way it shapes our belonging, being and becoming, we can educate children in the new literacy practices for participating as active agents in their socio-cultural world.

The Epiliterate Child Developing Literacy with the Mindset of the Artist

Studies of how older children employ digital affordances in their self-initiated and self-directed practices—inevitably outside the school environment—have led to observations about the characteristics of creative engagement and the way meaning is constructed in a globally connected, dispersed, open-access, digital world (Dezuanni, 2018; Huber et al., 2015; Jenkins et al., 2006; Kress, 2003; Marsh, 2006). While this is an evolving landscape, many of these identified practices correlate with the way artists engage in meaning making (Orr & Shreeve, 2017).

Kress (2003), arguing that being literate in the digital age involves being able to think and express ideas in different modalities or combinations e.g. text, visual and audio, proposes that a *design* approach to literacy suits the multimodal environment of a digital world. Huber et al. (2015) observe that a design approach invokes the mindset of the artist, "weaving together linguistic, visual, aural, gestural and spatial features to form coherent compositions" (p. 45). They go on to argue that, like the artist, individuals exercise agency to curate personalised multimodal compositions according to their requirements and preferences. The authors coin the term epiliteracy ('epi' from Latin meaning 'after' or 'post') to describe these emerging authorial practices.

Jenkins et al. (2006) lodge such practices within a 'participatory culture'. They observe how children, when left to their own devices, gravitate towards creating and communicating in collaborative and inclusive ways. In dispersed, fluid social networks formed across different cultural territories, the skills they activate are those of play-based learning, performance, appropriation, teamwork, transmedia navigation, networking, mentoring and negotiation. These practices resonate with the practices of the artist, and underscore the way the arts provide a bridge to the epiliterate practices required for participating in the new socio-cultural dynamics of an interconnected, digital world.

One of the reasons the expressive practices of artistic engagement align to those adopted in a digital world is that the Arts eschew prescriptive solutions. While 'well-structured' learning domains such as Maths are characteristically based on theorems and laws which are applied in a variety of situations with predictability, the Arts is an 'ill-structured' learning domain (Efland, 1995; Short, 1995) because meaning is not fixed: concepts vary in meaning, depending on the context.

The artist also typically has a navigational approach to learning (Bourriaud, 2006), assembling information, and linking up with collaborators, to meet specific needs, skill requirements and aspirations in a specific context. Perkins (2007) refers to this as an 'import' learning paradigm. He contrasts this to an 'export' learning paradigm readily seen in classrooms where information is learned so it can be applied at some later stage if required. Artists are drawn to researching and investigating topics, or learning arts-making processes, as required. They assemble information, or learn processes, or seek out experts, around the requirements they have in a particular context and at a particular time.

The multimodalities of the digital world with multiple points of entry and different ways of participating, mean that epiliterate people—like artists—are sourcing and combining different elements of experience and expression in different ways. They are actively involved in their own learning trajectory while pursuing their own response to challenges, problems or opportunities. They seek out information or skills to address their particular expressive desires in particular contexts. In the digital world, expertise does not reside in set edifices and locations, but is dispersed, and often localised. Seeking out and employing expertise in a 'just in time' way characterises the processes of meaning-making in both the artistic and digital worlds.

Being present in digital spaces is a growing phenomenon; and storytelling through video, podcasts, and games involves participation as a performer—or avatar. Increasingly the embodiment of characters in digital games and learning spaces invites children into scenarios where they view the world from their character's perspective. The 'performing arts' (dance, drama and music), offer familiar forms for children's learning through embodied expression and communication. In many online games, an immersive experience involves role playing and improvisation. These practices, whereby children learn by inhabiting or becoming a character, are traditionally found in drama activities. In immersive digital spaces children are often engaged with other participants and "collaboratively theorise about manipulating entirely new worlds" (Jenkins et al., 2006, p. 30). Again, in dance, drama and music, performers work collectively to realise a shared vision within a world they are creating and bringing into being.

The Nature of Arts Learning Experiences

In the Australian education context, the Arts Learning Area (Dance, Drama, Media Arts, Music and Visual Arts) is a mandated, educationally-rich curriculum from the Foundation year of schooling. Arts learning is also an important dimension of well-designed early childhood programmes. Despite this, the arts are likely to be present in the classroom in the form of 'directed production' activities such as holiday decorations and cards (Narey, 2009). This misunderstands and misrepresents the arts and their significant role in children's learning, whereby the arts are understood as semiotic systems for making ideas manifest in symbolic form (Gardner, 1999; Lowenfeld, 1947; Lowenfeld & Brittain, 1987; McArdle & Wright, 2014). In authentic arts learning experiences, children will construct knowledge within social and cultural contexts and make meaning through their explorations (Dinham, 2020; Dinham & Chalk, 2018; Narey, 2009; Wright, 2012).

Authentic arts learning is quintessentially play-based learning in that it is active, experiential and participatory; challenging and intellectually stimulating; open-ended and exploratory; supportive of imaginative engagement and interpretation; intrinsically motivating; pleasurable; and meaningful to the children involved (Dinham & Chalk, 2018). This reinforces its value in young children's learning.

In the Foundational year of the Media Arts curriculum, children are expected to use media technologies to capture images, sound and text. Working individually or collaboratively they are involved in telling their own stories in visual and multimedia forms. The curriculum asks that children learn the conventions of multimedia authorial, social and ethical practices, along with how to use equipment correctly and safely; and to reflect on the nature of media arts products and their cultural significance. Children's reflections may relate to their own creations, those of their peers, or those of experienced arts practitioners.

Learning activities for children include telling their stories in forms such as posters, comic strips, photo stories, digital storybooks, stop-motion animations, podcasts and videos. As children progress through the Media Arts curriculum, they are learning to be *creators* in the digital realm, as well as informed consumers.

Children's Storytelling in Media Arts

For the early childhood educator, an arts-integrated curriculum is a practical and play-based way of introducing children to many of the literacy practices of a digital world, because of the nature of arts learning and the emphasis placed on children telling their own stories in different modes. Some arts practices don't necessarily employ digital technologies but they help develop the epiliterate mindset. In Media Arts for example, the comic strip, which has been around for 400 years, introduces children to core concepts for telling stories in other multimodal forms such as digital storybooks and videos. Furthermore, since drawing is a significant 'first language' thereby enabling pre-literate children to make their ideas manifest (Lowenfeld, 1947; Lowenfeld & Brittain, 1987; McArdle & Wright, 2014)—the image-based nature of comics means they make an ideal entry point for young children.

The comic strip format and the causal or chronological sequence of events introduces the developmental challenges of thinking multimodally (image and text), episodically, and temporally. Since the narrative arc and plot of a comic strip is analogous with written narratives, the creation of comic strips lays foundational concepts for reading and writing (McArdle & Wright, 2014) as well as for digital storybooks and video creations. The comic-strip has its codes and conventions such as the iconic speech bubble, first used by Richard Outcault in *Yellow Kid* (1895), so learning how to use these, (the language), prefigures the codes and conventions children will learn to use with other communication formats such as the digital storybook or video.

In the same way that educators introduce children to planning for written storytelling (beginning, middle and end), so they can introduce children to planning for multimodal storytelling using a storyboard. Storyboards are a planning staple of the movie industry. Each shot in a film production is pictorially described in sequence, with relevant production notes, so for children it assists in refining a story's arc (for a comic strip) and telling it within a limited number of slides (for a digital storybook); and it underpins the organisation required when creating videos where there are separate production roles (cameraperson, audio, props).

Digital storytelling (using familiar PowerPoint® software) extends children's written storytelling through the use of images, text and a range of affordances. Children can curate images from the internet, photograph their own drawn, painted or collaged pictures, take photographs or copy existing ones. As with poster-making, they are learning many principles of graphic design as they consider how to arrange the components of image, graphics (such as lines, arrows) and text (font, size, colour and placement) to convey the message. Digital storytelling also introduces the possibilities of using different transitions between slides, sound effects, music, voice-over, staged release of components, and timing in the story's telling.

Stop-motion animations in the vein of *Wallace and Gromit* are created by stringing together a sequence of still-images that capture the progressive stages of physical actions. For young children these are more cognitively demanding than videos because the sequence of still-images required to create the appearance of movement extends outside the narrative arc of the story. In a study of 4 years old children's storytelling using stop-motion animation techniques (Marsh, 2006), it was observed that while a number of children used the technology to photograph sequences of images in an episodic comic-book style, one child demonstrated a nascent understanding of the animation concept in her creation. When the educator demonstrated an animation sequence, the child was able to independently create 23 more still-images, which, when sequenced, produced the effect of movement.

In other examples, Year Two children working in pairs demonstrated successful mastery of stop-motion animation concepts and processes using Claymation. Creating the images using three-dimensional objects such as matchbox cars and figurines in a self-constructed stage is conceptually easiest for children to master. When figures are created in clay or plasticine (Claymation) and remodelled by the child for each still image, the challenge is greater, but easier than drawing images to create the set of stills because the physical manipulation of the figurines mimics the way children manipulate toys in the physical world to animate them, such as walking a doll by manipulating its legs.

An important aspect of arts education is children's participation as audience members—including being an audience for their peers. Children's comic strips can be photocopied and added to the class library. In the same way educators use story-time to introduce young children to the conventions of children's books and story structures, they can introduce children to the conventions of comics and digital stories. Besides reviewing professional examples, the considered review of children's own creations acknowledges children as competent and valued participants in the world of ideas.

The Educator's Role

A recent survey by Early Childhood Australia revealed that only 13% of respondents believed early childhood centres were satisfactorily integrating technologies into play-based learning; and that many early childhood educators struggled with the concept (Edwards et al., 2018). However, as is argued in this chapter, children's rights to be competent and active participants in a digital world—with agency, voice and influence (Broström, 2012)—can be addressed within an arts-integrated programme where children's exploration, self-expression and storytelling are in focus; and where using the multimodalities offered by digital technology are easily incorporated.

The ArtsIN Framework (Hartle et al., 2015) conceptualises the educator's primary roles in an arts-integrated model of learning. There are five roles the educator plays: artist, researcher, designer, co-constructor, and advocate. Configuring this model specifically for integrating media arts in early childhood education provides a meaningful pedagogical approach for educators to adopt.

Artist: The concept of being co-learners with children highlights the importance of educators developing their own artistic capabilities and habits of mind. Undertaking professional development with the guidance of experienced artists or embarking on individual projects are valuable ways for educators to position themselves in the arts learning space. All educators are familiar with PowerPoint software, so a personal project to develop a children's digital storybook (as a teaching resource) means the educator is positioned within the experiential and creative space where learning is gained from direct experience. This develops the educator's skills, confidence and mindset in ways that align to epiliteracy principles.

Researcher: By being an engaged creative practitioner, educators develop experientially-driven understandings of the nature of artistic expression. This leads to a new appreciation of children's artistic and creative engagement; and the way it narrates their belonging, being and becoming. Through careful observation and reflection, educators can see how children are "experts on the subject of their own lives" (Broström, 2012, p. 261) and how children's voices and viewpoints enter the learning space in informative ways through their creative productions. This helps support the educator-researcher's recognition of the data that reveal the learning embedded in children's digital media arts creations.

Designer: As educators, through their own professional development and artistic engagement, become aware of the possibilities for creating and communicating in media arts contexts themselves, the possibilities for children-led explorations and creative inventions using the affordances of digital media become more apparent. Through well-designed media arts experiences, children will have opportunities to tell their own stories, develop interview and performance techniques for podcasts and videos, develop technical multimodal communication skills, and exercise critical thinking as they respond to the creations and stories of others.

Importantly, many digitally-based practices replicate or extend existing early childhood practices. For example, presenting children with a set of photographs that document the stages of a cooking process or the growth of a plant from seed, and asking children to sort them into chronological order, is a first step towards sequencing events in a narrative. It is also a step into a digital world where images have an enhanced role in storytelling. The parallels with picture books—and the way educators develop children's capacity to 'read' the pictures, make judgements from pictorial clues about the characters, and predict the next event—provide a pedagogical bridge into photographic storytelling. When young children are encouraged to tell their own stories on the topic of *My family* or *My favourite places in the garden,* by photographing and curating images, educators are supporting them to research, make decisions, organise their ideas, exercise judgement, and express their understandings, using media.

Co-constructor: With an emphasis on children expressing their own ideas and meaning-making, the educator's role in the learning process is one of a co-constructor who stimulates, challenges and stretches children's thinking. An important way of doing this is presenting open-ended challenges. This shifts the educator's role from directing the learning process towards a prescribed goal to one aimed at launching projects that invite creative thinking. This shifts the focus to brainstorming-type activities, research strategies, and experimentation; as well as exploring the inherent potential of different 'making' materials and processes; and developing understandings of the different forms of representation—the elements, codes and conventions. For example, in media arts, the point-of-view of the camera angle—looking down or looking up at the subject—conveys different messages about the power or significance of the subject.

The pedagogical shift towards supporting creative thinking places emphasis on the first two stages of the creative process, as identified by Wallas (1926). In the first of these, the Preparation stage, attention is given to expanding understanding of the problem or challenge. Brainstorming is a classic strategy but so too are strategies such as researching the concept, and viewing a range of visual stimuli. For example, a project about fish would include viewing images of different fish from colourful reef fish to deep sea monsters so that the initial idea of a fish is fully fleshed out. The second, Incubation stage of the creative process, is where ideas have time to percolate. Besides allocating time for ideas to resolve, this stage also includes active experimenting undertaken in a free-flowing way. In this stage of the creative process, artists play with different materials and different configurations. They generate testers, samples and models. Nothing is fixed but playing with the materials generates its own insights. As with the emergent curriculum model familiar to early childhood educators, the educator's role is to be responsive to the needs that emerge, or the possibilities that arise, from children's investigations.

Advocate: Engagement in authentic personal learning provides educators with a foundation for research and data-driven evaluations of children's learning. It promotes the design of meaningful learning opportunities that engage the educator as a co-constructor of learning through the adoption of appropriate pedagogical strategies for supporting children's construction of their own learning. This experience enables educators to become advocates when their "strong communication skills in multiple literacies provides courage of conviction and ability to take risks in new venues and environments" (Hartle et al., 2015, p. 296).

Conclusion

The significance of the transformational effects of the digital world cannot be underestimated. As with the transformation from oracy to a text-based world, new ways of constructing meaning and participating in the world are unfolding before us. The term epiliteracy is used here to conjure new ways of thinking, learning and participating in the multimodal digital world across interpersonal, social and cultural networks. The propositions made in this chapter are that the essence of epiliteracy can be found in the artist's habits of mind and creative practices; and that media arts experiences, as part of the Arts Learning Area in the school curriculum, provide a pedagogical bridge into the digital world for the early childhood educator.

Importantly, the principles of authentic arts education support the agency of children as competent communicators, whose ideas and views of the world are sought and valued, and whose rights to be creative participants in a digital world, are axiomatic. For early childhood educators, the five roles nominated in the ArtsIN Framework show how to support children's creative learning and active participation in digital spaces; and their development as epiliterate individuals.

References

Barr, R. (2019). Growing up in the digital age: Early learning and family media ecology. *Current Directions in Psychological Science*. https://doi.org/10.1177/0963721419838245.

Bourriaud, N. (2006). *Relational aesthetics* (rev. ed.) (S. Pleasance and F. Woods, Trans.). Adriana Hidalgo Editoria.

Broström, S. (2012). Children's participation in research. *International Journal of Early Years Education, 20*(3), 257–269.

Dezuanni, M. (2018). Minecraft and children's digital making: Implications for media literacy education. *Learning, Media and Technology, 43*(3), 236–249.

Dinham, J. (2020). *Delivering authentic arts education* (4th ed.). Cengage.

Dinham, J., & Chalk, B. (2018). *It's arts play: Young children belonging, being and becoming through the arts*. Oxford University Press.

Edwards, S., Straker, L., & Oakey, H. (2018). *Discussion Paper: Towards an Early Childhood Australia statement on young children and digital technology*. Early Childhood Australia Digital Policy Group. http://www.earlychildhoodaustralia.org.au/wp-content/uploads/2017/08/ECA-DPG-Disussion-Paper-April-including-appendices_FINAL-2.pdf.

Efland, A. (1995). The spiral and the lattice: Changes in cognitive learning theory and their implications for teaching and learning in the arts. *Studies in Art Education, 36*(3), 134–153.

Gardner, H. (1999). The happy meeting of multiple intelligences and the arts. *Harvard Education Letter, 15*(6), 1.

Gibson, J. J. (1986). *The ecological approach to visual perception*. Psychology Press.

Hartle, L. C., Pinciotti, P., & Gorton, R. L. (2015). ArtsIN: Arts integration and infusion framework. *Early Childhood Education, 43*, 289–298. https://doi.org/10.1007/s10643-014-0636-7.

Hilbert, M. (2011). The end justifies the definition: The manifold outlooks on the digital divide and their practical usefulness for policy-making. *Telecommunications Policy, 35*(8), 715–736. https://doi.org/10.1016/jtelpol.2011.06.012.

Hodge, R., & Tripp, D. (1986). *Children and television*. Polity Press.

Hopkins, J. (2016). The concept of affordances in digital media. In H. Friese, G. Rebane, M. Nolden, & M. Schreiter (Eds.), *Handbuch soziale praktiken und digitale alltagswelten* (pp. 1–8). Springer. https://doi.org/10.1007/978-3-658-08460-8_67-1.

Huber, A., Dinham, J., & Chalk, B. (2015). Responding to the call: Arts methodologies informing 21st century literacies. *Literacy, 49*(1), 45–54.

Institute for Culture and Society, UWS. (2014). *United Nations rights of the child in the digital age*. https://www.westernsydney.edu.au/ics/news_and_media/news/2014/united_nations_right_of_the_child_in_the_digital_age.

Jenkins, H., Clinton, K., Purushotma, R., Robison, A. J., & Weigel, M. (2006). *Confronting the challenges of participatory culture: Media education for the 21st century*. MacArthur Foundation.

Kress, G. (2003). *Literacy in the new media age*. Routledge.

Lowenfeld, V. (1947). *Creative and mental growth*. Macmillan.

Lowenfeld, V., & Brittain, W. L. (1987). *Creative and mental growth* (8th ed.). Pearson.

Marsh, J. (2006). Emergent media literacy: Digital animation in early childhood. *Language and Education, 20*(6), 493–506. https://doi.org/10.2167/le6660.0.
McArdle, F., & Wright, S. M. (2014). First literacies: Art, creativity, play, constructive meaning-making. In G. Barton (Ed.), *Literacy in the arts: Retheorising learning and teaching* (pp. 21–37). Springer.
Narey, M. J. (2009). Introduction. In M. J. Narey (Ed.), *Making meaning: Constructing multimodal perspectives of language, literacy, and learning through arts-based early childhood education* (pp. 1–2). Springer.
Ohler, J. (2008). *Storytelling and new media narrative.* http://www.jasonohler.com/storytelling/storymaking.cfm.
Ohler, J. (2010). *Digital community, digital citizen.* Corwin.
Orr, S., & Shreeve, A. (2017). *Art and design pedagogy in higher education: Knowledge, values and ambiguity in the creative curriculum.* Routledge.
Peers, C., & Fleer, M. (2014). The theory of 'belonging': Defining concepts used with belonging, being and becoming – the Australian early years learning framework. *Educational Philosophy and Theory, 46*, 914–928. https://doi.org/10.1080/00131857.2013.781495.
Perkins, D. (2007). Foreword. In L. Hetland, E. Winner, S. Veenema, & K. M. Sheridan (Eds.), *Studio thinking: The real benefits of visual arts education* (p. v). Teachers College Press.
Postill, J., & Pink, S. (2012). Social media ethnography: The digital researcher in a messy web. *Media International Australia, Incorporating Culture & Policy, 145*, 123–134. http://search.informit.com.au/documentSummary;dn=992474822836114;res=IELLCC.
Ribble, M. (2015). *Digital citizenship in schools: Nine elements all students should know* (3rd ed.). International Society for Technology in Education.
Royal Children's Hospital, Melbourne. (2017). *Australian child health poll: Young children owning smartphones is the new normal* [Press release]. www.rchpoll.org.au/media-centre/.
Short, G. (1995). Understanding domain knowledge for teaching: Higher order thinking in pre-service art teacher specialists. *Studies in Art Education, 36*(3), 154–169.
Singer, D. G., Singer, J. L., & Zuckerman, D. M. (1981). *Teaching television: How to use TV to your child's advantage.* Dial Press.
Stokol, D. (2018). *Social ecology in the digital age: Solving complex problems in a globalized world.* Academic Press.
Third, A., Bellerose, D., Dawkins, U., Keltie, E., & Pihl, K. (2014). *Children's rights in the digital age: A download from children around the world.* Young and Well Cooperative Research Centre.
Wallas, G. (1926). *The art of thought.* Harcourt Brace.
Wright, S. (Ed.). (2012). *Children, meaning-making and the arts* (2nd ed.). Pearson.
Zuckerman, D. M., & Zuckerman, B. S. (1985). Television's impact on children. *Design for Arts in Education, 86*(6), 39–45. https://doi.org/10.1080/07320973.1985.9940727.

Judith Dinham is Associate Professor, School of Education, Curtin University. Her discipline area is the arts. Her widely adopted university textbook written with Beryl Chalk, It's Arts Play: Young children belonging, being and becoming through the arts (OUP, 2018), positions the arts, including media arts, in contemporary early childhood education.

Chapter 10 The Impact of Digital Technologies on the Role of the Early Childhood Teacher

Vicki Schriever

Introduction

It is acknowledged by Ernest et al. (2014) that "…few topics change as quickly and elicit as strong an emotional response as a consideration of the role and influence of media, technology and screen time on young children" (p. 183). Uncertainties and tensions about the place of digital technologies in the life of the young child persist. This is despite the ubiquitous presence of digital technologies in society and the increasing prevalence of digital technologies in educational contexts, including kindergarten. The inclusion of digitally-focused learning outcomes in Australian early childhood curriculum documents, including *Belonging, being and becoming: The early years learning framework for Australia* (EYLF) (Department of Education, Employment and Workplace Relations [DEEWR], 2009), places an expectation on early childhood teachers to develop their own digital competencies and to facilitate children's learning with information communication technologies (ICT).

Transforming the intended curriculum, into the enacted curriculum and lived experience for the early childhood teacher and children is however, neither universal nor uniform. The findings reported in this chapter are drawn from research investigating how early childhood teachers working in kindergarten settings understand and manage their changing roles regarding digital technologies. This study found that the role of the early childhood teacher has been, and continues to be, impacted by digital technologies.

Young children's experiences with digital technologies can challenge early childhood professionals as they encounter new and uncharted situations in their work with children and families (Edwards et al., 2018). Lindahl and Folkesson

V. Schriever (✉)
University of the Sunshine Coast, Sunshine Coast, Australia
e-mail: vschriev@usc.edu.au

D. Holloway et al. (eds.), *Young Children's Rights in a Digital World*, Children's Well-Being: Indicators and Research 23,
https://doi.org/10.1007/978-3-030-65916-5_10

(2012) acknowledge that the introduction of digital technologies into kindergarten is expected to bring attention to demands on teachers' roles. This study acknowledged and exposed the significant impact digital technologies can have on the role of the early childhood teacher and revealed the actions taken by early childhood teachers to manage digital technologies within their kindergarten.

Literature Review

A prevailing narrative in our society is that "we live in rapidly changing times" (Gibbons, 2016, p. 367), with these changing times bringing forth new challenges not previously faced by past generations. A significant, contemporary challenge facing early childhood teachers is the need to manage the emergence of digital technologies in the context of their kindergarten. The EYLF (DEEWR, 2009) articulates an expectation that ICT will be actively and effectively used by both children and educators to support and facilitate learning. Outcome 4: Children are confident and involved learners states that "children resource their own learning connecting with people, places, technologies and natural and processed materials" (DEEWR, 2009, p. 37) and Outcome 5: Children are effective communicators declares that "children use ICT to access information, investigate ideas and represent their thinking" (DEEWR, 2009, p. 44).

Children's learning and engagement with ICT, and the early childhood educator's responsibility to support children to become effective communicators using a range of ICT are detailed within these learning outcomes. Learning Outcome Five is located within the broader sphere of communication and sits alongside literacy and numeracy learning outcomes. The inclusion of learning outcomes focused on the development of digital technology confidence, skills, and knowledge for young children is reflective of the needs of twenty-first century learners and citizens (Schriever, 2018).

While the curriculum details an expectation that early childhood teachers will integrate digital technologies and support children to attain ICT-focused learning outcomes, in practice, managing the emergence of ICT, facilitating digitally-enabled learning experiences and meeting the digital rights and needs of children, is a task fraught with tensions and uncertainties for early childhood teachers. Edwards (2016) acknowledges the difficulty of this task stating that "integrating technologies, digital media and popular culture with the provision of play-based learning experiences in early childhood education has not been easy for teachers to achieve in practice" (p. 513). Anxieties abound for early childhood teachers, as digital technologies appropriateness and place in young children's lives continues to be a contentious issue within society. Integrating digital technologies into early learning programs remains one of the most fiercely debated issues in the field of early childhood education and care (Ihmeideh & Al-Maadadi, 2018).

Early Childhood Australia [ECA] aimed to alleviate some of the uncertainties that exist surrounding young children and ICT by developing the *Statement on young children and digital technology* (ECA, 2018). The *Statement on young children and digital technology* was developed in recognition of the growing role of digital and interactive technologies in young children's lives and the need for greater guidance on technology use to inform the early childhood sector, families of young children and decision-makers about using interactive technology with, by, and for young children (ECA, 2018). Early childhood teachers seek to manage children's rights in a digital world and endeavor to balance the need between children's empowerment and protection with digital technologies.

Methodology and Methods

The methodology of grounded theory was used to investigate the following research question:

> **How do early childhood teachers working in kindergarten settings understand and manage their changing roles with digital technologies?**

Grounded theory moves beyond description, towards the development of an inductively derived substantive theory about a phenomenon, one which is discovered, rigorously developed and provisionally verified through the means of systematic data collection and analysis (Strauss & Corbin, 1990). A zig-zag, iterative approach to data collection and data analysis was used, which is unique to this methodology (Creswell, 2007). Four iterations of data collection occurred, interspersed with periods of data analysis. The constant comparison, constant interrogation method was used, whereby new data was analysed in light of the findings gained from previously collected and analysed data.

Interviewing was the principal mechanism used for the generation of data (Birks & Mills, 2015; Creswell, 2007). Semi-structured, in-depth interviews were undertaken in a face-to-face manner with 19 practising early childhood teachers. Through this process, the researcher was able to hear about their lived experiences with ICT, explore how they perceive their role, investigate how their role had been impacted by the inclusion of digitally-focused learning outcomes, and gain insights into their understanding and management of digital technologies within their kindergarten. There were numerous strengths associated with using semi-structured, in-depth interviews including its ability to gather rich information, follow interesting points, to include material which was not previously anticipated, and it enabled the researcher and the participant the opportunity to go into greater detail (Curtis & Curtis, 2011).

Theoretical sampling was used to purposefully select participants who were able to contribute to the development of the theory (Creswell, 2007). Theoretical sampling is unique to grounded theory research and is an essential element in making the process an emergent one (Birks & Mills, 2015). When sampling theoretically, the author went to places, persons and situations that provided information about the concepts she was endeavouring to learn more about (Corbin & Strauss, 2008). This

Table 10.1 Roles held by participants

Early Childhood Teacher: n = 19		
Additional roles	Centre director or co-director: n = 12	Centre Owner: n = 1

Table 10.2 Kindergarten settings

Total number of kindergarten settings: n = 19		
Community Kindergarten: n = 8	Kindergarten within a long day care centre: n = 7	Kindergarten co-located on school site: n = 4

process led the author to interview 19 early childhood teachers in regional Queensland, Australia who were responsible for planning and implementing the kindergarten program. Some participants held additional roles within their kindergarten such as centre director or co-director, and in one instance the participant was the part-time kindergarten teacher, director, owner and operator of a two-room long day care centre. The roles held by the participants is presented in Table 10.1.

Each of the 19 early childhood teachers interviewed for this study were employed at a different kindergarten setting. These settings were inclusive of community kindergartens, kindergartens within a long day care centre and kindergartens co-located on a state or private school site. Representation of the different kindergarten settings is detailed in Table 10.2.

Data analysis and data collection represent a tightly interwoven process within this study, as data analysis guided and informed future data collection (Creswell, 2007; O'Donoghue & Haynes, 1997). The iterative approach to data collection and data analysis occurred until the study reached a point of theoretical saturation, when no new data was being unearthed and the theory was well developed in terms of density and variation (Strauss & Corbin, 1998).

The processes of data analysis are a core feature of grounded theory, giving the methodology its distinctive nature and ensuring the final conclusions are robust, reliable and valid. All codes were derived and developed from the data, with no pre-determined codes used (Corbin & Strauss, 2008). The act of coding took the raw data and raised it to a conceptually higher level and enabled each concept to be developed in terms of their properties and dimensions (Corbin & Strauss, 2008). Each category that emerged as a result of coding was inductively derived and grounded in the data (Berg, 1998).

Open coding was the first step in data analysis and served to name and categorise the phenomenon through close examination of the data (Strauss & Corbin, 1990). The interviews were analysed individually and collectively, with comparisons occurring within a single interview transcript and across different interview transcripts to find similarities and differences (Charmaz, 2014). While open coding cracked the data apart, the second coding process, axial coding, put the categories together in conceptually different ways and established connections between a category and its subcategories (Punch, 2009; Strauss & Corbin, 1990). The third coding process, selective coding, was developed on the basis of axial coding (Strauss

& Corbin, 1990) and served to integrate and bring together the developing analysis (Punch, 2009). Through the process of undertaking open, axial and selective coding the significant findings from this research were determined.

Research Findings

This study revealed that the digitisation of early childhood education and care impacted the role of the early childhood teacher. Key findings from the research exposed three distinct ways early childhood teachers viewed their role and responded to the emergence of digital technologies within their kindergarten. Early childhood teachers acted to:

- Regulate digital technologies and privileged traditional pedagogies; or
- Valued digital technologies and facilitated children's digital experiences; or,
- Endeavoured to strike a balance by incorporating some digitally-enabled experiences within a play-focused environment.

The voices and lived experiences of early childhood teachers are shared to reveal how early childhood teachers perceive their role and how they understand and manage their role with digital technologies in the context of their kindergarten.

Regulate Digital Technologies and Privilege Traditional Pedagogies

There were early childhood teachers in this study who acknowledged the caring elements of their role, privileged traditional pedagogies that fostered play and socialisation and restricted the presence of digital technologies within their kindergarten. Aster described her kindergarten environment, remarking:

> I want this to be a place where children get their hands in mud and get dirty and play and have agency and talk to one another, and I would never I think be happy for children to be sitting in a line of computers playing games (early childhood teacher and director).

The pervasive presence of digital technologies within society and its subsequent emergence into kindergarten contexts led some early childhood teachers to take on the role of being a protector of children and childhood, and this included protecting children from engaging with digital technologies. This perspective was evident in Heather's statement, "it is not a necessary tool for us to have at this point in time when we are only three-and-a-half to four-and-a-half" (early childhood teacher and director). It was reflected in the views shared by Cedar who said, "the fact that they exist would cause some people to think that therefore they must be used. I don't subscribe to that view. I think about how it can be used—not to use it" (early childhood teacher) and in Bryony's remark, "I just don't like the philosophy of it with children" (early childhood teacher and director). This study found that some early childhood teachers marginalised ICT and endeavoured to create a kindergarten environment for young children that was free from the presence of working digital technologies.

The strategy of actively avoiding using digital technologies was employed by some early childhood teachers and traditional pedagogical practices were privileged. This approach was revealed in Holly's statement,

> I think if I had more ICT in the room I'd probably still not use them every day, even though they're there for us to access. I'd prefer to have the one-on-one conversations or do research with the group (early childhood teacher).

Holly retained a focus on traditional pedagogical practices including teacher to child interactions, conversations, and collaborative research and stated that she would choose not to use ICT to a greater extent, even if they were provided to her.

Value Digital Technologies and Facilitate Children's Digital Experiences

When asked to define their role, some early childhood teachers in this study focused on the educational aspects of their role. Zinnia declared, "first and foremost, I'm an early childhood teacher. So, I'm here to provide an educational program, a play-based educational program for the children" (early childhood teacher and director). When as educationally focused perspective was held, the early childhood teachers perceived it to be a responsibility of their role to provide children with opportunities to access ICT and to engage in digital experiences. These early childhood teachers recognised children as being agential citizens participating in a digital world and empowered children to engage with digital technologies. Their actions were driven by the early childhood teacher's desire to support children to be prepared to achieve ICT-focused school curricular expectations and to be futures-focused and capable of meeting the demands of a digital society. This perspective was evident in Laurel's statement, "it's not just school readiness, it's life readiness" (early childhood teacher and director). It was also evident in Daisy's justification for providing digital technologies within her kindergarten. She said,

> Children are going to be working with technologies. I never think it's too young for them to feel comfortable and confident using them. I feel it's important to be helping them, to be prepared as much as possible. If we don't start here we're really not keeping up (early childhood teacher and director).

Some early childhood teachers in this study acknowledged the digital landscape of children's lives, valued children's digital rights and cultural capital, and adapted their pedagogical practices to include digital technologies. Fern declared,

> They're digital citizens and we've got to respect that. We are in an era that we didn't grow up in, we have no knowledge of and we've got to respect that they are our future... They come

in with a wealth of knowledge. You've got to take that, you've got to embrace that and run with it (early childhood teacher).

These early childhood teachers acknowledged children's rights to access digital devices and recognised that digital technologies play a significant role in the child's life, now and into the future.

Striking a Balance

Within this study there were early childhood teachers that spoke of both the caring and educational aspects of their role. Lily remarked, "the main thing is to provide a learning environment for the children that's safe and secure, where they feel that they can come and learn without worrying about other things" (early childhood teacher and director). When Lily defined her role, she spoke of traditional caregiving, including providing social and emotional support, along with identifying educationally focused elements such as creating a learning environment where children can learn free from concerns.

Situated between active avoidance and marginalisation of digital technologies, and confident and enthusiastic provisioning of digitally-enabled learning experiences, were the early childhood teachers who viewed it as a feature of their role to strike a balance between traditional pedagogical practices and digitally-enabled learning experiences.

Zinnia shared, that, "it's really all about balance" (early childhood teacher and director), and Senna said, "it's just another way of extending children's play, which is what we do. I definitely see its role and I definitely see that we do need to show the children that we're using it" (early childhood teacher and director).

Some early childhood teachers in this study perceived that they had achieved the appropriate balance when they included some digitally-enabled experiences alongside traditional, play-focused pedagogical practices. Calla shared her perspective stating, "I think there needs to be a balance of ICT in the classroom. I would probably say maybe 10% ICT and 90% hands-on, real-life learning" (early childhood teacher).

The act of incorporating digital devices and digital practices into the kindergarten program was undertaken to meet the future learning needs of children, however, there remained a primary focus on play and socialisation. This was reflected in Lily's statement, "there's a balance here, a little bit of technology and then there's free play" (early childhood teacher and director). The role of the early childhood teacher was impacted by the emergence of ICT as these participants sought to modify their pedagogical practices to strike a balance that they were comfortable with. This included acknowledging children's digital rights and developing their digital competencies, alongside valuing traditional pedagogies.

Findings Summary

Findings from the data revealed that the emergence of ICT has impacted the role of the early childhood teacher, with actions taken by participants to manage digital technologies within their kindergarten. The presence of ICT within society has led some early childhood teachers to insulate their kindergartens from digital technologies and to privilege traditional and historically accepted pedagogies centred on play and socialisation. In contrast, there were early childhood teachers who embraced ICT and adapted their pedagogical practices to include a variety of digitally-enabled learning experiences. Other early childhood teachers were attempting to strike a balance by integrating aspects of ICT into a traditional early childhood program. Early childhood teachers made decisions about ICT and took action to manage the impact of digital technologies within their kindergarten.

Discussion

Jonsdottir and Coleman (2014) reported on a study into the professional identity construction of Icelandic preschool teachers and found that the preschool teachers appeared to be insecure about whether to emphasise the caring or teaching aspect of their role. Fueling this insecurity is the care/education dichotomy which persists within the profession globally. Research conducted by Einarsodottir (2006, 2008) revealed the role of being an early childhood teacher could be envisioned in three distinct ways, with each perspective influencing how the role was enacted. The three perceptions of the role were:

1. **Traditional perception of the role**: Focus is on providing care, emotional and social support and early childhood is the golden age of free play and development.
2. **Educationally focused perception of the role**: Adults are teachers, not caregivers and the role of the early childhood teacher is to ensure children learn.
3. **Dual care/education perception of the role:** Caregiving and teaching in a play-based learning environment are mutually inclusive where both are necessary to ensure high quality experiences and outcomes (Einarsodottir 2006, 2008).

The findings of Einarsodottir (2006, 2008) have been applied to the empirical data gathered in this study to provide a lens through which to understand how the early childhood teacher's role perception and the emergence of ICT into their kindergarten environment impacts their decision-making and management of digital technologies. Lightfoot and Frost (2015) propose that professional identity is not simply a matter of a role being adopted for instrumental reasons in the context of an occupation; it is about who we are, rather than the part we are playing. The early childhood teacher's professional identity construct has a significant impact on the perspectives they hold and their beliefs, actions and management of digital technologies in the context of their kindergarten setting. How a role is defined and viewed is significant, as roles are the perspectives from which people act (Hewitt, 2003). The participants in this study shared how they envisioned and enacted their roles as early childhood teachers, strengthening or diminishing aspects of the role depending on the degree to which it was valued and contributed to their role perception.

Digital Technologies and a Traditionally Focused Perception of the Role

Findings from this study revealed that when the role of being an early childhood teacher was focused on the social and emotional care and development of young children within an environment of free play and exploration, they prioritised non-digital experiences for children and avoided providing opportunities for the children to use digital technologies. The use of digital technologies with, for, and by, young children did not 'fit' within the early childhood teacher's perception of their role and facilitating children's digital experiences was situated outside the scope of their responsibilities. Zabatiero et al.'s (2018) research revealed that early childhood educators were concerned that children engaged excessively with digital technologies and that technologies displaced children's traditional play time engaging in physical activities and promoted sedentary behavior. The emergence of ICT impacted on the role of the early childhood teacher, as they took on the responsibility of sheltering children from further exposure to digital technologies and sought to protect children's experiences of play. The research literature and data from this study revealed that many early childhood teachers do not view ICT as associated with play but rather fear technology as a threat to traditional free play (Edwards et al., 2016; Lindahl & Folkesson, 2012; Moir, 2014).

Digital Technologies and an Educationally Focused Perception of the Role

When the early childhood teacher's perception of the role was educationally focused, they emphasised the need to prepare children to become school-ready and viewed their role as laying the foundation for future learning. These early childhood teachers viewed it as their responsibility to facilitate children's engagement with ICT. This was undertaken to support children's educational attainment by developing ICT-focused knowledge, skills and dispositions needed to meet school-based curricular expectations. Furthermore, digital technologies were incorporated into the children's early learning experiences to enable the development of technological competencies that would serve them in their future beyond school. The early childhood teacher's motivation to use technology came from a strong value belief centred on supporting children to learn and preparing children for their future (Ottenbreit-Leftwich et al., 2010).

Digital Technologies and a Dual Care/Education Focused Perception of the Role

The early childhood teachers that perceived their role to encapsulate both caring and educational responsibilities endeavoured to find a balance that they were comfortable with, whereby some ICT were used within their kindergarten program, although traditional pedagogical practices continued to take precedence. A key tension experienced by some of the early childhood teachers was mediating the purpose of early childhood education and care as being futures-focused for the purpose of school readiness, with the need to provide a space and time in which childhood and learning can be experienced at its natural pace (Gibbons, 2016). Early childhood teachers at times experienced tensions regarding digital technologies and were uncertain about how to appropriately incorporate ICT within their kindergarten. Nuttall et al. (2015) acknowledge that children's contemporary lives give rise to the development of tensions which are keenly felt by early childhood teachers as they consider how to foster children's learning through digital technologies. These uncertainties and tensions arose as it was deemed necessary to facilitate children's digital experiences to fulfil the educational aspects of the role and to prepare children to meet school-based expectations, however, they also wanted to ensure ample opportunities were provided for children to engage in traditional pedagogies centred on play and socialisation.

Conclusion

This chapter has reported a grounded theory study undertaken to investigate how early childhood teachers employed in kindergarten settings understand and manage their changing roles regarding digital technologies. Semi-structured, in-depth interviews were undertaken with 19 practicing early childhood teachers to hear their voices and to gain insights into their lived experiences with ICT. The rigorous processes of grounded theory data collection and analysis generated a wealth of data and revealed three distinct ways the early childhood teacher's role was perceived and impacted by digital technologies.

The emergence of ICT led some early childhood teachers to take on the role of protector as they endeavoured to shield children from ICT by creating a digitally-free or heavily restricted environment and traditional early childhood pedagogies dominated. An alternative approach taken by some early childhood teachers was to facilitate children's digitally-enabled learning experiences to support the societal agendas of school readiness and meeting the needs of a technologically-mediated society. Other early childhood teachers sought to strike a balance, by incorporating aspects of digital technology within a play-focused environment. This approach was taken to acknowledge children's digital rights and lives, and to support their attainment of future-focused learning goals while continuing to value traditional pedagogical practices. This research exposed three distinct ways early childhood teachers perceive their role, revealed the impact ICT has on the role of the early childhood teacher, and uncovered the actions taken by early childhood teachers to manage digital technologies within their kindergarten.

References

Berg, B. L. (1998). *Qualitative research methods for the social sciences* (3rd ed.). Allyn and Bacon.

Birks, M., & Mills, J. (2015). *Grounded theory: A practical guide* (2nd ed.). Sage.

Charmaz, K. (2014). *Constructing grounded theory* (2nd ed.). Sage.

Corbin, J., & Strauss, A. (2008). *Basics of qualitative research: Techniques and procedures for developing grounded theory* (3rd ed.). Sage.

Creswell, J. (2007). *Qualitative inquiry and research design: Choosing among five approaches.* Sage.

Curtis, B., & Curtis, C. (2011). *Social research: A practical introduction.* Sage.

Department of Education, Employment and Workplace Relations. (2009). *Belonging, being & becoming: The early years learning framework for Australia.* Commonwealth of Australia.

Early Childhood Australia. (2018). *Statement on young children and digital technology.* http://www.earlychildhoodaustralia.org.au/wp-content/uploads/2018/10/Digital-policy-statement.pdf.

Edwards, S. (2016). New concepts of play and the problem of technology, digital media and popular-culture integration with play-based learning in early childhood education. *Technology, Pedagogy and Education, 25*(4), 513–532. https://doi.org/10.1080/1475939X.2015.1108929.

Edwards, S., Henderson, M., Gronn, D., Scott, A., & Mirkhil, M. (2016). Digital disconnect or digital difference? A socio-ecological perspective on young children's technology use in the home and the early childhood centre. *Technology, Pedagogy and Education, 26*(1), 1–17. https://doi.org/10.1080/1475939X.2016.1152291.

Edwards, S., Straker, L., & Oakey, H. (2018). *Discussion paper: Towards an Early Childhood Australia statement on young children and digital technology.* Early Childhood Australia. http://www.earlychildhoodaustralia.org.au/wp-content/uploads/2017/08/ECA-DPG-Disussion-Paper-April-including-appendices_FINAL-2.pdf.

Einarsdóttir, J. (2006). Between two continents, between two traditions: Education and care in Icelandic preschools. In J. Einarsdóttir & J. Wagner (Eds.), *Nordic childhoods and early education: Philosophy, research, policy and practice in Denmark, Finland, Iceland, Norway, and Sweden* (pp. 159–183). Information Age.

Einarsdóttir, J. (2008). Children's and parents' perspectives on the purposes of playschool in Iceland. *International Journal of Educational Research, 47*(5), 283–291. https://doi.org/10.1016/j.ijer.2008.12.007.

Ernest, J. M., Causey, C., Newton, A. B., Sharkins, K., Summerlin, J., & Albaiz, N. (2014). Extending the global dialogue about media, technology, screen time and young children. *Childhood Education, 90*(3), 182–191. https://doi.org/10.1080/00094056.2014.910046.

Gibbons, A. (2016). Do 'we' really live in rapidly changing times? Questions concerning time, childhood, technology and education. *Contemporary Issues in Early Childhood, 17*(4), 367–376. https://doi.org/10.1177/1463949116677921.

Hewitt, J. (2003). Symbols, objects and meanings. In L. T. Reynolds & N. J. Herman-Kinney (Eds.), *Handbook of symbolic interactionism* (pp. 307–326). AltaMira Press.

Ihmeideh, F., & Al-Maadadi, F. (2018). Towards improving kindergarten teachers' practice regarding the integration of ICT into early years settings. *The Asia-Pacific Researcher, 27*(1), 65–78. https://doi.org/10.1007/s40299-017-0366-x.

Jonsdottir, A. H., & Coleman, M. (2014). Professional role and identity of Icelandic preschool teachers: Effects of stakeholders' views. *Early Years, 34*(3), 210–225. https://doi.org/10.1080/09575146.2014.919574.

Lightfoot, S., & Frost, D. (2015). The professional identity of early years educators in England: Implications for a transformative approach to continuing professional development. *Professional Development in Education, 41*(2), 401–418. https://doi.org/10.1080/19415257.2014.989256.

Lindahl, M. G., & Folkesson, A. M. (2012). Can we let computers change practice? Educators' interpretations of preschool tradition. *Computers in Human Behaviour, 28*(5), 1728–1737. https://doi.org/10.1016/j.chb.2012.04.012.

Moir, T. (2014). Getting in touch with technology without losing touch with early childhood pedagogy. *Educating Young Children, 20*(1), 34–37. search.informit.com.au/documentSummary;dn=280044741645353;res=IELHSS.

Nuttall, J., Edwards, S., Mantilla, A., Grieshaber, S., & Wood, E. (2015). The role of motive objects in early childhood teacher development concerning children's digital play and play-based learning in early childhood curricula. *Professional Development in Education, 41*(2), 222–235. https://doi.org/10.1080/19415257.2014.990579.

O'Donoghue, T., & Haynes, F. (1997). *Preparing your thesis/dissertation in education.* Social Science Press.

Ottenbreit-Leftwich, A. T., Glazewski, K. D., Newby, T. J., & Ertmer, P. A. (2010). Teacher value beliefs associated with using technology: Addressing professional and student needs. *Computers & Education, 55*(3), 1321–1335. https://doi.org/10.1016/j.compedu.2010.06.002.

Punch, K. (2009). *Introduction to research methods in education.* Sage.

Schriever, V. (2018). Digital technology in kindergarten: Challenges and opportunities. In A. Khan & S. Umair (Eds.), *The handbook of research on mobile devices and smart gadgets in K-12 education* (pp. 57–76). IGI Global. https://doi.org/10.4018/978-1-5225-2706-0.ch005.

Strauss, A., & Corbin, J. (1990). *Basics of qualitative research: Grounded theory procedures and techniques.* Sage.

Strauss, A., & Corbin, J. (1998). *Basics of qualitative research: Techniques and procedures for developing grounded theory* (2nd ed.). Sage.

Zabatiero, J., Straker, L., Mantilla, A., Edwards, S., & Danby, S. (2018). Young children and digital technology: Australian early childhood education and care sector adults' perspectives. *Australasian Journal of Early Childhood, 43*(2), 14–22. https://doi.org/10.23965/AJEC.43.2.02.

Vicki Schriever is a Lecturer in Early Childhood Education at the University of the Sunshine Coast. Vicki's research interests include understanding how early childhood teachers manage their changing roles with digital technologies, pedagogical practices and the role of digital technologies, and using digital technologies to foster relationships with families.

Chapter 11
Bridging Communities: Developing Digital Literacies and Introducing Digital Technologies in the Montessori Early Childhood Education Classroom

Samantha Owen, Sharon Davies, and Sarah Iles

Introduction

In 2018, River Montessori School (River)[1] in Australia commissioned a whole school action research project to identify whether there were authentic methods responsive to Montessori pedagogy to develop digital literacies and to introduce digital technologies into the Montessori classroom. The project was initiated by River in response to a change in 2017 in the local authority's mandated curriculum, which, if adopted, required the school to introduce digital technologies in the Early Childhood Education (ECE) spaces. Doing so presented a particular challenge because there was no pre-existing or agreed internal or official Montessori methodology for how to either introduce digital technologies in the ECE areas in the school or develop digital literacies throughout River. Working with the school leadership, we developed a whole school action research project which was guided by three research questions. First, did River educators need to adapt their teaching practices to respond to the new curriculum or was the prepared environment providing the groundwork for children to become digitally literate, which meant that they would meet the new curriculum demands at their own pace? Second, would the ECE curriculum benefit from focused teaching of digital literacies? Third, if so, how? The final question is important as at the heart of the considerations was the concern voiced by Sarah Jones in her study of effective technology integration in four U.S. Elementary (Primary) Montessori schools: "Can technology integration truly

[1] All School and Educator names have been changed for the purposes of this chapter.

S. Owen (✉) · S. Davies · S. Iles
Curtin University, Perth, Australia
e-mail: Samantha.owen@curtin.edu.au; Sharon.davies@curtin.edu.au; Sarah.Thomas1@curtin.edu.au

D. Holloway et al. (eds.), *Young Children's Rights in a Digital World*, Children's Well-Being: Indicators and Research 23,
https://doi.org/10.1007/978-3-030-65916-5_11

exhibit the spirit of a Montessori classroom, or will it simply become the obtrusive interference?" (Jones, 2017, p. 28).

To investigate these questions our action research project at River was implemented in three phases. Phase one comprised two professional development days during which we introduced the project and built capacity in the areas of digital literacies and digital technologies. We worked closely with the leadership team to better define the project and the outputs. Phase two was facilitation of three action research cycles with the critical friend. A critical friend is a "trusted person who asks provocative questions and provides data to be examined through another lens and offers critique of a person's work as a friend" (Costa & Kallick, 1993, p. 50). With a focus on collecting evidence, the critical friend provides support, challenges assumptions and assists in creating a vision for advanced practice (Eliott, 1985; Stenhouse, 1975; Timperley, 2015). Phase three commenced with further professional development for the ECE educators followed by a digital technologies implementation trial, which was informed by phase two. The project concluded with a whole school critical reflection workshops and delivery of the findings to the school community.

Drawing on the result of the research project, it is our contention in this chapter that for digital technologies and digital literacies to be successfully integrated and taught, a shared language and a shared approach for educators, students and families in the school community has to be identified. We argue that doing so is pertinent at a time in which children are "growing up at ease with digital devices that are rapidly becoming the tools of the culture at home, at school, at work, and in the community" (NAEYC, 2012, p. 2). Their world encompasses the virtual, digital and material or physical. They need a language that transfers across these spaces and anchors their understandings as well as their identity and community. Our chapter opens with a picture of the Montessori classroom followed by a review of the Montessori literature on digitisation of Montessori classrooms and lives. We then turn to a discussion of the immediate school context which gave rise to the project and introduce what we observed at the school as the potentially greater challenge they were facing. Finally we focus on the River project design and implementation to explain and examine our findings.

The Montessori Classroom

Montessori schools follow the teachings and pedagogy of Maria Montessori and are distinguished by their focus on working with and following the interests of the child (Montessori, 1912, 1949, 1963). Children in Montessori schools are taught in 3-hour learning cycles in which they complete sequenced jobs. The educator works in this space as a guide and may provide lessons and instructions, other children may also take on a teacher role and children may work independently or in groups (Montessori, 1963). The focus in the classroom is on the prepared environment, which allows for the children to independently engage and explore and master new tasks. The role of the educator is to be responsive and to allow for experiential learning (Montessori,

1912). Children formally enter the Montessori classroom at age 3 and their education follows the Planes of Development, the four stages of growth Maria Montessori identified: Infancy, the Absorbent Mind, birth to age 6; Childhood, Conscious Imagination, ages 6–12; Adolescence, Social Independence, ages 12–18; and Maturity. Each stage corresponds to a particular developmental stage and are translated as the age groupings for the Montessori classrooms (Montessori, 2004).

At River, which is regarded as a best-practice Montessori school, the prepared environment in the ECE areas for First Plane children is home-like. These classrooms are beautifully presented and contain a child-height kitchen, a bathroom, cleaning implements, tables and chairs for eating, spaces for self-care, reading corners, work spaces, mat areas and the educational materials. The environment is deliberately screen-free. As children progress through the Planes, the classrooms change to meet their new neurological and developmental needs and—accordingly—screen and digital technologies are slowly introduced until fully integrated when they enter the Third Plane. These decisions had been made in line with the personal beliefs of the leadership team, in response to some parent pressure and in keeping with what was regarded as best Montessori practice.

Digitisation and Montessori Classrooms

In the Montessori community, literature discussing the place of digital technologies in the classroom first emerged in the U.S. and in the early 1990s. Presenting a view representative of her community, Jennifer Turner, ECE Montessori educator, asserted that when it came to digital technologies the guiding pedagogical principle, to 'follow the child', should be ignored as, while children would want to use computers, their presence in the classroom would restrict creative thinking and disengage children from reality (Turner, 1992, p. 32). She also suggested parents should be instructed to limit the use of digital technologies at home (Turner, 1992). In 1999 Peter Montminy, an educator living in Eastern Pennsylvania, questioned the practicality of Turner's directive, especially as he worked in a school where he had no choice but to use technology in the ECE classroom. Thus, he acknowledged that "many Montessorians are apprehensive about technology in the classroom" and hold "concern that the purity of the philosophy/method will be compromised" but in the case they were forced to do so, he offered "Best Practice Guidelines" (Montminy, 1999, p. 30). These were: the materials should be sequential, stimulating, not overwhelming, 'aesthetically pleasing' and nonviolent; the use must be meaningful, useful, age-appropriate and not reward focused; "the materials must have built-in flexibility" for the child to be "self-directing, self-pacing, and self-correcting"; the technology should allow the child to work independently, creatively and be differentiated; and, finally, "materials should enhance or complement the teaching . . . be well integrated within the learning environment" (Montminy, 1999, pp. 30–31). If these conditions could be satisfied with the suggested digital technology then Montminy's view was that it could be integrated into the classroom.

Montminy's work raised the important question of how Maria Montessori would respond and was influential for Montessori educators seeking an approach. Following Montminy, Arlene Love and Pat Sikorski (2000), elementary school educator Elizabeth Ross Hubbell (2003), the Chair of the American Montessori Schools (AMS) Heads Section (Cichuki, 2006) and the parent representative to their Board (Moore, 2006) agreed that 'meaningful' use was key. Hubbell defined meaningful digital learning experiences as those which connected students to the real world and honoured the Montessorian directive to prepare children for future life (Hubbell, 2003, p. 40). She suggested that WebQuest and digital storytelling fulfilled these requirements (Hubbell, 2003, p. 41). Montessori Educator and Professor of Education at Radford University, Barbara Foulks Boyd agreed and controversially suggested that if the tenant that the Montessori classroom should be "a reflection of the home, the community, and the world" was taken as paramount then there must be full classroom integration of digital technologies (2008, p. 30).

However, an ambivalence about the place of digital technologies in Montessori classrooms, especially ECE, remained and was raised in Kathy Carey's 2012 *Montessori Life* editorial. Carey questioned what preparation for the future actually looked like and whether being literate in digital technologies would be an essential skill set. She referred educators to *The Secret of Childhood* (Montessori, 1963) and rejected the need for education on digital technologies. She instead suggested children needed to be taught ways of thinking (Carey, 2012). Carey's concerns connected to wider and varied discussions over children, screen time and restriction and the future of work (Jenkins, 2015; Lumby, 2018; Richtel, 2011). Carey's dismissal of digital technologies in the classroom was moderated by Montessori educator, Jana Morgan Herman's guidelines for 'Balanced Use' (Herman, 2012, p. 36). Herman did not agree with complete restriction as characterised herself as "an immigrant in a brave new digital world" in which the children are "digital natives" (Herman, 2012, p. 36). She also made a distinction between the screens, media and digital technologies and endorsed the previous directives for meaningful use. The philosophy behind her approach was that schools should model behaviour so that families know how and when digital technologies should be used. Her guidelines for schools were: (i) they "should emphasise the importance of play" when using digital technologies; (ii) only use computers to extend established creative and productive processes; and (iii) not interrupt the 3-hour work cycle (Herman, 2012, pp. 37–40). With these habits established at school, Herman suggested that parents can support their children by limiting technology, increasing bonding time and experiences, eating dinner together, observing digital bans, having technology-free spaces and ensuring that there is "No media, ever, for children under 2" (Herman, 2012, p. 41).

These debates informed the 2013 American Montessori Society (AMS) Position Statement on Information Technology. The Position Statement gave the directive that the introduction of digital technologies to education spaces should not be to the detriment of, or to replace, the existing curriculum. Instead digital technologies were to be adopted as tools which could add to or complement the Montessori curriculum: "Intelligently integrated into the Montessori environment, technology can be a valuable communications tool and may promote traditional Montessori tenets"

(AMS, 2013, para 6). With respect to how and when digital technologies should be used, the AMS recommended practitioners make implementation decisions which were authentic to Montessori pedagogy and which respected children's neural development: "Using Montessori's methods of observing will help us determine the place of technology in the Montessori world" (AMS, 2013, para 11). Following the Position Statement, in 2014 the first academic study of robotics in the ECE Montessori environment was published by Tufts University researchers. They responded to research highlighting the need for early exposure to STEM and they wanted to develop "an effective approach for foundational programming and engineering concepts into Montessori education" (Elkin et al., 2014, p. 154). The researchers selected the STEM toy, LegoWeDo (https://education.lego.com/en-au/product/wedo-2) to work with because they have many of the qualities Montessori educators look for in resources: self-correcting; sensory; collaborative; promotes respectful learning; and differentiated. After observing the classroom and interviewing the teacher, the researchers found that simply introducing the new technology was not enough, educators had to be trained in their use and the new technologies needed to be thoughtfully integrated with classroom routines to be meaningful and effective (Elkin et al., 2014, p. 166). The Tufts University study coincided with the launch of Cubetto, a small wooden robot for 3- to 6-year-children designed with Montessori principles in mind by UK-based toy company Primo Toys (https://www.primotoys.com/). The company claims that children can learn to code and program using Cubetto by placing instruction blocks on a wooden interface board which sends Cubetto off on an adventure through wireless technology. A research project sponsored by Primo toys to trial Cubetto in the ECE Montessori classroom concluded it "integrate[d] perfectly into the Montessori learning environment" (Stockdale, 2016) because it met the criteria for a Montessori resource or toy.

Despite these advances, the extent to which digital technology should be integrated was still under question, and in 2015 the Executive Director of the American Montessori Society clarified some of the open-ended elements in the 2013 Position Statement. Richard A. Ungerer advised that children in the First Plane of Development should have limited to no engagement with digital technologies and that when and if they were introduced in the Second Plane the interaction should be limited to meaningful and authentic use consistent with Montessori pedagogy (Ungerer, 2015). Elaborating this position, the Association Montessori Internationale (AMI) gave the advice that if an educator were contemplating introducing digital technologies into Montessori classrooms, they should first ask the questions: "Is there a practical purpose" and "Does it offer an alternative approach that exists in no other format?" (MacDonald, 2015). It is these questions which we took to the research project as we tried to ascertain the link between digital literacies and digital technology use—and how they translated across the home and school environments.

The Global Meets the Immediate School Context

The first research phase of our project commenced with professional development with the teaching staff and leadership team to establish current educator knowledge and confidence in the areas of digital technologies and digital literacies and to chart their responses to the mandated curriculum changes. As an opening activity we asked educators to form groups to complete an anonymous survey. They were unanimous in their understanding that the school had to respond to the curriculum changes but they did not all have the same reasoning. All groups but one gave reasons which resonated with the early literature and which implied an openness to the use of digital technologies across all Planes of Development: "Children are already exposed at home, so we cannot ignore existence and development of digital technology;" "It's the way of the future so we ought to prepare them." The outlier group returned to the point made by Carey (2012) and privileged the development of digital literacies and "critical problem-solving skills" as the necessary skills to "interact effectively with the ever changing digital world." The survey also asked the educators about use. Responses to "How do digital technologies sound in your classroom?" picked up on classroom chatter in two ways. First, as discussion of use at home: "Children talk about the games they play on iPad" and "There is a fair amount of conversation between the children about technologies used at home". Second, "The computers are silent but students tend to hover around them and chat, discuss, giggle, laugh etc."; "Discussions, trial and error—robots crashing, videos playing"; "Noisy because the children using them tend to talk"; "It is often collaborative and the students are verbally sharing ideas, skills and knowledge." Similarly, to the question "How do digital literacies and digital technologies feel in your classroom?" the answers crossed home and school. Responses ranged from "Not necessary," "Not missed by the children" and "Unfamiliar and is not part of the daily experiences in the classroom" to "Not available" and "Absent." The absence was clarified in further comments: "Conversational: a couple of children refer to games" and "Restricted—the children know that computers or iPads are not available in class, as they may be at home" or "Limited but very in demand." Some gave examples of use: "Children feel privileged when they use the computer or camera" and "Enjoyable: children have access to a CD player. They enjoy using it to listen to stories and songs." One group directed the discussion to meaningful use: "Appropriate use feels ok, there is a time, age and reason where the use of digital technologies are valuable and warranted."

The survey concluded with questions related to knowledge acquisition. First, "How do children learn to use digital technologies?" The educators overwhelmingly responded that children learned "at home." The answers also picked up on the passage of the knowledge from the home to the school—"Through interaction with entertainment devices initially then with guidance/lessons and further time to experiment and explore"—and vice versa: "Their critical thinking and problem solving skills are developed so that they have the capabilities to explore technology outside of the schooling environment." Educators also commented on *how* learning happened: "In the context of other subjects"; "Planned units of work and also on an as needed basis"; "When it is presented in a meaningful way." They also explained the learning with digital technologies needed to happen within the Three-Period lesson structure: "They learn by seeing and exploring it. Trying something and seeing how it works." To the question "How do children learn digital literacies?" the educators again referred back to Montessori pedagogy: to the prepared environment and "Experiences with the Montessori materials would prepare them to use and understand digital technologies." Thus, educators indicated their perception that even if they were not dealing with digital technologies in their classrooms, their curriculum supported the development of digital literacies. When asked to name digital teaching strategies the educators specifically referred to the ECE classrooms stating: "children are learning many skills that will later help them with their digital skills, such as decision making, analysing, questioning, analytical thinking skills, concentration, intrinsic motivation, perseverance, fine motor co-ordination."

Educators at River were secure in their pedagogy and they saw it was possible that the routines demanded by the prepared environment and the Montessori curriculum provided for the use of digital technologies in their classrooms and development of digital literacies by their students. The responses also pointed to something else: the premise for the professional development day and for the research project was to discuss the new mandated curriculum. However, their cited motivation was more frequently around the need for consistency between the school environment and the students' wider world. There was something of an acknowledgement that the school was out of step with the everyday realities for children and it became clear that there was a—perhaps unacknowledged—reliance on the link between school and home learning: that when children came to use digital technologies in their classrooms there was an expectation that they had already had exposure at home and learned how to use them from their parents, siblings or peers.

The responses highlighted a larger problem the school needed to address: for Montessori education to be effective and children to have the most developmental potential, there must be continuity between the home and classroom spaces (Bronfenbrenner, 1979). What emerged from the survey results was a perceived disconnect between the Montessori philosophy which drove and defined the school pedagogy and the attitudes to screens and digital technology use of some educators at the school. The conflictual interaction had the potential to create problems as it suggested that—simultaneously—the school relied on the digital immersion occurring outside of the school space but by doing so parents did not respect the need for continuity between the home and classroom spaces which the school requested. The latter is an important point as the school is fee-paying: the parents of River school had chosen the school, gone through the interview process and made an agreement with the school to accept the guiding ethos. The school was firm in its belief that children under the age of 8 should not be using screen based digital technologies and for children younger than 6 years old learning should be sensorial. However, the River educators recognised that to remain as 'authentic' Montessori they needed a set and consistent policy for the use of digital technologies. It was this we identified was actually driving the research project and which redefined our research questions:

How did the prepared environment and the Montessori curriculum provide for the development of digital literacies and the use of digital technologies?

What Are Digital Literacies in the Montessori Classroom?

At the second professional development day we facilitated a workshop with the educators to identify what digital literacies looked like at River. The ten aspects they selected were: critical thinking; navigation skills; communication skills; collaboration skills; problem solving; participation in and contribution to civil society; self-regulation and independence; global citizenship; multimodality and innovation; and creativity. The educators were then asked to collaboratively define each of these and to provide examples of what they looked like or could look like in their classrooms. For communication skills the identified embedded skills and knowledge were social and emotional competence; empathy; and extensive vocabulary knowledge. The observable behaviours in daily life included: correct use of associated vocabulary in context; ability to detect social and emotional cues; and able to disseminate information gathered and analysed. The educators saw that this transferred into observable behaviours in digital technologies use as the correct application of associated vocabulary in a digital context and ability to detect social and emotional cues in a digital world. The selection process and finding the examples helped the educators to map the prepared Montessori environment and the Montessori curriculum to the digital lives of their students and their school.

The digital literacies aspects, embedded skills and knowledges and observable skills and behaviours provided the educators with a shared language and understanding. We asked them to complete the planning and reflection templates for their classrooms (See Table 11.1). This tool enabled educators to identify a lesson within their current Montessori Curriculum which contained the pre-conditions for the development of digital literacies and which could be extended through the use of digital technologies. Educators were asked to teach the lesson with a focus on intentional teaching of digital literacies and then journal what children did, said and produced to capture demonstrated opportunities for, and the development of, digital literacies and children's learning.

The educators' journals provided a window into understanding how the educators taught their lessons as they began to think consciously about digital literacies and introducing the new curriculum. One educator, Serena, focused on the 3-year-old job of flower arranging (Fig. 11.1). She noted in her journal:

> This activity involves a sequence of steps:
>
> [. . .]
>
> Pouring water and pouring through a funnel
>
> Choosing vase and flowers from a selection
>
> Making a judgment about how much stem needs to be cut [. . .]

Table 11.1 Planning and reflection template—Serena

1. PRE—Example from planning [Experience/lesson/activity—exploring manipulation, repetition, frequency]	2. Links to Curriculum	3. Digital Literacy competencies	4. Ideas for extension [Progression]	5. Ideas for extension [links to other Digital Literacy competencies]	6. Identify how DT can be incorporated	7. POST Implementation—reflection based on observation related to Digital Literacies identified in 3
Flower arranging 3 year old job	• 'People produce familiar products to meet personal and community needs' • Engineering principles and systems • Food and fibre production • Designing • Producing and implementing • Evaluating	• 1—Critical thinking • 5—Problem solving • 6—Participate and contribute to civil society • 7—Self-regulation and independence • 8—Innovation and creativity	• Growing own flowers to cut and arrange • Arranging flowers to give to other classes	• 8. Global citizenship	• Digital photography to map seasons • Growth and life cycle • Needs of plants • Pressing of flowers for preservation	• Steps help to prepare the brain for sequencing/coding? • Enhances the children's awareness of the products made and used in society: decorative/symbolic/emotional? • Collaboration • Navigation skills

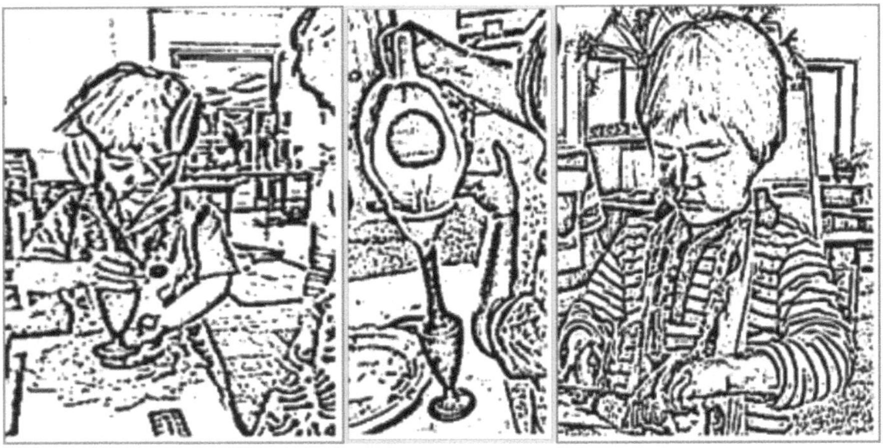

Fig. 11.1 Step 3—Pouring water and pouring through a funnel, Step 5—Making a judgment about how much stem needs to be cut and Step 7—Arranging in the vase

Cutting the stem

Arranging in the vase (designing)

Choosing a place in the classroom to put the finished flower arrangement. Placing the vase on a doily to beautify the room

[...]

Drying everything on the table, hanging up wet towel, replacing with a dry towel and making the workspace ready for the next person to use.

When the critical friend met with Serena she encouraged her to think critically and reflectively in the next cycle about how each of the steps of the lesson demonstrated could be linked to digital literacy classifiers and provide direct preparation for the use of digital technologies in her critical reflection journal. Serena then photographed the activity and underneath the photographs her notes read.

Serena, can I put a flower on your table?

([Digital literacy] Communication skills)

I see:
- repetition
- social participation
- Concentration
- judgement
- self-chosen work
- following multiple steps

I wonder: if this is supporting collaboration? Design?

Her journal concluded with a critical reflection.

I wonder:

... if the process of remembering and following the many steps helps prepare the brain for sequencing/coding?

... if the activity enhances the children's awareness of the products made and used in society [...]

... if this activity supports collaboration? [...] often conversation is opened up between the giver and receiver.

... Navigation skills [...]

She used the school's new shared digital literacies language in her journal, which enabled her to recognise how, with new emphases placed, the prepared environment of the ECE classrooms could provide indirect preparation for developing digital literacies to use digital technologies.

When Digital Literacies Meet Digital Technologies in the Montessori Classroom

In the third research phase, River introduced one Cubetto to each ECE classroom. The choice was guided by a decision that the ECE classrooms would remain screen-free. We used the reflective journals and individual educator and leadership meetings to design the approach, which had two essential elements. First, digital technologies would be introduced in the third period of the Three-Period Lesson, which is when the educator asks the child to recall the knowledge or lesson by naming the new concept or idea. Digital technologies could be introduced and integrated here if their use represented a meaningful intervention consistent with Montessori Pedagogy. Second, we agreed ECE educators should decide how Cubetto would be introduced into the prepared environment and that educators should develop Digital Technology lessons which complemented the current curriculum. We used the third Professional Development to both familiarise the ECE educators with Cubetto and to workshop a method of introducing Cubetto to the prepared environment and a Three-Period Lesson for coding in the ECE classroom. For Period One they repurposed a pattern identification exercise which asked children to stick coloured pieces of paper onto a squiggly line. Then, once the children had displayed competency, they moved to Period Two where they were introduced to a more targeted coding activity which required them to use the command blocks for Cubetto and to draw a command sequence which would have Cubetto reproduce the squiggly line from Period One. Finally, in Period Three the children were encouraged to work collaboratively to use the story they had written in Period Two to program Cubetto and send it on a journey.

We asked the educators to critically reflect and at the concluding whole school Professional Development workshop they reported on the success of the trial. However, to improve the classroom experience, they also identified a need to further develop the shared digital literacies language and to standardise language related to digital technologies for the whole school. Finally, we recommended that they develop a whole school digital resource reference library for digital literacies, tools, technologies and approaches, complete with materials and lesson cards for the classrooms. The assumption was that once the educators had a consistent language for digital literacies and approach to digital technologies use it would filter into the wider school community and open channels of communication with caregivers.

Conclusion

Our Whole School Action Research project provided professional development to River educators in the area of digital literacies and digital technologies. The argument we have made in this chapter is that the fundamental problem the school experienced with implementation of digital technologies was not only that there was no agreed approach for implementation in the ECE space but that there was a disconnect in the school community between the educators and the families who attended the school. Hence, the children attending the school lived—to various extents—digitised lives and so the school needed to actively engage with these lives not through restriction but through explicit guidance which understood digital technology use as a "normal part of a child's social development" (Holloway et al., 2013, p. 25). During the three action research cycles, educators were supported by researchers who redirected the lenses they used to critically reflect on their practice. The project supported a reflective process and located a methodology for authentically integrating technologies into the River ECE classrooms, which we modelled with Cubetto. Cubetto was not "an obtrusive interference" (Jones, 2017, p. 28) but an example of how digital technologies could be introduced and used in the ECE Montessori classroom in a manner which is authentic and responsive to Montessori pedagogy and contiguous with the River's philosophy. With a common understanding and language for digital literacies to inform focused teaching, it was possible to develop digital literacies in the prepared Montessori environment, even with the initial absence of digital technologies. By defining digital literacies in the context of the school, the educators could begin to catalogue a shared language for digital literacies and digital technologies use which can be referred to and used by educators, children and parents and caregivers. Hence, the language provided the discursive continuity for the microsystems to be mutually supporting and compatible in the demands they are making of children (Bronfenbrenner, 1979). Furthermore, a shared language resolved the internal contradiction in the school attitude to digital technologies use as educators were confident that the approach was consistent with Montessori pedagogy and the school's philosophy and curriculum, which opened up a channel for communication between educators, children and caregivers. It was at this point that digital technologies could be introduced to the ECE Montessori classroom in meaningful and authentic manner.

References

American Montessori Society. (2013). *AMS position statement on information technology*. http://www.tjleone.com/ams_technology_position_statement.pdf.

Boyd, B. F. (2008). Assistive technology for every child. *Montessori Life, 20*(1), 30.

Bronfenbrenner, U. (1979). *The ecology of human development*. Harvard University Press.

Carey, K. (2012). Montessori, children, and the digital age. *Montessori Life, 24*(1), 2–7.

Cichuki, P. H. (2006). Technology: We have come such a long way. *Montessori Life, 18*(2), 9.

Costa, A. L., & Kallick, B. (1993). Through the lens of a critical friend. *Educational Leadership, 51*(2), 49–51.

Elkin, M., Sullivan, A., & Bers, M. U. (2014). Implementing a robotics curriculum in an early childhood Montessori classroom. *Journal of Information Technology Education: Innovations in Practice, 13*, 153–169. http://www.jite.org/documents/Vol13/JITEv13IIPvp153-169Elkin882.pdf.

Elliott, J. (1985). Facilitating educational action-research: some dilemmas. In R. Burgess (Ed.), *Field methods in the study of education*. Falmer Press.

Herman, J. M. (2012). Creating balance in the new age of technology. *Montessori Life, 24*(3), 36–43.

Holloway, D., Green, L., & Livingstone, S. (2013). *Zero to eight. Young children and their internet use*. LSE: EU Kids Online. http://eprints.lse.ac.uk/52630/.

Hubbell, E. R. (2003). Integrating technology in a Montessori classroom. *Montessori Life, 15*(2), 40–41.

Jenkins, M. (2015, December 2). Tablets out, imagination in: the schools that shun technology. *The Guardian*. https://www.theguardian.com/educator-network/2015/dec/02/schools-that-ban-tablets-traditional-education-silicon-valley-london.

Jones, S. J. (2017). Technology in the Montessori classroom: educators' beliefs and technology use. *Journal of Montessori Research, 3*(1), 16–28.

Love, A., & Sikorski, P. (2000). *Integrating technology in a Montessori classroom*. Educational Resources Information Center.

Lumby, C. (2018). Screentime panics. In C. Nelson & R. Robertson (Eds.), *Dangerous ideas about mothers* (pp. 49–58). UWA Press.

MacDonald, G. (2015). Technology in the Montessori classroom. Benefits, hazards and preparation for life. *Association Montessori Internationale – USA Journal, 3*, 4–7.

Montessori, M. (1912). *The Montessori method* (A. E. George, Trans.). Frederick A. Stokes Company.

Montessori, M. (1949). *The absorbent mind* (C. A. Claremont, Trans.). Kalakshetra Publications.

Montessori, M. (1963). *The secret of childhood* (J. M. Costelloe, Trans.). Random House.

Montessori, M. (2004). *The four planes of education*. AMI Pamphlet, Montessori Books.

Montminy, P. (1999). Best practice guidelines for computer technology in the Montessori early childhood classroom. *Montessori Life, 11*(4), 30–31.

Moore, A. M. (2006). A parents look at technology. *Montessori Life, 18*(2), 11.

National Association for the Education of Young Children (NAEYC). (2012). *Technology and interactive media as tools in early childhood programs serving children from birth through age 8 [Position Statement]*. National Association for the Education of Young Children, the Fred Rogers Center for Early Learning and Children's Media at Saint Vincent College. https://www.naeyc.org/sites/default/files/globally-shared/downloads/PDFs/resources/topics/PS_technology_WEB.pdf.

Richtel, M. A. (2011, October 22). Silicon Valley school that doesn't compute. *The New York Times*. https://www.nytimes.com/2011/10/23/technology/at-waldorf-school-in-silicon-valley-technology-can-wait.html.

Stenhouse, L. (1975). *An introduction to curriculum research and development*. Heinemann.

Stockdale, M. (2016). Introducing "Cubetto". *Montessori International, 119*, 32–33.
Timperley, H. (2015). *Professional conversations and improvement-focused feedback. A review of the research literature and the impact on practice and student outcomes*. AITSL.
Turner, J. (1992). Technical literacy. *Montessori Life, 4*(4), 32–33.
Ungerer, R. A. (2015). Montessori education, technology, and meaning in our lives. *Montessori Life, 27*(1), 3.

Samantha Owen is an Early Career Academic at Curtin University where she is the Humanities and Social Sciences Education Lecturer. She is Editor in Chief of Cultural Science Journal and National Convenor, Australian Women's History Network. Samantha works on 'alternative' education, policy and impact assessment.

Sharon Davies specialises in Early Childhood and STEM Education. She has a particular interest in how action research and professional development can produce meaningful change and determine rich educational experiences for all children. She is a Senior Lecturer and the Master of Teaching Course Coordinator at Curtin University in Western Australia.

Sarah Iles is an experienced educational leader, qualified to work in Primary and Early Childhood Mainstream and Educational Support settings. She has extensive experience in leading whole school change. Sarah has a particular interest in developing digital literacies and the use of digital technology with children in diverse abilities and needs.

Chapter 12
Understanding the Mutuality of Play and Media Literacy in Young Children

Helle Hovgaard Jørgensen and Helle Marie Skovbjerg

Introduction

Article 31.1 of the Convention on the Rights of the Child states that 'States Parties recognise the right of the child to rest and leisure, to engage in play and recreational activities appropriate to the age of the child and to participate freely in cultural life and the arts' (UNHCR, 1989). This assertion of the right to play can be understood to refer to spontaneous, self-organised play, with no external purpose and where the participating children themselves decide what is meaningful, without external interference.

In Scandinavian countries, there is a strong tradition of taking children's play seriously, as exemplified by the characters in the stories of Astrid Lindgren. In late 1960s Denmark, youth television positioned the child at the centre of a world of their own. *Super Carla,* a famous children's programme from 1968 about a girl with super powers, put it like this: "Well, this is how real children live: they do exactly what they like, they don't go to school. Because it is more important to play than to write and calculate" (Super Carla, 1968). This is a microcosm of contemporary ideas in childhood sociology and a broader international movement that acknowledges children as active participants in their own lives and gives them a voice of their own, both in research (Corsaro, 2003; James et al., 1998) and in society. Today, when discussing children's play, their participation and their right to play, their use of digital technologies must form part of that discussion.

H. H. Jørgensen (✉)
University College Lillebælt, Odense, Denmark
e-mail: Hhjo1@ucl.dk

H. M. Skovbjerg
Design School Denmark, Kolding, Denmark
e-mail: Skovbjerg@dskd.dk

D. Holloway et al. (eds.), *Young Children's Rights in a Digital World*, Children's Well-Being: Indicators and Research 23,
https://doi.org/10.1007/978-3-030-65916-5_12

To gain some insight into that changing landscape, this chapter addresses the following research question: *What can be learned about media literacy among pre-primary children by studying in-school play activities from the child's perspective?*

In that light, the concept of media literacy is revisited to understand the nature of play with digital technologies and the importance of children's participation, based on a field study of children aged 5–6 years in two Danish public pre-primary schools. Our ambition here is to enhance existing understanding of the links between play and media literacy by including children's perspectives on the matter, encompassing all the wild, crazy, frivolous and silly behaviours that characterise play.

A conceptual framework is therefore needed that combines play and media literacy—both in the context of everyday life, where playful practices are part of young children's way of being together, and in a broader aesthetic context that acknowledges the importance of the child's perspective and participation. To that end, we draw on Bateson's concept of *framing* (Bateson, 1972), along with the concept of *participation* (Mouritsen, 1996), the idea of *play as heteroglot* (Bakhtin, 1981), Bogost's (2016) understanding of materials in play and the idea of play as a way of being and exploring the world based on the concept of mood (Skovbjerg, 2018, 2021). Additionally, we understand media literacy as vernacular, social, affective, material and embodied. In so doing, we consider play as more than just a developmental tool (as in Piagetian or Vygotskyian accounts). Similarly, we consider media literacy as more than a matter of acquiring skills to avoid harm or exposure to inappropriate material.

After addressing the theoretical understanding of media literacy in relation to play, we consider the methodological implications of investigating these as related issues. We go on to analyse two interviews by and with children. One concerns drawings of the infamous computer game GTA and the fun of it from a 6-year-old's perspective; the other explores two children's understanding of how to make a game. Finally, we discuss the results and implications of looking at media literacy in this way from the child's perspective.

Media Literacy and Participation

In approaching media literacy by looking at children's play, we acknowledge the importance of the child's perspective. Understanding media literacy is not easy. In general, literacy is used to define standards for the acquisition of skills as practised in the contemporary testing culture in schools worldwide (PISA, ICIL). From the alternative perspective of *new literacy studies* (Gee, 2010), literacy is thought of as a social practice (Barton & Hamilton, 1998). This approach looks at reading and writing in the broader context of communicative practices as 'vernacular, networked and embodied' (Marsh & Hallet, 2008; Rowsell & Pahl, 2015). In other words, children in preschool are embedded in a literacy 'of their own', entangled in playful activities and other social practices that influences the meaning making processes in which they participate.

In moving from an understanding of literacy in terms of standards that can be trained and measured to a broader account encompassing social practices and communication, it becomes clear that the understanding of the concept has evolved. This shift of perspective moves away from an emphasis on protection—focused mainly on cognitive skills for avoiding harm and exposure to inappropriate material—to an emphasis on participation, focusing on cultural issues of how to get things done by creative and productive use of digital media (Erstad & Amdam, 2013). Adopting this socio-cultural approach, we view media literacy as the trichotomous ability to access, understand and create/express oneself through media (Carlsson, 2013), and as a social and communicative practice embodied in children's playful ways.

To date, most of the literature on media literacy acknowledges play as young children's primary activity but treats it as secondary to the more obvious learning-related aspects of media literacy. For example, when Henry Jenkins mentioned play as one of three core skills linked to media literacy (Jenkins, 2009, p. 40), it was as a means of achieving 'clearly defined' learning goals, even though learning outcomes are difficult to measure in relation to new media literacy. Marsh characterised the virtual worlds in which children participate as part of their 'everyday landscape of play' (Marsh, 2012, p. 193), and most investigations of children's ways with media refer to 'playfulness' (Burnett et al., 2014) and the importance of 'play' (Wohlwend, 2015, 2018). However, a deeper understanding of play and playful is needed, as well as an elaboration of what a play perspective might offer or to elaborate what a play perspective might offer (Marsh, 2015). Nevertheless, most play researchers acknowledge the participatory nature of children's play (Caillois, 2001 [1958]; Huizinga, 1963 [1938]; Mouritsen, 1996; Sutton-Smith, 1997). The following section explores that participatory position in terms of five dimensions of play that link it to media literacy within an overall conceptual framework.

Five Dimensions of Play

In *The Ambiguity of Play*, Sutton-Smith (1997) noted the difficulty of exploring play, either as an empirical object or as an analytical concept. For present purposes, we address this difficulty by using what might be called 'gentle' language to convey the ontological dimension of play. Imagine putting your hands around a jelly cake—not squeezing too hard but still holding it, showing it to others, maybe eating it with good friends. That jelly cake is the type of language required here. Epistemologically, to become knowledgeable practitioners of play, we need to find methods and techniques that enable us to explore, try out, participate in, and play, in all sorts of ways. To understand play, we have to be close to the actual emergence and sharing of play (Skovbjerg, 2018).

In our understanding of play, we emphasise the following five dimensions. First, based on Bateson's (1972) concept of *framing*, we stress the need to look at practice within play as a valuable situation in itself, where meaning may have different connotations than in non-play situations. Following Bateson, we can illuminate how these frames for play are created. Following Mouritsen (1996), we stress the importance of *participation, skills and performance* in play. According to Bakhtin, the chaos between children, spaces and objects is both *dialogical* and *heteroglot*—not predictable but emerging and re-emerging from situation to situation as centrifugal and centripetal forces influence language and meaning. Finally, we consider the relevance of *mood* in play (Karoff, 2013; Skovbjerg, 2018, 2021), emphasising the importance of the being aspects of play—how I relate to the world and to other people in different ways, sometimes within a specific practice and sometimes breaking the rules to explore further. As described by Bogost (2016), this emergence can involve the circumscription of objects. Taken together, these five dimensions capture the complex experience of play without sacrificing its poetics.

The Research Context

The cases reported here form part of a large field-based study involving 120 pre-primary school children aged 5–6 years in two Danish public schools in Odense. This qualitative study employed participatory methods that included fieldwork, participatory observations, interviews and interventions (Burn & Richards, 2014; Clark et al., 2014; Gulløv & Skreland, 2016; Marsh, 2012). The study focused on how children use digital media during playful activities involving both discursive and interactive practices (Couldry, 2004; Reckwitz, 2002; Schatzki, 2001; Swidler, 2001). The fieldwork began by asking *what is going on here*, in the first instance through conversations with the gatekeepers and the children, as well as through planned interventions. Data collection also included focus and semi-structured interviews with the children to explore the narrative structure of experiences and artfulness (Holstein & Gubrium, 1995, p. 18). Parents provided full consent to publish the children's pictures and names.

Grand Theft Auto or 'The Car Game': Driving Fast and Crashing Cars

As part of this process, the children made drawings of their 'favourite games' and interviewed each other about the games represented, allowing us to observe the actual emergence of digital play. The initial idea was simply to bring digital play into the children's everyday school life through stories and drawings. However, it transpired that the interviews themselves were used as a 'thing' with great play potential (Bogost, 2016)—that is, the Q&A mechanism of the interview became a formula on which to improvise (Mouritsen, 1996), making it a 'thing' circumscribed in play. Most of the themes of the 88 drawings were predictable and age-related. The representations were interesting—for example, the 'squareness' of *Minecraft* and the 'greenness' of *FIFA* (Kress & Van Leeuwen, 2006)—as were the high levels of activity around the drawings. During the process, the images were often performed (e.g. a 'scissor kick' in *FIFA*, jumping in *SuperMario*) as narratives connected to the games, and words were invented in the knowledge sharing processes around the classroom tables. All of the 88 children moved effortlessly from one medium (a game) to another (a drawing) without questioning the task. The process of drawing and interviewing became playful, partly because of the open-endedness of the situation in terms of the play material (games), and because playmates (other children) were invited to join in. The interview framing created a situational invitation to be curious (ask questions) and to tell stories, perform skills (the artfulness of the interview) and be social. With participation as a precondition and performing skills as a practice, this openendedness created moments of devotion, intensity and euphoria, echoing Skovbjerg's play moods (Karoff, 2013). Among the drawings, the infamous *Grand Theft Auto* (GTA) seemed a deviant case. Because it is rated for 18-year-olds and was therefore considered age-inappropriate, the game created a certain fascination precisely because of its illegality. It obviously worries adults that 6-year-olds are fascinated by such a game, but in a play context, the interesting question is of course *what is the fun of it?*

Of the 120 children, only about 10 made drawings of GTA, calling it 'the car game' or just 'GTA'. It was unclear whether they all actually played the game. When the researcher asked more specific questions (e.g. "Do you have it at home? or "When do you play the game?"), it became obvious that they had probably not really played the game. What was clear, however, was that certain games had a particular status among the children, which had to do with age and elder brothers or fathers who played the game.

'The Car Game': '... and you can do all kinds of strange stuff'

According to the children, GTA is 'real fun' for a number of reasons: "You can drive in cars very fast"; "You crash cars, steal cars, kill people, escape the police and so on"; and "You can do all kinds of strange stuff". In other words, the game is mainly about death and destruction, which is a play paradox. Because the practice is to drive really fast and crash cars, it is simultaneously about destruction (the cars really do crash in the game) and not (no cars are really destroyed). The children understood this, pointing out that the fun of GTA is that "you can do all kinds of *strange* stuff". They also understood that this is not the normal way of doing things; the authorities (police) are powerless while the outlaws are powerful. The children are in control in a world without adults, where they can drive recklessly, shoot and steal in a lawless gameworld. In short, the world is upside down, and the Bakhtinian concept of *carnivalism* seems relevant—what is generally considered strange becomes fun and out of the ordinary. The drawings express the dynamics of GTA. The squiggles

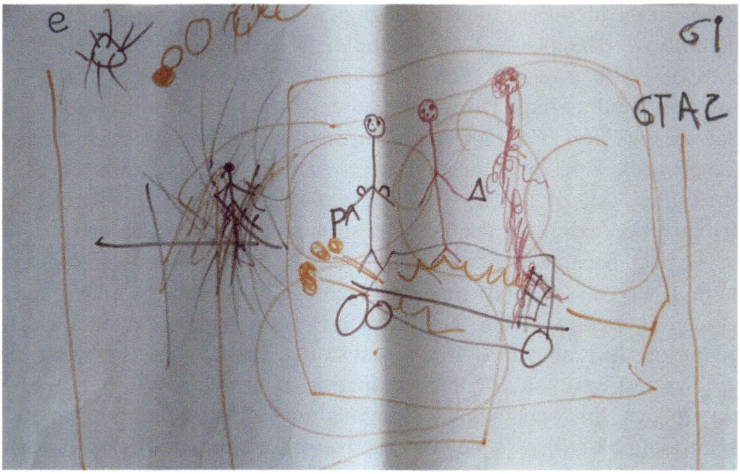

Fig. 12.1 A 6-year-old's drawing of the game GTA2

are more squiggly (less 'nice' or look-alike) than in Minecraft and other drawings, perhaps referring to the game's complexity (or reflecting these children's lack of normative drawing skills, which makes their great efforts to complete the task all the more interesting) (See Fig. 12.1).

A Semiotic Reading of the Drawing

In Fig. 12.2, GTA2 is represented by a stick figure of a strong man and his car. The man is at the centre of the drawing, and the two half circles on his upper arm signify muscles. Below, the analogical representation of wheels as circles creates the sign for a car (Kress & Van Leeuwen, 2006). In the mess on the left side of the drawing, there is another stick man, and the orange dots near the car suggest an analogy to speed. As the car is driving away from the stripes, it can be assumed in the context of the game that the man has been run over. The analogies are constitutive principles for the metaphoric processes that creates the signs, and the signs depend specifically on the context, rooted in a micro game culture that preoccupies the child.

Child-to-Child Interview

During the interview about the GTA drawings, it becomes obvious that children who do not play the game lack specific knowledge about it. The illustrator (D) explains what the game is about:

Fig. 12.2 A 6-year-old's GTA drawing of what "looks a lot like a helicopter"

"Well, you can chase cars and you can shoot, and you can steal buses and drive into people, and then you can also make a code bmx-bike, and then you can make a super code".

The interviewer (Q) says "*mmm*" and directs attention to the inscription (*GTA2*) in the drawing, asking "*Why does it say Gustav?*" (pointing to the letters GTA). "*It doesn't*", replies D, surprised by Q's ignorance. Nevertheless, the conversation continues; Q is obviously not alarmed by the content of the game, and later, when D talks about *Franklin* (the main figure in GTA), Q does not know who he is.

What seems important is to find common ground. When D shows Q another drawing of GTA and says: "*This is a helicopter*", Q immediately says: "*I know*", and D replies: "M*aybe you figured that out*". Q then says: "*Yes, because it looks a lot like a helicopter*" (See Fig. 12.2). Clearly, it does not, but that seems unimportant. D is acknowledged as the expert reporting from a micro-game-culture while Q knows about games; what matters is that they generally play in children's culture.

This meeting and exchange of knowledge primarily reveals a peer-to-peer culture in which children's communicative practices and participation are play-related, and the main goal is mutual understanding. The dialogical dimension of the interview creates artful opportunities for reworking language and meaning; the letters G-T-A become 'Gustav', and the drawing of something that most resembles a car becomes a helicopter in the dialogical practice of language's centripetal and centrifugal forces as the Q&A formula of the interview allows the children to improvise and reframe the situation as play.

The practice of *sliding* engenders a mood of devotion, where room is made for agreement. This changes towards the end of the interview, when D exclaims a satisfied "*Ha!*", and both children sense the thrill of having avoided misunderstandings, conflicts and appearing unknowledgeable. The practice turns to *shifting,* and the mood changes momentarily to intensity (Karoff, 2013). Framing the situation as playful makes it possible to perform dialogical skills, and within that framing, non-sense becomes part of the meaning-making processes that emerge in an unpredictable manner.

From the children's perspective, GTA is 'fun' because you can do all kinds of 'strange stuff' linked to destruction, frivolity and an upside-down world, and the framing is play. In that sense, and according to the trichometry of media literacy, the children are perfectly aware of the gameplay mechanisms as distinct from 'the real world'. Among the children, knowledge about games is considered valuable. At a micro level, this game-related knowledge and expertise is quite specific and is known only to members and participants in a certain game culture. Being able to join in and participate in other children's micro cultures is fruitful and is made possible by a fundamental acceptance of the importance of playing games.

The GTA drawings highlight the importance of being close to the actual emergence of play, which was made possible by inviting the children to interview each other, using a handy camera for documentation, circumscribing the interview as a playful situation.

Making a Game

In the final phase of the field work and based on the findings in the two other phases, interviews were conducted in relation to the trichometry of media literacy. Six focus group interviews with 12 children addressed the issue. In pre-primary classes, *access* to digital media was strictly regulated and most often forbidden. However, as soon as the researcher created access by inviting the children to tell stories about their digital media worlds and introducing a digital camera to the field, the children interacted to create, invent and express themselves in ambiguous ways without breaking the rules.

As indicated above, understanding of digital media was demonstrated through creativity, communication and aesthetic expression. The following interview between two of children (A and J) further clarifies these characteristics. Beginning as an actual interview, with questions and answers, the following question changed the situation by introducing issues of creative expression and level of media literacy:

"What if you were supposed to make your own game—what kind of game would you make?"

The first answer is "I have no idea", but then, as A scans the room, she begins to find things (which happen to be kitchen items), and she brings furniture to set up a scene for the game that she says she wants to build. The things at hand are now circumscribed for play purposes—in this case, for a restaurant game. A did not simply talk about what she imagined; she *made* the game by building it and acting it out at the same time, establishing a social community of cooperation (and communication).She then gives her game a name: *Restaurant.* It is both a quiz game and a food game, she says; initially, it is about "soups and coffees". The action is framed as play when A says "Then, let's say, that it was him who ordered this soup and coffee". In so doing, she is explicitly addressing the action as *make-believe* ("let's say"). Moreover, she shifts to past tense, which is a typical indicator of a play situation (Mouritsen, 2003). In Bateson's terms, the situation is framed in a meta-communicative manner as play. The action relates to her individual game, but as J is automatically co-playing, 'the game' is elaborated in a cooperative process.

A then moves on and talks about how to win hearts and reach the "toughest levels"—where, she says, you end up with something "soft and slimy". As she describes the toughest level, the climax of the game, her voice and gestures shift dramatically (Fig. 12.3, left).

When the researcher asks a question about the toughest level, she does not answer; she is in a mood of devotion, absorbed in her associations. She continues, surprised by her own ideas: "Oat meal is soft and slimy!"

Meanwhile, J is playing with a plastic plate. A directs her gaze at him while saying: "And monsters can also come—then you can kill them and make meat balls of them". She rises to demonstrate that she is a monster; she does not need to ask J to follow her, as he rises in reaction to her (Fig. 12.3, centre).

It is important to recognise the collective aspect of this communication. To keep the 'game' going, the children needed to cooperate, with and without words, devoted to the process of making the game. They demonstrated a solid understanding of 'the game' by referring to levels, rewards, actions, conflicts, monsters and killing; narrative aesthetic techniques were demonstrated, and scenes were set and performed. In this way, they made room for play, and media literacy emerged through the processes of play and communication as ideas, modes, moods, sounds and words came up and were made up.

The children's approach to the making of a game was explorative, seeking possibilities in the things at hand. At the same time, the things define the possibilities, so constituting the game, as Bogost also mentions. The presence of the camera also contributes to defining the situation; A addresses the camera with her eyes as a communicative act—an interaction that frames and maintains the situation.

These practices are on many levels dialogic; media texts, language, participants (J, the researcher and the camera) are all in dialogue to make this happen, as are the things in the room. Inspired by Bakhtin, the concept of dialogical practices indicates

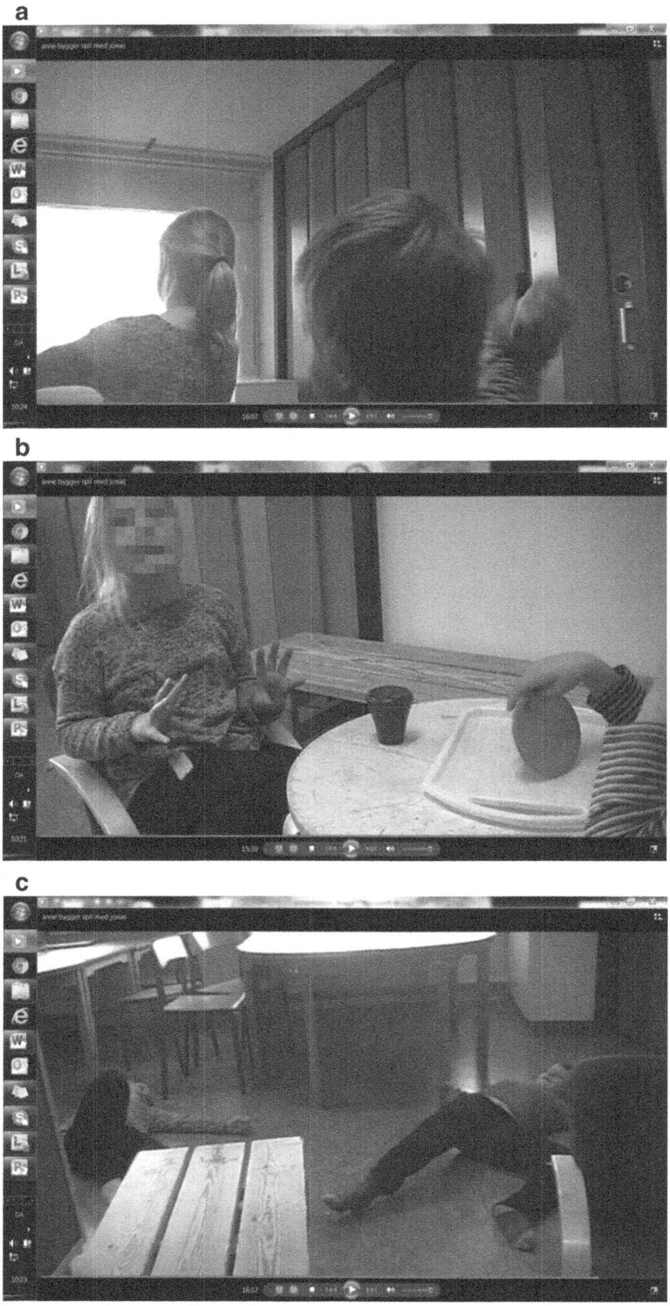

Fig. 12.3 Video stills: A and J making a game

the processes going on, circumscribing the here and now for play. At the threshold between the interview and the initiative to play, a lot of things happen in setting up a play action. The explorative and entrepreneurial approach to making and building a game reveals the playfulness of media literacy.

Conclusion

Connecting levels of media literacy and the five dimensions of play creates more ways of understanding children's practices and how these can be explored through participation in play. By observing how children use play practices to explore the possibilities associated with media literacy, we can identify what they find important in shared play. Understanding children's playful use of media literacy necessarily involves access to digital media, and participation is also a precondition. That means understanding creative improvisation of practices circumscribing the now, introducing language (in its broadest sense), narratives and other aesthetic techniques to situations that make expression and communication possible, framed as play. Media literacy can be seen in having fun, being together, practising certain skills and using things, play moods and play practices. To understand play, self-expression and communication in such contexts, it seems important to acknowledge the role of media literacy. In the present case, worries about GTA are replaced by insights from being close to the actual emergence of play, inviting a broader and more nuanced view of play that helps to explain the fascination and fun of crashing and turning the world upside down. Through this common language, we can hope to find other ways to create 'spielraum' for its silliness, nonsense and wildness of play. The Convention on the Rights of the Child does not exclude any form of play, and the present analysis confirms that digital play is an ambiguous matter, in which curiosity, exploration and knowledge open new pathways for school pedagogical practice. As Bogost claims, play is in everything, but not everything conveys play. When planning media literacy events as part of everyday school life, it is essential to take account of children's perspectives in finding ways to value, guide and build on their ideas.

References

Bakhtin, M. (1981). *The dialogic imagination. Four essays*. University of Texas Press.

Barton, D., & Hamilton, M. (1998). *Local Literacies. Reading and writing in one community*. Routledge.

Bateson, G. (1972). *A theory of play and fantasy. Steps to an ecology of mind*. Ballantine Books.

Bogost, I. (2016). *Play anything. The pleasure of limits, the uses of boredom, and the secret of games*. Basic Books.

Burn, A., & Richards, C. (2014). *Children's games in the new media age. Childlore, media and playground*. Ashgate.

Burnett, C., Merchant, G., Pahl, K., & Rowsell, J. (2014). The (im)materiality of literacy: The significance of subjectivity to new literacies research. *Discourse: Studies in the Cultural Politics of Education, 35*(1), 90–103. https://doi.org/10.1080/01596306.2012.739469.
Caillois, R. (2001 [1958]). *Man, play and games*. University of Illinois Press.
Carlsson, U. (2013). *What is media literacy*. http://nordicom.statsbiblioteket.dk/mld/en/medialiteracy.html.
Clark, A., Flewitt, R., Hammersley, M., & Robb, M. (2014). *Understanding research with children and young people*. SAGE.
Corsaro, W. A. (2003). *We're friends, right? Inside kids' culture*. Joseph Henry Press.
Couldry, N. (2004). Theorising media as practice. *Social Semiotics, 14*(2), 115–132. https://doi.org/10.1080/1035033042000238295.
Erstad, O., & Amdam, S. (2013). From protection to public participation. *Javnost – The Public, 20* (2), 83–98.
Gee, J. P. (2010). *New digital media and learning as an emerging area and "worked examples" as one way forward*. The MIT Press.
Gulløv, E., & Skreland, L. L. (2016). Ethnographic studies of young children. In J. Prior & J. Van Herwegen (Eds.), *Practical research with children*. Routledge.
Holstein, J. A., & Gubrium, J. F. (1995). *The active interview*. SAGE.
Huizinga, J. (1963 [1938]). *Homo ludens*. Gyldendal.
James, A., Jenks, C., & Prout, A. (1998). *Theorizing childhood*. Polity Press.
Jenkins, H. (2009). *Confronting the challenges of participatory culture: Media education for the 21st Century*. MIT Press.
Karoff, H. S. (2013). Play practices and play moods. *International Journal of Play, 2*(2), 76–86. https://doi.org/10.1080/21594937.2013.805650.
Kress, G., & Van Leeuwen, T. (2006). *Reading images: The grammar of visual design* (2nd ed.). Routledge.
Marsh, J. (2012). Children as knowledge brokers of playground games and rhymes in the new media age. *Childhood, 19*(4), 508–522. https://doi.org/10.1177/0907568212437190.
Marsh, J. (2015). *Exploring play and creativity in pre-schoolers' use of apps: Final project report*. http://www.techandplay.org/reports/TAP_Final_Report.pdf.
Marsh, J., & Hallet, E. (2008). *Desirable literacies. Approaches to language & literacy in the early years*. SAGE.
Mouritsen, F. (1996). *Legekultur*. Syddansk Universitetsforlag.
Mouritsen, F. (2003). Æstetisk refleksivitet og refleksionstyper i børns former for narrative udtryk og leg. In J. W. Gleerup & F. Wiedemann (Eds.), *Pædagogisk forskning og udvikling*. Syddansk Universitetsforlag.
Reckwitz, A. (2002). Toward a theory of social practices. A development in culturalist theorizing. *European Journal of Social Theory, 5*(2), 245–265.
Rowsell, J., & Pahl, K. (2015). *The Routledge handbook of literacy studies*. Routledge.
Schatzki, T. (2001). Introduction: Practice theory. In K. K. C. T. Schatzki & E. von Savigny (Eds.), *The practice turn in contemporary theory*. Routledge.
Skovbjerg, H. M. (2018). *The value of play. Design School Kolding*.
Skovbjerg, H. M. (2021). *On play*. Copenhagen: Samfundslitteratur.
Super Carla. (1968). Written by Thomas Winding and Lulu Gauguin. Produced by DR. Retrieved March 29, 2020, from https://www.dr.dk/bonanza/serie/487/boernetv/52913/1960erne-super-carla.
Sutton-Smith, B. (1997). *The ambiguity of play*. MIT Press.
Swidler, A. (2001). What anchors cultural practices. In K. K. C. T. Schatzki & E. von Savigny (Eds.), *The practice turn in contemporary theory*. Routledge.
UNHCR, U. N. o. H. R. (1989). *Convention on the rights of the child*. https://www.ohchr.org/en/professionalinterest/pages/crc.aspx.
Wohlwend, K. E. (2015). One screen, many fingers: Young children's collaborative literacy play with digital puppetry apps and touchscreen technologies. *Theory Into Practice, 54*(2), 154–162.

Wohlwend, K. E. (2018). Play as the literacy of children: Imagining otherwise in contemporary early childhood education. In D. E. Alvermann, N. J. Unrau, & M. Sailors (Eds.), *Theoretical models and processes of literacy* (7th ed.). New York, NY: Routledge.

Helle Hovgaard Jørgensen, works at the Centre for Applied Research in School and Education at University College Lillebaelt, Denmark. Her field of research is play and media literacy in children's perspectives.

Helle Marie Skovbjerg is Professor at Design School Denmark and Head of Research for Lab Design for Play. She has been researching play for several years and has published in several books and journals on the concept of 'mood perspective'.

Chapter 13
Digital Technologies and Children: Does more Digital Interactivity Make for Better Learning?

Jo Li Marie-Joelle Tay

Introduction

In the last decade, there has been a shift toward the digitisation of learning and education. This has partly been fuelled by the introduction of new digital technologies such as smartphones and tablets, which has resulted in the proliferation of digital media use across generations, from babies to the elderly. The shift has also partly been prompted by a belief that digital technologies are more interactive than non-digital ones, and that traditional methods of education are no longer effective. In this context, more interactivity is seen to be better—it is associated with being more engaging, more entertaining, and more fun. This belief has led both parents and educators to adopt the use of digital technologies and techniques (such as the gamification of learning) to engage young learners. The question is, does more digital interactivity really make for better learning?

To address this question, the concept of cognitive load was used to understand the results of visually mapping the dimensions of interactivity in three different Learning Experience Scenarios (LES). This mapping was done using the model of interactivity created for the author's doctoral dissertation (Tay, 2018). The model was developed using Zimmerman, Forlizzi, and Evenson's (2007) research through design methodology. This involved the ideation and testing of five iterations of the model, and the application of the final model by the researcher on the three LES presented here. The three different LES selected for comparison were a printed worksheet, Mathletics (an online learning mathematics website used by many Australian primary schools), and Minecraft (a popular multi-player online video game that has been used by some educators to gamify learning). Since the author was

J. L. M.-J. Tay (✉)
Curtin University, Perth, Australia
e-mail: joli.tay@curtin.edu.au

D. Holloway et al. (eds.), *Young Children's Rights in a Digital World*, Children's Well-Being: Indicators and Research 23,
https://doi.org/10.1007/978-3-030-65916-5_13

both researcher and participant, there are obvious limitations to this research. However, it is still important to introduce the model of interactivity as a useful tool and to present these preliminary findings here, if only to justify the need for additional research to be done with large sample sets of actual students, across a wider range of learning experiences. Such research would contribute significantly to our understanding of interactivity and its role in learning.

This chapter starts by examining the notion of interactivity in relation to learning, experience, technology, education, and cognitive load. The model of interactivity is then briefly described, followed by an overview of the visual mapping of the three chosen LES. Finally, this chapter concludes with a discussion that uses cognitive load to understand what the visual mapping of the three experiences suggest about digital technologies, interactivity, and learning.

Learning, Interactivity, and Experience

Much of the learning we do as human beings comes *from* real-world experience. We learn that ice cream is cold and sweet from eating it, and that sandpaper is rough from touching it. As Pine and Boy stated, "Learning is a consequence of experience" (as cited by Boy & Pine, 1999, p. 221). Pine and Boy also described learning as "an experience that occurs inside the learner and is activated by the learner" (as cited by Boy & Pine, 1999, p. 221). Thus, it can also be said that *the act of learning itself constitutes an experience*; in this chapter this is referred to as a learning experience. Just as real-world experiences come in many different forms, so do learning experiences. Crucially, it is proposed here that what makes each real-world experience different from the next, and what makes one learning experience different from another, is the difference in the interactivity present in the experience. The problem, however, is that the general interpretation of interactivity today is much more limited than our understanding of experience.

At present, interactivity is typically viewed as a characteristic of technology. Davis (2012, p. 5), for example, cited Jenkins et al., who declared that "interactivity is a property of the technology, while participation is a property of culture". Although there is some merit in relating interactivity to technology, the view that interactivity is *solely* a characteristic of technology is restrictive as it suggests that interactivity is unique to digital technologies. This is not necessarily the case. Dewey's pragmatism considered experiences to be the means by which humans interact with their environment (Tay, 2018). These "interactions are *what* human organisms experience" (Garrison, 1994, p. 9) and interactivity is what happens during an interaction. This perspective underlies the broader understanding of interactivity proposed by Tay (2018), including the notion that *all experiences involve interactivity*. This means the simulated experiences more generally associated with digital technologies are not the only experiences that are interactive; real-world experiences should be considered interactive as well.

Digital Technologies and Interactivity in Education

The use of digital technologies in education is contentious. Educators and researchers have both resisted and embraced the digitisation of learning. Li (2007, p. 390) observed that "computers have been installed in almost all schools but many teachers have not used them". Davis (2012) highlighted the benefits of using simulations in an educational context, however she simultaneously noted that simulations were also an area of concern for others. Dewey, for instance, cautioned against using simulations to replace real-life experiences (as cited by Davis, 2012, p. 7). Jenkins et al. (as cited by Davis, 2012, p. 10) shared this apprehension, stating that excessive use of technology and simulated environments (e.g. in the form of video games) could result in health problems as well as diminish access, opportunities, and time for other forms of learning and play. Davis distinguished between real-life and simulated experiences, declaring that real-life experiences are "active" and simulated experiences are "interactive, though ultimately passive" (p. 10). She questioned whether technology has made online experiences more social or if it has reduced real-life social connection instead: "While interactivity and connectedness can lead to socialization [sic], it can also lead to isolation." (Davis, 2012, p. 3).

Conversely, others have argued against these negative views of technology in education (Marsh, 2005; Smith, 2002; Yelland, 2007). Marsh (2005), for example, suggested that media technology can help with the social and personal development of children. Smith (2002) observed higher levels of engagement and learning when technology was used during play. Yelland (2007) suggested that using technology would give learners more access to learning experiences and expand the way they think.

Both sides offer valid and important insights, however these opposing views suggest that there continues to be uncertainty around the use of digital technologies in education. What this does ascertain, though, is the need for a better understanding of what interactivity is and its role in the use of digital technologies for educational purposes.

Understanding Cognitive Load

Before discussing the relationship between interactivity and cognitive load, it is necessary to first explain the concept of cognitive load. To do this, we must begin with Sweller's (1994, p. 298) proposition that one function of learning is "to store automated schemas in long-term memory". A schema refers to the "cognitive construct" we create to organise information in our brain. With practice and time, we are able to process schemas automatically, with little to no conscious effort. This automation is useful because it allows schemas to be stored in our long-term memory while also reducing demands on our working memory. The concept of working memory used here is the one popularised by Atkinson and Shriffin (1968)–it acts as short-term mental storage for sensory data, while at the same time drawing from long-term memory to process (or make sense of) this data.

According to Sweller, the limitations of our working memory mean that we can only process a small number of different items at any one time. This becomes problematic when there is "high element interactivity", which occurs when "interactions between many elements must be learned" at the same time. All this simultaneous interactivity results in a "high cognitive load". (Sweller, 1994, p. 295) Conversely, when only a small amount of information needs to be processed by our working memory, then there is a low cognitive load.

Cognitive load refers to the strain that the brain experiences when trying to process information. Sweller (2010) identified three types of cognitive load: intrinsic, extraneous, and germane. Intrinsic cognitive load directly relates to the inherent difficulty and complexity of the content that must be learned, and the existing knowledge that the learner has. Extraneous cognitive load relates to the way in which the learning content is delivered or taught. It is considered to be a bad form of cognitive load because it is thought to negatively impact learning, which is why instructional designers seek to minimise it. Whether or not the cognitive load is intrinsic or extraneous depends on what needs to be learned. (Sweller, 2010) For instance, if the learner needs to learn a mathematical concept, using a computer to do so may produce an extraneous cognitive load. However, if the learner needs to use the same mathematical concept to write a piece of computer programming code, then using the computer may constitute an intrinsic cognitive load.

Germane cognitive load refers to cognitive load that results from the construction of schemas (i.e. models generated by a learner to organise knowledge) and is necessary for knowledge to enter into the learner's long-term memory. Because of this, it is dependent on intrinsic cognitive load. It is considered to be a good form of cognitive load and to be necessary for learning to occur. Thus, instructional designers usually seek to increase germane cognitive load. (Sweller, 2010)

In addition to these three types of cognitive load, there are a variety of cognitive load effects. Only three of these are of relevance here and thus only these will be examined. These are: the redundancy effect, the split attention effect, and the modality effect. The redundancy effect refers to the inclusion of redundant or unnecessary information that is not directly relevant to learning and is thought to have a negative impact on learning. The split attention effect refers to instances where learners have to cognitively process multiple sources of information at the same time in order to integrate them. The modality effect is partly related to the split attention effect. It refers to the positive effect that can result from integrating some modes of information (e.g. visual and auditory) in a mutually supportive way so that extraneous load is reduced. (Sweller, 2010).

The notion of cognitive load, together with the three cognitive load effects outlined above, suggest that perhaps the interactivity in digital technologies may not be suitable for all learning contexts. The question is, how can we determine the kinds of interactivity suitable for each given learning context? This is where the model of interactivity can be useful.

The Model of Interactivity

Even though digital technologies offer numerous benefits, there is still concern (e.g. Davis, 2012) that excessive technology use to overwhelm our real-world lives. However, no one has offered a concrete solution that can help us balance our use of technology. The model of interactivity was created to help find a balance between real-world and simulated experiences, by allowing the different dimensions of interactivity that exist to be identified. It is based on the assumption that interactivity is present in all experiences. Thus, it was designed to be able to visually map the different dimensions of interactivity present in any kind of experience, particularly so that real-world experiences could be compared with virtual, simulated ones. To achieve these goals, the model had to be complex. This complexity can be seen in Fig. 13.1.

The model of interactivity lists seven dimensions of interactivity (see the key on the right of Fig. 13.1). These dimensions are categorised as internalist or externalist. The internalist dimensions of interactivity refer to those dimensions that are directly related to the body of the individual who is engaged in an experience. There are four internalist dimensions of interactivity: affective, sensory, mental, and motor. The externalist dimensions of interactivity refer to those dimensions that are external to the body of the individual who is engaged in an experience. It can also be thought of

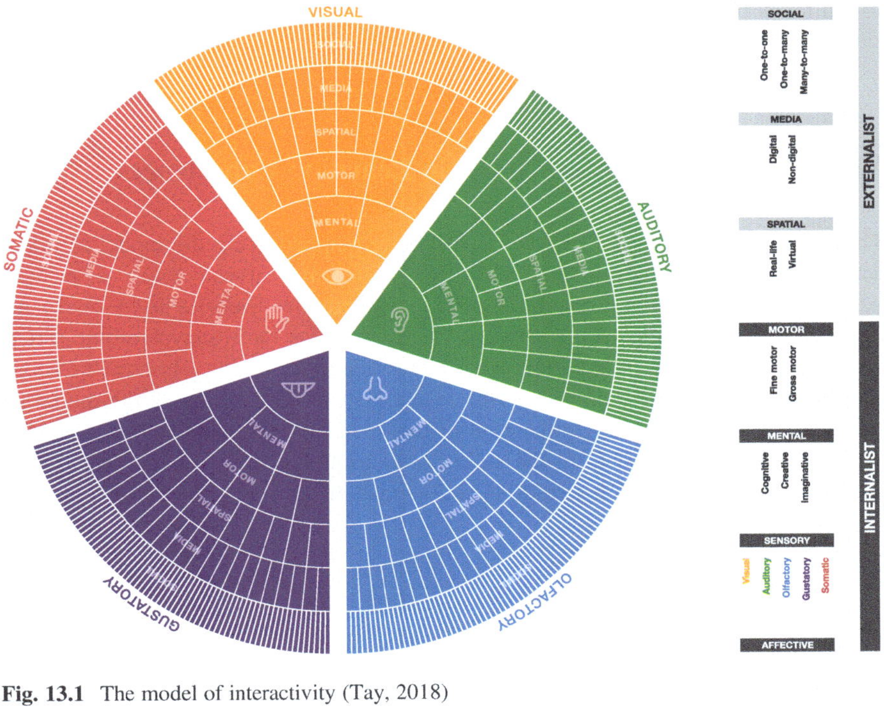

Fig. 13.1 The model of interactivity (Tay, 2018)

as the context of the experience. There are three dimensions of externalist interactivity: spatial, media, and social. Detailed explanations of each of the dimensions can be found in Tay (2018).

The key on the right of the model in Fig. 13.1 outlines the arrangement and names of each dimension and sub-dimension. The three inner rings in the model are the internalist dimensions of interactivity. The ring in the centre represents sensory interactivity; it has five sub-dimensions: visual, auditory, gustatory, olfactory, and somatic. The second ring is mental interactivity and has three sub-dimensions: cognitive, creative, and imaginative. The third ring is motor interactivity and has two sub-dimensions: fine motor and gross motor. The outer three rings are the externalist dimensions of interactivity. The fourth ring is the first externalist dimension, spatial interactivity. It has two sub-dimensions: real-life and virtual. The fifth ring is media interactivity and has two sub-dimensions: digital and non-digital. The sixth and outermost ring is social interactivity and has three sub-dimensions: one-to-one, one-to-many, and many-to-many.

It is necessary to note that although affective interactivity is listed as an internalist dimension in the key, it is not actually mapped in the model of interactivity. Since affective interactivity is present all the time, showing the relationship between the affective dimensions (emotions) and other dimensions of interactivity with a single visual map would be too complex, because it is often possible for several emotions to be present at the same time. The intention was to address affective interactivity in future research, where the model of interactivity could be used to generate separate visual maps of the dimensions of interactivity for each individual affective dimension occurring simultaneously.

Visual Mapping of Interactivity in Learning Experiences

Three learning experience scenarios (LES) were mapped using the model of interactivity. These three LES involved three different learning tools: a printed worksheet, Mathletics (an online learning mathematics website used by many Australian primary schools), and Minecraft (a popular multi-player online video game that has been used by some educators to gamify learning). For each LES, a description is provided followed by the visual map of the LES using the model of interactivity along with an explanation of the visual mapping. The content in this section has been adapted from the author's doctoral dissertation (Tay, 2018). For ease of explanation, the descriptions and explanations below are written in first person.

Description of Learning Experience Scenario 1

In Learning Experience Scenario 1 (LES 1), I did a printed math worksheet individually, not in a group. This is typical of how worksheets are used in classrooms. A printed math worksheet typically consists of a series of sums printed in black on white paper. Since colour printing is significantly more expensive than black-and-white printing, worksheets are usually black-and-white photocopies or printouts. A math addition worksheet was selected for LES 1 so it can be compared with Learning Experience Scenario 2 (i.e. the World Challenge in the Live Mathletics learning option), which also involves math addition. LES 1 is different from the other learning experience scenarios because the learner works alone and does not have a choice of learning options. This is common in many classroom environments where the teacher controls the learning activities.

Application of Model to Learning Experience Scenario 1

Figure 13.2 shows the model as applied to LES 1. Here, only two sensory interactivity sub-dimensions are present: visual interactivity and somatic interactivity.

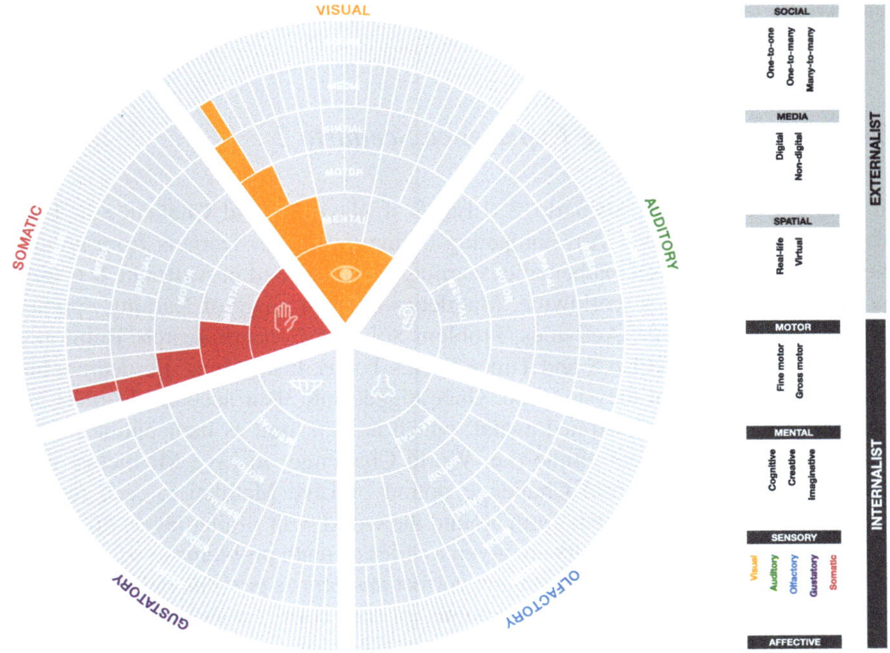

Fig. 13.2 Visual map of interactivity in LES 1 (Tay, 2018)

Visual interactivity is present because I have to look at the printed symbols (i.e. numbers, plus sign, etc.) on the worksheet to interpret them, and also to see what I am writing. Somatic interactivity is present because I use my sense of touch to interact with the worksheet and my pencil.

The mental interactivity that is linked to visual interactivity is cognitive, since cognitive skills are needed to complete the worksheet. These include using memory (i.e. remembering what the symbols represent), constructing mental models (i.e. understanding how addition works), and problem-solving (i.e. figuring out which numbers added together produce which numbers). The mental interactivity linked to somatic interactivity is also cognitive. However, the cognitive skills required are somewhat different. Even though memory is also required here, for instance, the task is different (i.e. remembering the shape of the numbers and how to move the pencil to produce them on paper).

The motor interactivity linked to both the visual and somatic interactivity is fine motor. In relation to visual interactivity, it involves the movement of my eyeball to read what is on the worksheet and to write, as well as the movement of my fingers and hand, which must sync with what I see so I can write legibly. In relation to somatic interactivity, it involves the movement of the fingers and hand as necessitated by the act of writing. The spatial interactivity linked to both visual-motor and somatic-motor interactivity is real-life, since they both involve interacting in the real world with real-life objects rather than with virtual ones. The media interactivity is non-digital, since a non-digital print medium is being used. There is no social interactivity here since the worksheet is being completed individually.

Description of Learning Experience Scenario 2

In Learning Experience Scenario 2 (LES 2), I played World Challenge in Live Mathletics on a Macbook Air, competing online with other learners from around the world. Mathletics is a mathematics learning website based in Australia. It is used by schools and families worldwide. Mathletics offers a range of learning options, such as Live Mathletics, Activities, Problem Solving, Concept Search, Rainforest Maths, Prepare for NAPLAN and Times Tables Toons. Each Mathletics learner has a personal account that they log into. Once logged into their account, learners are asked to customise an illustrated avatar, which is then used as their public profile image when they compete online (i.e. World Challenge) in Live Mathletics with other learners. In the World Challenge option of Live Mathletics, each learner competes against other learners from around the world with similar levels of ability, online and in real-time. Learners are given a series of math problems to solve within 60 seconds. The goal is to give as many correct answers as possible before the time runs out. Each learner is then awarded points for each correct answer, and additional points are given if the learner has the most points (i.e. comes in first) or if the learner gets a new high score (i.e. exceeds than their existing high score).

Application of Model to Learning Experience Scenario 2

Figure 13.3 depicts the model as applied to LES 2. In LES 2, there are three sensory interactivity sub-dimensions present: visual interactivity, auditory interactivity, and somatic interactivity. Visual interactivity is present because I have to look at the images and text on the screen, and also at the keyboard. Auditory interactivity is present because I use the sound of the keyboard to track whether the keys have been pressed. Somatic interactivity is present because I use my sense of touch to interact with the keyboard.

In relation to visual interactivity, the mental interactivity is cognitive; cognitive skills are needed to come up with the answers to the World Challenge questions. Because speed is a factor here, the main skill required is memory (i.e. remembering which numbers added together produce which numbers). In relation to auditory interactivity, the mental interactivity is also cognitive, since cognitive activity is required to process the sounds I hear. In relation to somatic interactivity, the mental interactivity is again cognitive. However, the cognitive skills required are somewhat different. Even though memory is also required here, for instance, the task is different (i.e. remembering where the numbers are located on the keyboard).

The motor interactivity linked to all sensory interactivity sub-dimensions is fine motor. In relation to visual, auditory and somatic interactivity, I engage in finger and

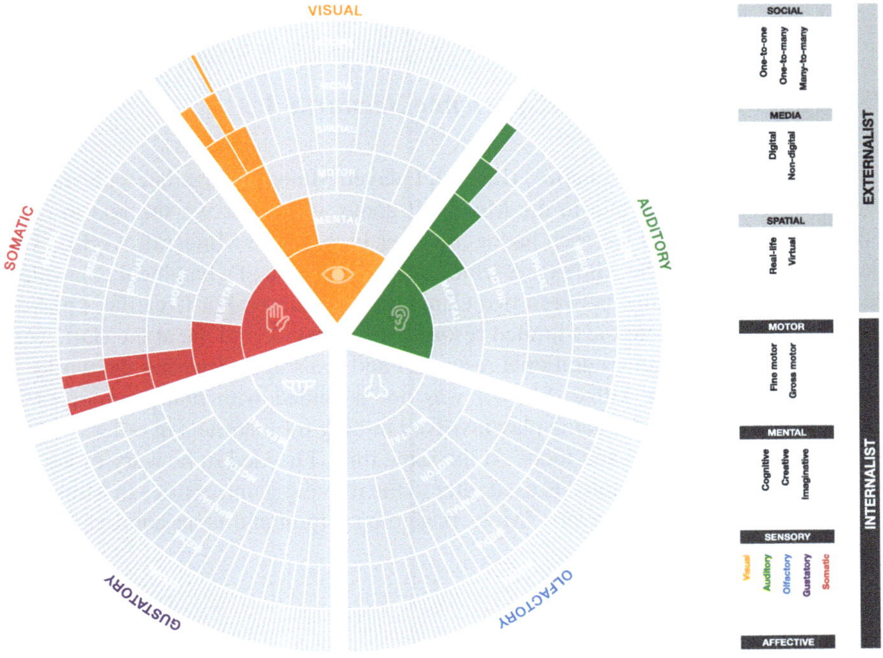

Fig. 13.3 Visual map of interactivity in LES 2 (Tay, 2018)

hand movements based on the sensory feedback I get (e.g. the images I see on screen, the sound from typing on the keyboard, and the feeling of the keyboard buttons). The spatial interactivity linked to both visual-motor and somatic-motor interactivity is both real-life and virtual. This because I interact with both objects in real-life (i.e. keyboard and computer screen), as well as people and objects in virtual space (i.e. the people, images and text in Mathletics). The spatial interactivity linked to auditory-motor interactivity is real-life only because the only sound I hear is from the tapping of the keys on the keyboard; the World Challenge does not have any sound. The media interactivity is digital, since digital media (an Apple Macbook Air laptop) is being used in this instance.

There is many-to-many social interactivity, but only in relation to the visual-motor-virtual-digital interactivity. Although I can see the other learners—in the form of the profile images and personal information (name, country, school)—and they can see me, I cannot chat to them, either via text messages or speech, nor can I touch them, virtually or in real-life. All I can do is compete with them in solving math problems. Despite this, the limited interaction is enjoyable, primarily because I am able to compete with many other learners at the same time, and vice versa. With the exception of this form of interaction, Live Mathletics is a solo activity, especially in terms of auditory and somatic interactivity. This is because I am not interacting with others in real life, and the only way I am interacting with others is using my visual sense in a virtual space (i.e. I can see their names, schools/locations, and how many questions they have answered, but only on screen).

Description of Learning Experience Scenario 3

In Learning Experience Scenario 3 (LES 3), I played Minecraft together with another player, both in real-life and in the Minecraft game world (i.e. we sat next to each other on the couch and we each had a character in the same Minecraft world). The device used here was the Xbox One, which required the use of a handheld controller. Minecraft offers a choice between two Game Modes: Survival or Creative. I played in Survival mode, where only limited resources are provided to me at the start. This meant I needed to acquire additional resources and tools upon entering the world. It also meant I could be killed. Since I was not provided with many resources in Survival mode, part of the game play involved mining for materials in order to build more elaborate buildings and tools. At the same time, I needed to be wary of hostile mobs. The term 'hostile mobs' is used refer to the moving game entities in Minecraft that will chase or attack players. 'Mob' is gaming jargon and is an abbreviation of 'mobile'.

Application of Model to Learning Experience Scenario 3

Figure 13.4 shows the model as applied to the LES 3 scenario. In relation to cognitive-gross motor-virtual-digital media streams in visual, auditory, and somatic interactivity, one-to-one social interactivity was coloured. This is because I am seeing, hearing, and touching my friend's avatar in the virtual Minecraft world. However, in real life, I am only seeing and hearing my friend without the use of media, and this relates to my friend's fine motor movement (e.g. his use of the game controller). Thus, the one-to-one social interactivity segments relating to the cognitive-fine motor-real-life-(no digital media) streams of visual and auditory interactivity are coloured in. The non-digital media segment is not coloured in as no media was used. With regard to imaginative interactivity, however, I am imagining seeing, hearing, and touching my friend inside the virtual Minecraft world, so the one-to-one social interactivity segments linked to the imaginative-gross motor-virtual-digital streams in visual, auditory, and somatic interactivity are all coloured in.

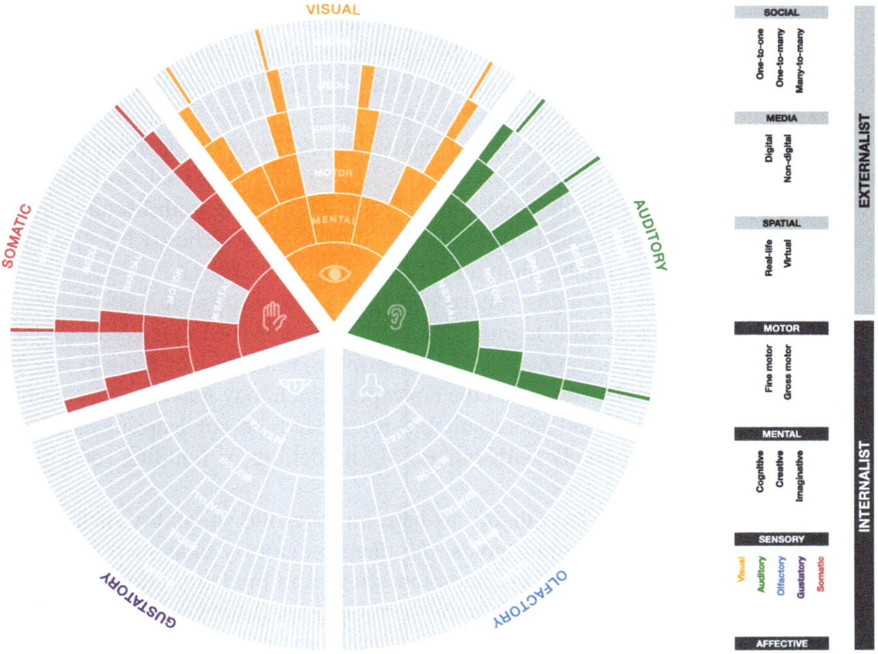

Fig. 13.4 Visual map of interactivity in LES 3 (Tay, 2018)

Discussion: Cognitive Load and Interactivity

Comparing Figs. 13.2, 13.3, and 13.4, it is clear that LES 3 (Minecraft) has the most interactivity, while LES 1 (printed worksheet) has the least interactivity. From this, one could argue that Mathletics is better than a printed worksheet since it is more interactive (and therefore more likely to attract students' interest), but once we examine this in relation to cognitive load, a different picture emerges.

As explained earlier, the construction of a new schema involves learning many pieces of new and different information at same time (i.e. high element interactivity), resulting in a high cognitive load. When schema cannot be processed automatically, working memory is taken up with trying to process the interactions between the learner's existing knowledge and the new information. Thus, if a learner is trying to learn a new mathematical concept, much of the working memory is required, resulting in high intrinsic and germane cognitive loads. In this situation, where the goal of learning is to learn and understand a new mathematical concept, having a high extraneous cognitive load is likely to make it more difficult for the student to learn this new concept. In this context, adding digital interactivity (e.g. in the form of an app or a computer game) will potentially increase the extraneous cognitive load, since this interactivity is not necessary for the new mathematical concept to be learned. If we relate this situation to the Learning Experience Scenarios that were mapped, using a printed worksheet (as in LES 1) is likely to be more appropriate than using Mathletics (as in LES 2).

If a learner has already learned and understood a mathematical concept, but wants to get faster at using the concept, then the goal of learning is now different —it is learning to be faster at using the concept. Here the intrinsic cognitive load and germane cognitive load are both low, so one can assume that an increase in the extraneous cognitive load is unlikely to negatively impact on the learner. In other words, it will not matter too much whether the learner does a printed worksheet or uses Mathletics. However, if the goal is to have fun competing with friends on the speed and accuracy of using mathematical concepts, then Mathletics (LES 2) would probably be much more suitable than a printed worksheet (LES 1).

Given the comparatively higher interactivity that Minecraft (as in LES 3) has, it may not be ideal to use it with learners who have not achieved mastery or proper understanding of mathematical concepts. Furthermore, it has less of a direct link to mathematical concepts (depending on the concept), which may make it less suitable for learning mathematics, and more suitable for other types of learning, such as learning to work cooperatively with others, 3D modelling or building, or virtual orienteering.

Besides using cognitive load theory to help interpret the suitability of learning experiences for particular learning goals, the three previously examined cognitive load effects (redundancy effect, split attention effect, and modality effect) can help us to understand more clearly why some learning experiences work better than others. The redundancy effect is particularly relevant to the use of digital technologies in learning, since it refers to the inclusion of redundant or unnecessary information that is not directly relevant to learning. Taking into consideration the redundancy effect can help with one of the difficulties parents and educators face today—determining whether a learning app is effective. Assumptions about the level of cognitive load can be made based on the number of redundant elements identified in a learning app. However, further research is needed to establish a more precise method for doing so.

When considering the use of digital technologies in learning, it is also important to take into account the split attention effect. This occurs when learners have to cognitively process multiple sources of information at the same time in order to integrate them. Many learning apps, for example, require the use to process visual animations, as well as sound effects, in addition to the learning content. These animations and sound effects may actually be distractions that result in the learner splitting their attention between them and the learning content. In a sense, these animations and sound effects can also be considered redundant if they do not directly contribute to or support the content being learned.

Finally, it is also important to consider the modality effect in conjunction with the split attention effect. If there is a positive effect resulting from the integration of some modes of information (e.g. visual and auditory) and this is done in a mutually supportive way so that extraneous load is reduced, then this is actually advantageous. What the modality effect points out is that animations or sound effects are not necessarily bad, but rather, that they must be fit for purpose.

Conclusion

The model of interactivity has the potential to be a useful tool to help educators and instructional designers establish whether or not digital technologies should be used and, if they are necessary, how, when, and where particular digital technologies should be used. With further research, the model can also contribute to the limited body of research involving the gustatory and olfactory dimensions in the context of learning (e.g. Classen, 1999; Di Fuccio, Ponticorvo, Ferrara, & Miglino, 2016; Jonsson, Ekström, & Gustafsson, 2005; Mustonen, Rantanen, & Tuorila, 2009; Tijou, Richard, & Richard, 2006).

More importantly, it illustrates the usefulness of cognitive load theory in interpreting the visual maps of the interactivity in learning experiences. Used together, cognitive load theory and the model of interactivity can help to provide a better understanding of the suitability of learning experiences (including whether or not the use of digital technologies is appropriate), so they can be selected and designed more appropriately for the learning goals each learner has. More research is needed to build on the preliminary findings presented in this chapter, to validate the findings with a larger sample size of actual students.

References

Atkinson, R. C., & Shriffin, R. M. (1968). Human memory: A proposed system and its control processes. *Psychology of Learning and Motivation, 2*(1968), 89–195. https://doi.org/10.1016/S0079-7421(08)60422-3.

Boy, A. V., & Pine, G. J. (1999). *A person-centred foundation for counseling and psychotherapy*. Springfield, IL: Charles C. Thomas Publisher.

Classen, C. (1999). Other ways to wisdom: Learning through the senses across cultures. *International Review of Education, 45*(1999), 269–280. https://doi.org/10.1023/A:1003894610869.

Davis, H. E. (2012). Technology in the classroom: A deweyan perspective. *Kentucky Journal of Higher Education Policy and Practice, 1*(2), 2. Accessed from https://uknowledge.uky.edu/kjhepp/vol1/iss2/2.

Di Fuccio, R., Ponticorvo, M., Ferrara, F., & Miglino, O. (2016). Digital and multisensory storytelling: Narration with smell, taste and touch. In K. Verbert, M. Sharples, & T. Klobučar (Eds.), *Adaptive and Adaptable Learning (EC-TEL 2016, Lecture Notes in Computer Science, vol. 9891, pp. 509–512)*. https://doi.org/10.1007/978-3-319-45153-4_51.

Garrison, J. (1994). Realism, Deweyan pragramatism, and educational research. *Educational Researcher, 23*(1), 5–14. https://doi.org/10.3102/0013189X023001005.

Jonsson, I. M., Ekström, M. P., & Gustafsson, I. (2005). Appetizing learning in Swedish comprehensive schools: An attempt to employ food and tasting in a new form of experimental education. *International Journal of Consumer Studies, 29*(1), 78–85. https://doi.org/10.1111/j.1470-6431.2005.00382.x.

Li, Q. (2007). Student and teacher views about technology: A tale of two cities? *Journal of Research on Technology in Education, 39*(4), 377–397. Accessed from http://files.eric.ed.gov/fulltext/EJ768884.pdf.

Marsh, J. (Ed.). (2005). *Popular culture, new media and digital literacy in early childhood*. London: Routledge Falmer.

Mustonen, S., Rantanen, R., & Tuorila, H. (2009). Effect of sensory education on school children's food perception: A 2-year follow-up study. *Food Quality and Preference, 20*(3), 230–240. https://doi.org/10.1016/j.foodqual.2008.10.003.

Smith, C. R. (2002). Click on me! An example of how a toddler used technology in play. *Journal of Early Childhood Literacy, 2*(1), 5–20. https://doi.org/10.1177/14687984020021001.

Sweller, J. (1994). Cognitive load theory, learning difficulty, and instructional design. *Learning and Instruction, 4*(4), 295–312. https://doi.org/10.1016/0959-4752(94)90003-5.

Sweller, J. (2010). Element interactivity and intrinsic, extraneous, and germane cognitive load. *Educational Psychology Review, 22*(2), 123–138. https://doi.org/10.1007/s10648-010-9128-5.

Tay, J. L. M. (2018). *A model for mapping interactivity in learning experiences*. (Doctoral dissertation). Accessed from https://espace.curtin.edu.au/handle/20.500.11937/70568

Tijou, A., Richard, E., & Richard, P. (2006). Using olfactive virtual environments for learning organic molecules. In Z. Pan, R. Aylett, H. Diener, X. Jin, S. Göbel, & L. Li (Eds.), *Technologies for E-Learning and Digital Entertainment (Edutainment 2006, Lecture Notes in Computer Science, vol. 3942, 1223–1233)*. https://doi.org/10.1007/11736639_152.

Yelland, N. (2007). *Shift to the future: Rethinking learning with new technologies in education*. New York: Routledge.

Zimmerman, J., Forlizzi, J., & Evenson, S. (2007, April). Research through design as a method for interaction design research in HCI. In *Proceedings of the SIGCHI conference on Human factors in computing systems (pp. 493–502)*.

Jo Li Marie-Joelle Tay is a Lecturer in the School of Design and the Built Environment at Curtin University, where she teaches in the area of graphic design. Her research interests include aesthetics, interactivity, experience design, learning, and design thinking.

Chapter 14
Disability and Remote Learning during COVID-19

Catia Malaquias and Katie Ellis

Introduction

Public debate on human rights, inclusive education and the concept of digital rights has grown considerably since the closure of schools to prevent the spread of COVID-19. Disability activists and their allies point to the high stakes for children with disability being home-schooled during the pandemic. Impossible decisions between health, work and education were forced onto these children and their families (Duffy, 2020). In many cases, long-held disabling approaches to education were exacerbated in this new untested digital environment as the world grappled with both the health and economic implications of COVID-19.

While the disruption caused by the COVID-19 pandemic to education systems and the closure of schools for varying periods created an unexpected challenge to education stakeholders in Australia and across the world, it also provided education systems with an opportunity to innovate educational practices and to explore the potential to use digital technology to supplement and enhance traditional teaching. In addition, with parents in many countries faced with the shift to learning at home and teachers placed under significant pressure to adapt and develop content for children in isolation at home in a short period of time, this also brought into sharper focus the value of teachers and the importance of education.

This chapter begins with an introduction to the social model of disability before moving on to a discussion of the *United Nations Convention on the Rights of Persons with Disability* (CRPD) and the right to an inclusive education articulated in Article 24. The next section introduces the main case study of this chapter—a report prepared by the organization Children and Young People with Disability

C. Malaquias · K. Ellis (✉)
Curtin University, Perth, Australia
e-mail: katie.ellis@curtin.edu.au

D. Holloway et al. (eds.), *Young Children's Rights in a Digital World*, Children's Well-Being: Indicators and Research 23,
https://doi.org/10.1007/978-3-030-65916-5_14

Australia in July 2020 titled *Not even remotely fair: Experiences of students with disability during COVID-19*. The chapter then offers an analysis of and discussion about the findings in the context of both digital rights and the right to an inclusive education. It then considers further extending the social model of disability to a digital environment, examining the potential for inclusive education to address disabling attitudes while considering Article 8 of the CRPD.

COVID-19 has forced a redesign of our everyday lives to facilitate the digital delivery of education, work and telehealth. We acknowledge this in this chapter and conclude it with a reflection that highlights the importance of using the lessons learned in this pandemic to develop future approaches to learning that are inclusive.

Locating Disability in Society

This chapter proceeds from both a social and human rights approach to disability. While disability is typically located as a problem within a damaged body, the social model of disability instead highlights the way disability is the result of inflexible and inaccessible social practices, policies, and procedures. This model locates disability in society and impairment within the body. As Michael Oliver explains:

> Disability according to the social model, is all the things that impose restrictions on disabled people; ranging from individual prejudice to institutional discrimination, from inaccessible public buildings to unusable transport systems, from segregated education to excluding work arrangements, and so on. Further, the consequences of this failure do not simply and randomly fall on individuals but systematically upon disabled people as a group who experience this failure to discrimination institutionalised throughout society (Oliver, 1996)

A human rights approach extends the social model of disability and offers a framework for removing socially created barriers. A human rights approach to disability recognizes that people with disability have the same rights as the rest of the population; however, environmental, attitudinal and organizational barriers create obstacles to the enjoyment of those human rights. A rights model recognizes the effects of impairments but places responsibility on society and governments to ensure the full inclusion of people with disability. While the social model recognizes disability as a social construct, a rights approach understands that "human rights do not require the absence of impairment" (Degener, 2016, p. 35).

As COVID-19 was declared a pandemic, quarantine and social distancing requirements saw whole populations rapidly and effectively shift to online forms of participation such as remote learning, working from home and telehealth. Governments, businesses, and institutions moved quickly to redesign practices and workplaces, in many cases to offer digital flexibility.

Disability activists and their allies described the response as bittersweet as they witnessed and participated in a world suddenly willing to embrace flexibility. For Gerard Goggin and Katie Ellis, the digital social life under COVID emergency conditions was a new normal that continued to enforce a disabling normalcy:

> As people heeded the edict 'Stay home. Save lives', practices, discourses and shared meanings emerged. Many people with disabilities found this highly ironic. For a significant proportion of people with disabilities staying at home is a default. In their lives, if infrastructure, capacity, and preference exists, their reliance on digital participation modes has long been heightened. Accordingly, with nearly full spectrums of populations precipitating the switch to digital-by-default social life, across disability communities many people expressed the feeling of potential relief and shared solidarity. Now the rest of the population had the opportunity to gain a sense and even share the lived experience of what such necessary confinement to households would be like. Unfortunately, this feeling was quickly followed by the realisation that this long-run familiarity and expertise in the stay-at-home conditions of digitalised everyday life now shared across populations still did not spare disabled people from structural exclusion (Goggin & Ellis, 2020, pp. 168–169)

Inclusive education for children with disability younger than age eight was one area of persisting structural exclusion. South African-born British disability activist Victor Finkelstein's thought experiment 'the village' is useful in understanding the experiences of children with disability and their parents during COVID-19. The late scholar and pioneer of the social model of disability articulated a vision for a world designed to include people with disability focusing specifically on the needs of wheelchair users (Finkelstein, 1980). In his so-called village, the built environment, the media and health care all reinforced the social convention that to use a wheelchair was normal and species typical. This normalcy was enforced through building design to include this population, with wide door frames, low ceilings and no stairs. The media represented only the experiences of people with disability and the community existed thinking everyone was exactly like them. As a result, when some able-bodied visitors arrived in their village and wanted to settle there, they were pathologized by the majority population of wheelchair users. The visitors were issued helmets to protect their heads from the low doorframes and any evidence of the damaging impacts of the built environment such as bruises on their foreheads were heavily stigmatized and would result in exclusion from the workforce. This hypothetical society illustrates the way the environment and people's attitudes are created and therefore could be redesigned or changed to include and exclude certain bodies.

By positioning disability as the result of socially created exclusion and discrimination, the social model of disability is effective in bringing together a variety of human rights concerns. Recognition of disability as a human rights issue is key to solving the problem of disability in a number of contexts. This chapter focuses on education and proceeds from a human rights approach to disability, particularly for children younger than eight years old being home-schooled in Australia during the COVID-19 pandemic.

A Universal Right to Education

The universal right to education has been formally recognized under a range of international human rights instruments since the *Universal Declaration of Human Rights* (United Nations, 1948) was adopted in 1948. *The International Covenant on Economic, Social and Cultural Rights* (United Nations, 1966), which was adopted by the United Nations General Assembly almost two decades later in 1966 as a multilateral treaty, went further in outlining the content of the right more comprehensively, devoting its Articles 13 and 14 to education.

But while these and other early international instruments recognised the right to education as a fundamental human right and affirmed the important principle of universality, that the right to education applies regardless of individual characteristics or attributes, Article 24 of the United Nations *Convention on the Rights of Persons with Disabilities* (United Nations, 2008) specifically applies these rights and principles to the context of people with disability and the barriers they commonly face in accessing education. In light of this and the severe exclusion that people with disabilities have faced historically, including through State-endorsed policies of confinement and segregation (Appleman, 2018), it is not surprising that the concept of inclusion is at the core of the expression of the right to education in the *Convention on the Rights of Persons with Disabilities,* the CRPD.

Relevantly, Article 24.1 of the CRPD provides:

> State Parties recognize the right of persons with disabilities to education. With a view to realizing this right without discrimination and on the basis of equal opportunity, State Parties shall ensure an inclusive education system at all levels . . . .

Article 24.2 of the CRPD further states that State Parties shall ensure that:

1. people with disability are not excluded from the general education system on the basis of disability;
2. people with disability can access an inclusive, quality and free primary education and secondary education on an equal basis with others in the communities in which they live;
3. reasonable accommodation of the individual's requirements is provided;
4. required supports are given within the general education system; and
5. individualized support measures are provided in environments that maximize academic and social development, consistent with the goal of full inclusion.

In September 2016 the Committee on the Rights of Persons With Disabilities (CRPD Committee) issued General Comment No.4 (The Right to Inclusive Education) (GC4, United Nations, 2016), an interpretation of the normative content of Article 24 as well as policy guidance for its application. In essence, GC4 makes it clear that the right of people with disability to education is the right to accessible, quality and inclusive education. The commentary states "Inclusion involves a process of systemic reform" and "changes in content, teaching methods, approaches, structures and strategies in education to overcome barriers with a vision serving to provide all students ... with an equitable and participatory learning experience" (United Nations, 2016, para 9).

Importantly, GC4 also defines the distinct concepts of integration, segregation and exclusion that are not compatible with inclusive education and makes it clear that segregated or special schools and classes for students with disability are forms of segregation. General Comment No.6 (Article 5: Equality and Non-Discrimination) (GC6, United Nations, 2018), issued two years later and providing the interpretative guidance on Article 5 of the CRPD, expressly states that "segregated models of education, which exclude students with disabilities from mainstream and inclusive education on the basis of disability, contravene articles 5(2) and 24(1)(a)" of the CRPD (United Nations, 2018, para. 64(a)).

Australian governments regularly cite the Disability Discrimination Act, 1992 (DDA) to evidence the implementation of Australia's obligations under the CRDP, including Sect. 22 which makes unlawful some forms of disability discrimination by schools and education authorities and imposes on them a requirement to provide 'reasonable adjustments' to students with disability. The Disability Standards for Education, 2005 (the Standards) which have been established under the DDA purport to clarify the obligations of educational authorities, institutions and other education providers under the Act.

However, the DDA makes an exception where the school or education authority can establish that compliance with Sect. 22 would cause 'unjustifiable hardship'.

While the concepts of reasonable adjustment and unjustifiable hardship in the DDA seem to reference the concepts of reasonable accommodation and undue burden in the CRPD, they appear to be significantly more limited (Cukalevski & Malaquias, 2019, p.7). In this regard, the CRPD Committee has taken the view that reasonableness is to be "understood as the result of a contextual test that involves an analysis of the relevance and the effectiveness of the accommodation, and the expected goal of countering discrimination" (United Nations, 2016, para. 28; 2018, para. 25(a). In contrast, the Standards provide for a lower competing interests standard in that "an adjustment is reasonable in relation to a student with a disability if it balances the interests of all parties affected" (Disability Standards for Education, 2005, para. 3.4(1)). Similarly, the CRPD Committee has stated that "undue burden cannot be claimed to defend the failure to provide accessibility" (United Nations, 2016, para. 28).

The CRPD Committee has once again urged Australia to reform the DDA and the Standards to address their flaws (United Nations, 2019a, b). Speaking at the Constructive Dialogues held in Geneva on September 12, 2019 for the CRPD Committee's review of Australia's combined second and third periodic report on the implementation of its obligations under the CRPD, CRPD Committee chairperson Mr. Danlami Umaru Basharu raised with the Australian government delegation the issue of "reforming the Australian legal and policy framework, including the [Disability Discrimination Act, 1992], to ensure that the rights of students with disabilities to inclusive education are upheld, and there is immediate and progressive implementation of Article 24 and General Comment No.4, including specific measures to address cultural and attitudinal barriers within education departments and at school administration levels and ensure adequate training of and support to school administrators and educators, for the inclusion of students with disabilities." (United Nations, 2019b, at 1.31.34 minutes). In its Concluding Observations, the CRPD Committee specifically recommended the Australian government review the Disability Standards for Education, 2005 in line with GC4 (United Nations, 2019b, para. 46).

A 2019 analysis of segregation patterns in Australia in the period from 2009 to 2015, conducted by Kate De Bruin, corroborates long-standing concerns about the effectiveness of the DDA and the Standards (United Nations, 2013, para. 46(b)) and concluded as follows (De Bruin, 2019, p. 11):

> While the DDA and [Standards] reforms have both supported the general aim of increasing the access of students with a disability to the regular education system, the data presented here clearly demonstrates that these reforms have failed to bring about important improvements in access to regular schools and classes for students with a disability. Instead, special school placements have substantially increased following the introduction of the DDA and the [Standards], meaning that, paradoxically, more students are segregated into these settings following legislative reforms designed to reduce this. Although both the DDA and the [Standards] have led to a reduction in the rate of exclusion for students with disabilities, this rate remains concerningly high and the reforms cannot be considered to have successfully addressed the persistent issue of exclusion for this reason.

Further, a 2020 scoping review, to determine the extent to which the DDA and DSEs were achieving the elimination of disability discrimination against students with disability, identified a range of problems and concluded that "Australia has missed the mark in achieving non-discrimination in primary and secondary schools" (Duncan, Punch, Gauntlett, & Talbot-Stokes, 2020, p.15).

Remote Learning for Children with Disability during COVID-19

This context of uneven support and a patchy commitment to inclusive education is where students with disability found themselves in during COVID-19, as they learned remotely. The rush to online remote learning brought into sharp focus existing inequalities in a number of educational contexts (Ellis, Kao, & Pitman, 2020). In an analysis of the use of technology to support the wellbeing of children during the COVID-19 pandemic, Karen Goldschmidt explores the way technology became essential to the social, physical, emotional, intellectual and spiritual wellbeing of children at this time (Goldschmidt, 2020). While hopeful that "change can be useful" Goldschmidt recognizes the persistence of the digital divide. Her consideration of the experiences of children with disability in this context however focuses on therapeutic rather than pedagogical services.

In order to ensure the delivery of education for all students during the COVID-19 pandemic, a National Cabinet established by the Australian government agreed on April 16, 2020 to seven *Covid-19 National Principles for School Education* including principle 6 that "all students must continue to be supported by their school to ensure participation in quality education during the COVID-19 crisis" (Australian Government, 2020). However, despite this ostensible commitment of the Australian government, the challenges of accessing quality education and appropriate supports have not been equally experienced by all students and, notwithstanding the requirements of the DDA and the Standards which continued to apply during the COVID-19 pandemic, it would seem that education systems have not given equal priority to the learning of all students.

For children with disability, while experiences of learning at home varied, existing barriers experienced before the pandemic were further aggravated in the time that schools were closed to most students and in the period when transition back into classrooms occurred, as demonstrated by various surveys undertaken by disability advocacy organisations in this period (see for example Down Syndrome Australia, 2020). The reality for many children with disability was that the COVID-19 pandemic added another layer to the systemic neglect and exclusion they had already been experiencing.

Not Even Remotely Fair: Experiences of Children with Disability during COVID-19

This section of the chapter focuses on research undertaken by the young people's organization Children and Young People with Disability Australia during COVID-19 and published in a report titled *Not even remotely fair: Experiences of students with disability during COVID-19* in July 2020 (CYDA Report). CYDA is the national representative organization for children and young people with disability in Australia. The report focused specifically on the experience of students with disability in accessing education during the COVID-19 pandemic through an analysis of both quantitative and qualitative data collected nationally with an online survey of more than 700 family members of students with disability as well as students with disability themselves (Dickinson, Smith, Yates, & Bertuol, 2020, p. 2).

Key findings identified in the CYDA Report addressed a lack of support in areas including appropriate materials, and social support. The provision of supports for students with disability by schools was significantly reduced during the relevant period (Dickinson et al., 2020, p. 3), including in respect of modification of curriculum content (decreased by 18%), social support (decreased by 34%), supervision (decreased by 38%) and individual access to support workers (decreased by 44%) and specialist allied health (decreased by 23%), with parents sometimes using funding from the National Disability Insurance Scheme for education support purposes (although there was confusion about the use of funding in this way, see Dickinson et al., 2020, p. 25).

There was a lack of assertive and proactive support for students with disability and many students were left isolated without regular contact from school or connection to peers for significant periods and unable to access learning material and engage in online class lessons (Dickinson et al., 2020, p. 32–37). This lack of appropriate materials and adequate supports in many cases placed the onus on parents to attempt to adapt learning materials themselves without having the skills to do so, with some parents reporting that their child received exactly the same materials as provided to students without disability, without adjustments being made to support their access and participation (Dickinson et al., 2020, p. 26–33).

The provision of adequate social support in particular was critical to learning engagement and well-being during the remote learning phase, by ensuring that students remained connected to their learning community and peers and mitigating the risk of social isolation (Dickinson et al., 2020, p. 27–31. The most significant positive impacts were reported when schools planned and delivered two or more types of educational and social supports to students with disability, with those students being 88% more likely to feel part of a learning community, 109% more likely to receive adequate support in education, 48% more likely to be engaged in their learning and 18% less likely to feel socially isolated (Dickinson et al., 2020, p. 5 and p. 35–36).

The CYDA Report also noted that many parents had to undertake significant advocacy, to ensure their child's access to supports and resources (Dickinson et al., 2020, p. 39). These findings corroborated informal accounts being provided by many parents of students with disability and some teachers engaging in online groups and forums at this time. For example, teachers and families recounted that even for students who ordinarily attended regular classrooms in general education, in some cases the learning materials that were being provided to them were not the curriculum-aligned differentiated class lesson content that was being provided to peers, but were materials prepared by so-called special schools and in some cases sourced from outdated and previously discarded alternative curriculum developed for students with disability.

In an anonymous personal communication, one parent told us of their eight-year-old child receiving lesson material that was, not only unconnected to the material being provided to the non-disabled students in the class, it was also not age-appropriate. The parent noted that a 'kinder storybook' was provided, accompanied by worksheets for an activity that involved minimal literacy or numeracy content and had been downloaded from an online teaching resource website, and in their words "it was a cut out and colour activity for a kinder kid and my daughter was not interested" and "it was just busy work to keep them occupied."

Concerns of this nature were also reflected in the CYDA Report, which includes the following statement sourced from a parent:

> During COVID19 there was more 'alternative' material provided which risked disconnecting the student from class learning. It is as if there was failure to recognise in the same way that even in a home learning context the principles of inclusive education should be applied.

Drawing on Salend's extensive literature review on inclusive education, Garry Hornby identifies four key principles of inclusive education (Hornby, 2014):

> Firstly, providing all learners with challenging, engaging, and flexible general education curricula; secondly, embracing diversity and responsiveness to individual strengths and challenges; thirdly, using reflective practices and differentiated instruction; and fourthly, establishing a community based on collaboration among students, teachers, families, other professionals, and community agencies.

Offering children learning materials below their age or intellectual abilities is in clear contradiction to these principles. Further, inclusion is already a key feature of early childhood educators, as a teacher cited in Cologon explains (Cologon, 2014, p. 12):

> Inclusion is really (when you think about it) what teaching is: Meet each child where they are at, build on their strengths and interests to move them along, and adapt your teaching style, resources and pace to each of them.

This teacher is drawing on the principles of universal design for learning—an approach that recognises that the education system must adapt to individual learners because there is no one typical way to learn. Even as these issues began to surface during COVID-19, advocacy organizations sought to raise these concerns directly with education ministers and their departments to develop resources to support students with disability and their families. For example, the Australian Coalition for Inclusive Education (ACIE), which comprises advocacy organizations from around Australia, developed a set of principles and recommendation to ensure the right of students with disability to inclusive education during the COVID-19 pandemic, which were provided to the Commonwealth and state and territory education ministers (Australian Coalition for Inclusive Education, 2020) and held a national roundtable with stakeholders including representatives from education departments around the country on May 20, 2020. The principles and recommendations document reflected experiences that had been reported to ACIE member organizations since the pandemic began but also expressed broader concerns about potential longer term effects of social and academic disconnection being experienced by students with disability beyond the pandemic, as well as the risk that stop-gap approaches and poor practices adopted during the crisis, such as the use of alternative learning materials in place of the curriculum-aligned and differentiated regular class lesson, could stay beyond it and after students transitioned back to learning at school.

Digital Rights and the Rights of People with Disability

The lack of support for families in key areas during the pandemic has contravened the best practice advice for parents (regardless of whether children had a disability) to maintain a routine and stay socially connected to schools during COVID-19. We began this chapter citing Finkelstein's thought experiment. In a later publication, the disability rights activist argued that disability would be redefined again following the information age. For Finkelstein, technologies such as braille keyboards would allow people with vision impairments to fully participate in the workforce (Finkelstein, 1981). These insights have influenced the *Convention on the Rights of Persons with Disabilities* which is recognized as the first UN convention to articulate digital rights (Goggin, 2015).

Digital technologies or modes of access are key to achieving many of the rights articulated throughout the CRPD. For example, people who are blind or vision impaired and people who are deaf can access television and other forms of cultural life through audio description and captions respectively (Ellis, 2019). Likewise, just as Finkelstein predicted, there are a number of technologies that students with disability can make use of to access education. There have been documented benefits of making these inclusive technologies available to the entire student population. For example, Microsoft's immersive learning tool assists students with dyslexia read and is effective in increasing sight recognition of words for all students.

Returning to Finkelstein's thought experiment with which we began this chapter, during COVID-19 the world was redesigned using digital technologies. Prior to the pandemic, disability theorists acknowledged the potential of digital technologies in facilitating a more inclusive environment for people with disability but cautioned that discriminatory attitudes would continue to disable people with impairments (Ellis & Kent, 2011; Goggin & Newell, 2003). Gerard Goggin and the late Christopher Newell in particular reflected on the ways the digital environment would never become fully accessible until disability was recognised as a marker of social exclusion in the same was as race, class and gender. They emphasised the way disabling attitudes continued to persist with negative consequences in the digital context. For students with disability participating in remote learning, negative attitudes and lack of support continued to affect them and in some cases exacerbated the lack of support they already received.

Improving representations of disability also features heavily in the CRPD. Article 8 of the CRPD extends long-argued concerns of social model and critical disability scholars such as Beth Haller, Gerard Goggin, Katie Ellis and Elizabeth Ellcessor (Duncan, Goggin, & Newell, 2005; Ellcessor, 2015; Ellis, 2012; Ellis & Goggin, 2015; Haller, 1995, 2000, 2010), mandating that state parties "adopt immediate, effective and appropriate measures:"

(a) To raise awareness throughout society, including at the family level, regarding persons with disabilities, and to foster respect for the rights and dignity of persons with disabilities;
(b) To combat stereotypes, prejudices and harmful practices relating to persons with disabilities, including those based on sex and age, in all areas of life;
(c) To promote awareness of the capabilities and contributions of persons with disabilities. (Article 8 United Nations, 2006)

Article 8 also highlights inclusive education as one avenue to achieving the goals of the article and the convention at large.

Conclusion

The closure of schools as governments around Australia attempted to contain the spread of COVID-19 and the delivery of education to students at home, which tended to be more protracted for students with underlying immunity or other health conditions, demonstrated that in the current climate, students with disability are at the high risk of being adversely affected by the COVID-19 pandemic and similar emergency situations, and need appropriate and planned support from their schools to ensure their engagement in learning and social and emotional well-being. It also highlighted the fears that existing practices in support of students' rights to inclusive education, such as the provision of curriculum-aligned differentiated content, could be weakened by the COVID pandemic, or another pandemic.

While the move to digital delivery modes in the COVID pandemic offered an opportunity to redesign the world to be more inclusive and flexible, the experience of the children being taught during this period, particularly those younger than the age of eight, reinforces Gerard Goggin and Christopher Newell's warning in *Digital Disability* that the same disabling attitudes created in the analogue world would simply be reproduced in digital environments if we did not address them as enforced and created socially. The solution, according to the *Convention on the Rights of Persons with Disabilities* is to recognise the role of these disabling attitudes and attempt to change them through the normalisation of inclusive education practices.

On a more encouraging note, some experiences also demonstrated that students with disability who receive adequate social and academic supports from their schools and stay connected to their learning communities, can progress their learning safely in their home environment in periods of emergency or crisis, and realise their right to inclusive education. For educators, this means ensuring online spaces are accessible to all students and underpinned by universal design for learning and providing students with disability with access to the regular class lesson content, differentiated, as may be appropriate, and delivered by their class teachers to the student along with their class peers. It also requires schools to deliver social supports to foster students' membership of, and meaningful connection to, their regular learning communities and maintain strong engagement with students learning at home to seek feedback and promptly identify what is and is not working for the student.

The evidence gathered during the COVID-19 pandemic must now be used to inform and guide future policy responses to the crisis and efforts to ensure the preparedness of education systems, to deliver inclusive education and support the learning of all students in situations of emergency that may result in the temporary closure of schools and the delivery of education to students at home.

References

Appleman, L. I. (2018). Deviancy, dependency, and disability: The forgotten history of eugenics and mass incarceration. *Duke Law Journal, 68*(3), 417.

Australian Coalition for Inclusive Education. (2020, April). *Providing inclusive education in a time of crisis: Principles and recommendations.* Accessed from acie.org.au/2020/04/27/providing-inclusive-education-at-a-time-of-crisis-principles-and-recommendation-from-acie

Australian Government. (2020, April). *COVID-19 national principles for school education.* Accessed from www.dese.gov.au/covid-19/schools/national-principles-for-school-education

Cologon, K. (2014). *Inclusive education in the early years: Right from the start.* South Melbourne, Victoria: Oxford University Press.

Cukalevski, E., & Malaquias, C. (2019). A CRPD analysis of NSW's policy on the education of students with disabilities–a retrogressive measure that must be halted. *Australian Journal of Human Rights, 25*(3), 1–16. https://doi.org/10.1080/1323238X.2019.1609720.

De Bruin, K. (2019). The impact of inclusive education reforms on students with disability: An international comparison. *International Journal of Inclusive Education, 23*(7–8), 811–826. https://doi.org/10.1080/13603116.2019.1623327.

Degener, T. (2016). Disability in a human rights context. *Laws, 5*(3), 35.

Dickinson, H., Smith, C., Yates, S. & Bertuol, M. (2020). *Not even remotely fair: Experiences of students with disability during COVID-19.* Report prepared for Children and Young People with Disability Australia (CYDA), Melbourne.

Disability Discrimination Act. (1992). (Cth). Accessed from www.legislation.gov.au/Details/C2016C00763

Disability Standards for Education. (2005). (Cth). Accessed from www.education.gov.au/disability-standards-education-2005

Down Syndrome Australia. (2020, July). *Submission to Disability Royal Commission Emergency Planning and Response.* Accessed from www.downsyndrome.org.au/blog/2020/07/08/dsas-submission-to-the-emergency-planning-and-response-issues-paper.

Duffy, C. (2020, 24 March). *Coronavirus opens education's digital divide, as COVID-19 forces schools into online learning.* Accessed from www.abc.net.au/news/2020-03-24/coronavirus-closes-schools-online-learning-no-internet/12082580?utm_source=abc_news&utm_medium=content_shared&utm_content=link&utm_campaign=abc_news

Duncan, K., Goggin, G., & Newell, C. (2005). Don't talk about me . . . like I'm not here: Disability in Australian national cinema. *Metro, 146/147,* 152–159.

Duncan, J., Punch, R., Gauntlett, M., & Talbot-Stokes, R. (2020). Missing the mark or scoring a goal? Achieving nondiscrimination for students with disability in primary and secondary education in Australia: A scoping review. *Australian Journal of Education, 64*(1), 54–72. https://doi.org/10.1177/0004944119896816.

Ellcessor, E. (2015). *Very special webisodes: Web series, disability, and cultural accessibility.* Accessed from www.flowjournal.org/2015/03/very-special-webisodes

Ellis, K. (2012). Because of rather than in spite of: Friday night lights' important cultural work of intersecting disability and masculinity. *Interactive Media E-Journal,* 8 (Masculin/Feminin). Accessed from http://nass.murdoch.edu.au/issue8/pdf/IM8-masculine-feminine-article-04-ellis.pdf

Ellis, K. (2019). *Disability and digital television cultures: Representation, access, and reception.* New York: Routledge.

Ellis, K., & Goggin, G. (2015). *Disability and the media.* New York: Palgrave Macmillan.

Ellis, K, Kao, K and Pitman, T (2020, forthcoming). The pandemic preferred user. *Fast Capitalism.*

Ellis, K., & Kent, M. (2011). *Disability and new media.* New York: Routledge.

Finkelstein, V. (1980). Attitudes and disabled people: Issues for discussion. *Disability archive UK.* Accessed from www.leeds.ac.uk/disability-studies/archiveuk/finkelstein/attitudes.pdf

Finkelstein, V. (1981). Disability and the helper/helped relationship: A historical view. *Disability archive UK.* Accessed from www.leeds.ac.uk/disability-studies/archiveuk/finkelstein/Helper-Helped%20Relationship.pdf

Goggin, G. (2015). Communication rights and disability online: Policy and technology after the world summit on the information society. *Information, Communication & Society, 18*(3), 327–341. https://doi.org/10.1080/1369118X.2014.989879.

Goggin, G., & Ellis, K. (2020). Disability, communication, and life itself in the COVID-19 pandemic. *Health Sociology Review, 29*(2), 168–176. https://doi.org/10.1080/14461242.2020.1784020.

Goggin, G., & Newell, C. (2003). *Digital disability: The social construction of disability in new media.* Lanham: Rowman and Littlefield Publishers.

Goldschmidt, K. (2020). The COVID-19 pandemic: Technology use to support the wellbeing of children. *Journal of Pediatric Nursing, 53*, 88–90. https://doi.org/10.1016/j.pedn.2020.04.013.

Haller, B. (1995). Rethinking models of media representation of disability. *Disability Studies Quarterly, 15*(2), 26–30.

Haller, B. (2000). If they limp, they lead? News representations and the hierarchy of disability images. In D. Braithwaite & T. Thompson (Eds.), *Handbook of communication and people with disabilities* (pp. 225–237). Mahwah, NJ: Lawrence Erlbaum.

Haller, B. (2010). *Representing disability in an ableist world: Essays on mass media.* Louisville, KY: Advocado Press.

Hornby, G. (2014). *Inclusive special education: Evidence-based practices for children with special needs and disabilities.* https://doi.org/10.1007/978-1-4939-1483-8.

Oliver, M. (1996). *Understanding disability: From theory to practice.* Houndsmill, Basingstoke: Macmillan.

United Nations. (1948). *Universal declaration of human rights*, Accessed from www.un.org/en/universal-declaration-human-rights

United Nations. (1966). *International covenant on economic, social and cultural rights.* Accessed from https://www.ohchr.org/en/professionalinterest/pages/cescr.aspx

United Nations. (2006). *Convention on the rights of persons with disabilities.* Accessed from www.un.org/development/desa/disabilities/convention-on-the-rights-of-persons-with-disabilities.html

United Nations. (2008). *Convention on the Rights of Persons with Disabilities* (CRPD). Accessed from www.un.org/development/desa/disabilities/convention-on-the-rights-of-persons-with-disabilities.html

United Nations. (2013). *Concluding observations on the Initial Report of Australia, Adopted by the Committee at its Tenth Session* (2–13 September 2013) (CRPD/C/AUS/CO/1). Accessed from https://tbinternet.ohchr.org/_layouts/15/treatybodyexternal/Download.aspx?symbolno=CRPD%2fC%2fAUS%2fCO%2f1&Lang=en

United Nations. (2016). *General Comment No. 4 (Article 24: Right to Inclusive Education)* (CRPD/C/GC/4. Accessed from https://tbinternet.ohchr.org/_layouts/15/treatybodyexternal/Download.aspx?symbolno=CRPD/C/GC/4&Lang=en

United Nations. (2018). *General Comment No. 6 (Article 5: Equality and Non-Discrimination)* (CRPD/C/GC/6). Accessed from https://tbinternet.ohchr.org/_layouts/15/treatybodyexternal/Download.aspx?symbolno=CRPD/C/GC/6&Lang=en

United Nations. (2019a). *Concluding observations on the Combined Second and Third Periodic Reports of Australia* (CRPD/C/AUS/CO/2–3). Accessed from https://tbinternet.ohchr.org/_layouts/15/treatybodyexternal/Download.aspx?symbolno=CRPD%2fC%2fAUS%2fCO%2f2-3&Lang=en

United Nations. (2019b). UN WebTV. *Consideration of Australia (Cont'd) – 500th Meeting 22nd Session Committee on Rights of Persons with Disabilities – 13 September.* Accessed from webtv.un.org/search/consideration-of-australia-contd-500th-meeting-22nd-session-committee-on-rights-of-persons-with-disabilities/6085975013001/?term=&lan=english&page=2

Cátia Malaquias is a lawyer, board director and award-winning human rights and inclusion advocate who works to promote recognition of the rights of people with disability, including the right of every child to inclusive education. Catia is currently undertaking a PhD at the Centre for Human Rights Education, Curtin University.

Katie Ellis is Professor in Internet Studies and Director of the Centre for Culture and Technology at Curtin University. Her research is located at the intersection of media access and representation. She has authored and edited 17 books on the topic of disability and digital media, including most recently Disability and Digital Television Cultures (Routledge, 2019).

Part III
Connected Toys at Home and School

Chapter 15
Internet of Toys and Forms of Play in Early Education: A Longitudinal Study of Preschoolers' Toy-Based Learning Experiences

Pirita Ihamäki and Katriina Heljakka

Introduction

This chapter centers around pre-schoolers' play and agency in a connected world. A fully connected world means a thoroughly "smart world" in which the relationship between objects, their environment, and people become tightly intertwined (Rose, Eldridge, & Chapin, 2015). During the past few years, connected technologies have entered the realm of early education. For example, the Internet of Toys (or, IoToys) have made their way to kindergartens and preschools. Magid (2019) questions, how in an increasingly digitally-based world, most educational systems are incorporating more digital resources into their curriculum. In our work, the interest is in the ways toy-based pedagogies connect with the learning possibilities that the IoToys offer.

Integrating new technologies to learning environments may bring some benefits for children (Plowman & Stephen, 2005). The current IoToys have some advantages as they integrate digital content in traditional toys and enhance possibilities for technologically-mediated interaction (Kara, Aydin, & Cagiltay, 2013). For example, they may enable the physical toy to be connected to digital apps (with iPads or smart phones).

We have sought to examine the relationship between IoToys, play and learning from the perspective of preschool-aged children and their teachers through qualitative research, which includes a 6-month longitudinal study conducted with 20 preschool-aged children and their two teachers. Our study presents, how a connected

P. Ihamäki (✉)
Prizztech Ltd, Pori, Finland
e-mail: pirita.ihamaki@prizz.fi

K. Heljakka
University of Turku, Pori, Finland
e-mail: katriina.heljakka@utu.fi

D. Holloway et al. (eds.), *Young Children's Rights in a Digital World*, Children's Well-Being: Indicators and Research 23,
https://doi.org/10.1007/978-3-030-65916-5_15

toy-based learning environment including IoToys can provide possibilities to both physical and digital interaction between toys and their players, enable physical toys to be connected to virtual content (Yelland, 1999) and enhance the educational value of children's play (Hinske, Langheinrich, & Lampe, 2008).

Our study focusing on Finnish early education argues that the toy-based learning experiences offered by IoToys involve many forms of object play, which support preschooler's learning, for example, through coding and other creative exercises. By playing with IoToys as the ones employed in our research, Wonder Workshop's Dash robot and Fisher-Price Junior Smart Bear, children may learn skills related to visual coding and languages (Heljakka & Ihamäki, 2019a; Heljakka, Ihamäki, Tuomi, & Saarikoski, 2019).

Our proposition was that a purposefully framed approach to play including connected toys will support children's learning outcomes more so than other forms of play. The subsequent findings of this study show how toy-based learning might be used in creation of a pedagogical model for thinking about play with IoToys in early childhood. We now turn to an examination of the literature associated with toy-based learning and pedagogy in early childhood education to frame the way we have been thinking about toy-based learning with internet connected toys.

Toy-Based Learning and Pedagogy in the Early Childhood Curriculum

Work in the area of toy-based learning and pedagogy seeks to consider the ways in which play with tangibles, such as physical toys is related to and/or used in early childhood as a basis for pedagogy. Core to this research and discussion in this area is the way in which toys (in this case, IoToys) are understood, positioned or defined in relation to pedagogy (Ihamäki & Heljakka, 2018a, b). A historically important perspective has been associated with the role of open-ended play as a basis for pedagogy. According to this perspective, open-ended play is important for young children because it provides opportunities for exploration and discovery, which are necessary for supporting learning (White et al., 2007).

The significance of play in learning is strongly supported by established pedagogical theory (Hutt, Tyler, Hutt, & Christopherson, 1989; Piaget, 1951; Vygotsky, 1978). Lampe and Hinske (2007) pointed out that an ideal learning experience comes from the combination of physical experience, digital content, and imagination of the child. Another interpretation of play in early childhood is to consider it from the perspective of there being different forms of play in association with different toys, each of which have distinct learning benefits. This play type approach was introduced by originally Hughes (2002) and then Marsh, Plowman, Yamada-Rice, Bishop, and Scott (2016) and this model has been extended by a toy-based learning approach summarised by the authors in Table 15.1.

Table 15.1 Forms of play with IoToys: A toy-based learning approach in early education

Exploratory play with the Internet of Toys: Using physical skills to learn about the properties of the toys, how to interact with them, and exploring how these toys can be played with.
Constructive play with the Internet of Toys: The personalisation of the toys (by recording one's own voice to toys) and building trails by programming/coding. This type of play involves creating, recognising and solving tasks with toys.
Creative play with the Internet of Toys: Engaging in open-ended play patterns related to the toys such as care-taking (e.g. playing house with the toys) or coding (e.g. programming the toys to move), and also employs other materials such as art supplies in ways that encourage fluency, flexibility, originality, imagination and the making of novel connections.
Pretend, fantasy and socio-dramatic play with the Internet of Toys: role-play, pretending with objects, pretend actions and situations with the toys, within the imaginary play frame to create a play episode or event. When it involves interaction and verbal communication with one or more players it is socio-dramatic play (Monighan Nourot, 2006).
Physical locomotor play with the Internet of Toys: activities that involve all kinds of physical movement for their own sake and enjoyment with the toys (Heljakka & Ihamäki, 2019a, b). In this type of play, a range of motor skills is mastered and children are employing the toys as a part of physical and socially shared play.
Language or wordplay with the Internet of Toys: Spontaneous manipulation of sounds (e.g. by recording one's own voice to toys), forming of words with rhythmic and repetitive elements (e.g. coding sounds for the toys and making the toys move with the sound). As children get older, this kind of play often incorporates rhyme, word play and humour.

So far, only few studies on how children play with smart toys have been published (Cagiltay, Kara, & Aydin, 2014, Holloway & Green, 2016, Heljakka & Ihamäki, 2018, 2019a, b, c Heljakka et al., 2019, Ihamäki & Heljakka, 2018a, b, c, Johnson & Christie 2009).

From this research and discourse is emerging a sense of what might constitute "toy-based learning". Toy-based learning involves considering many forms of object play (including play with contemporary smart toys) as possible informants to the early childhood curriculum, which contribute to children's capacities to engage in meaning making and develop an understanding of their worlds (Wood, 2010). A focus on enabling meaning and supporting understanding suggests that toy-based learning is understood as context specific and personally subjective so that a range of approaches can be adopted to meet diverse learning needs. The way of thinking about toy-based learning has informed our perspective on the learning and teaching in early childhood education. Rather than focusing only on how open-ended play intersects with early education, we have been able to draw on an understanding of toy-based learning, which provides reference to a range of experiences, including what might be considered structured play with IoToys.

Our connection with a toy-based learning perspective on play emerged from an early study completed in 2017 when we first started investigating how preschools in Finland used contemporary toys in their open-ended, free-time (unstructured) play, and how they have provided toy-based learning for the preschool children. Through this work, we came to understand six different types of play that emerge through the toy-based learning approach and that seemed to influence teacher planning for learning with toys and the ways in which children were able to share what they understood of the content associated with each play type with internet connected, or IoToys (see Table 15.1).

Our Study

Data Collection

In our study, we focus on preschool children's own perspectives. Our research represents a 6-month study on the IoToys Wonder Workshop's Dash and Fisher-Price's Smart Toy Bear conducted in the context of Finnish preschool. In order to collect a rich data set, we carried out three play-sessions, during which we interviewed (n = 20) children in a focus group interview situation. The duration of each play session was approximately 35–60 minutes. Additionally, the research material included interviews with two teachers with a university degree.

Moreover, the participating preschoolers played occasionally with the IoToys in their unstructured playtime during the 6-month period. During these times teachers observed and video-recorded children's play. Furthermore, the children themselves also video-recorded their own play sessions as they created stories around the IoToys. Overall, the video-data generation included the recordings of the three observed play-sessions, preschoolers own videos of their play, and video-material recorded by the teachers. The interviews were conducted as group interviews, in which the participants took turns to answer researcher's questions. All of the settings included preschool children aged five to seven years. The data was collected between October 2018 to March 2019.

Data Analysis

To analyse the data, we used both a thematic analysis on interview materials collected from teachers as well as observational data in association with the video-recorded play sessions in order to illustrate forms of play in a toy-based learning context.

Each set of teacher interviews and planning documents were examined and categorised to each play type (see Table 15.1) to determine the content and pedagogical strategies teachers identified prior to and after implementation of the play type. After this initial analysis, we identified key results in order to understand the ways the educators were making meaning about the relationship between the toy-based learning context and the IoToys under investigation.

A Case Study of IoToys

Wonder Workshop's Dash and Fisher-Price's Smart Junior Toy Bear were included in the repertoire of playthings available for the preschoolers to use during their free playtime. Both of these IoToys are marketed as educational toys (or *edutainment*), which according to the toy-makers offer possibilities for many kinds of toy-based learning, as explained in the following descriptions.

Wonder Workshop's Dash

As the website for Wonder Workshops' Dash mentions, "play is a powerful teaching tool". The website includes an educational part, on which benefits for using Wonder Workshop's Dash in educational programs are listed. It is supportive twenty-first century skills, by "encouraging students to continue building critical thinking, creativity, communication, and collaboration skills for tomorrow's job market" (Wonder Workshop website, Education, 2018). According to Kolodny, some 8500 schools are using Dash and Dot around the world today.

Fisher-Price Smart Toy Bear

The Fisher-Price Smart Toy Bear is described as "an interactive learning friend as unique as your child" that comes without a screen. The toy responds to what the player says, and remembers things. It takes cues from its player, then invites play, talk, movement, imagination and learning. The Fisher-Price Smart Toy Bear is also able to recognise images from the nine Smart Cards included, which the player can choose to play activities with, like listen to stories, play games and 'go on adventures'. As the Fisher-Price Smart Toy Bear is constantly updated, it is said to encourage socio-emotional development, imagination and creativity: "Your Smart Toy learns new activities every month" (Smart Toy Bear, Fisher-Price, Mattel, 2020).

Findings

In this study, our goal was to understand the toy-based learning approach in relation to forms of play that emerged in a series of play sessions with two IoToys, the Dash toy robot and Fisher-Price Smart Toy Bear. The findings of the case study reported in this chapter focus on researching the six types of play with the chosen IoToys, and in particular, with our hypothesis regarding the usefulness of purposefully framed toy-based learning.

Our case study shows that, for example, the Fisher-Price Smart Toy Bear has been used by the preschoolers as a part of *exploratory play*, where children have made contact with the Smart Bear by hugging and taking care of it. During their free time play, the preschoolers have used this toy for playing 'house' and 'nursery'. They have listened to the stories about what the smart toy has to tell, and the preschool teacher has translated the stories because the main language in the preschool is Finnish but some of the children's main language is English. In this multilingual learning environment, we see that IoToys like the Fisher-Price Smart Toy Bear give an opportunity for children to interact with each other in new ways. Thus, the toy becomes a communication tool for the children. Some preschoolers are shy, or they are not on the same language level because their main language is not Finnish. Therefore, the Fisher-Price Smart Toy Bear was considered as helpful in facilitating learning of *language* and *wordplay*. For example, this happened in association with the counting of English numbers as a part of social play. The preschoolers also repeated what the Smart Bear told them. In this way, the Fisher-Price Smart Toy Bear acted as a tool for learning languages. Based on our findings, the Fisher-Price Smart Toy Bear also supports *fantasy* and *socio-dramatic play*, where children are engaged in role-play and the toy has played a role in play patterns such as the playing of house and playing of nursery. In these play scenarios, the Fisher-Price Smart Toy Bear acted as a fellow player, or 'child', which was involved in an imaginary story, created and followed by the preschoolers.

When engaging with the Dash toy robot, preschoolers were involved in *constructive play*, like solving tasks related to coding together. For example, the preschoolers recorded their own voice with the Dash robot and programmed Dash to move with the help of an ipad-operated app. Dash involved the preschool children to move and thus, engage in *physical locomotor play*, and was in this way enjoyed as a part of collaborative play. The preschoolers' *creative play* happened when one child programmed the Dash robot to move and others tried to run away from the toy or to catch it. This means that the preschoolers came up with their own 'games' with the toy in collaborative ways (see Heljakka et al., 2019; Heljakka & Ihamäki, 2019a).

Our results describe how preschool-aged children received toy-based learning experiences by engaging with the two IoToys. The results of our study are summarised and presented in the *Dimensions of toy-based learning experience* chart (see Fig. 15.1.), and clarified in the following:

The *physical dimensions* of the toys (their aesthetics, form of a character) provide *wow* experiences, or experiences of surprise and delight for the players—the toys represent novel playthings with many affordances fitted to be used as a part of traditional *haptic-oriented object play*.

The *fictional dimensions* of the toys enable *flow* experiences, or, experiences of immersion, as players become engaged in the stories that the toys tell. Fictional experiences were enjoyed by the preschoolers when they listened intensively to the Fisher-Price Smart Toy Bear's stories, and even forgot about the presence of the researchers who were observing the play sessions. Fictional experiences with the

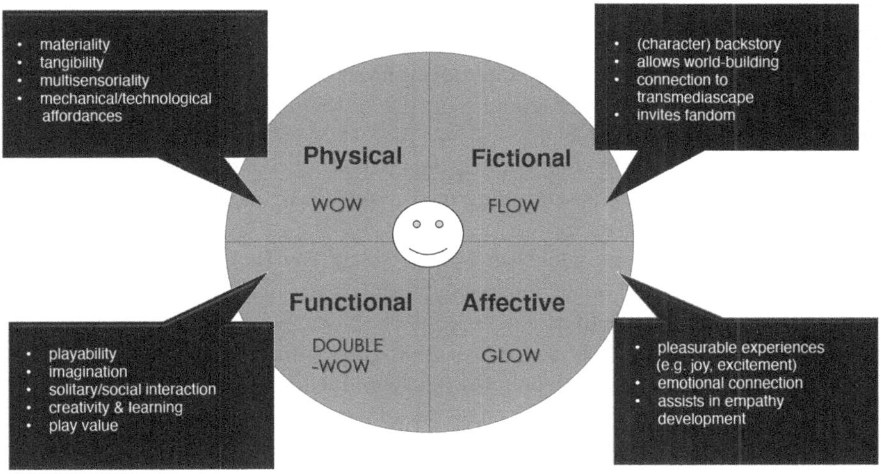

Fig. 15.1 New dimensions of the toy-based learning experience with IoToys

Smart Toy Bear happened through role-playing with the toy, for example, preschoolers listened to the toy's story and started to develop the story further, and play along with the story. In our study, one story involved the narrative of a boat trip affected by a storm. The preschool children who participated in our study extended the story by playing with the Fisher-Price Smart Toy Bear by moving around in the rooms of the preschool facility and pretended to be in a storm. In association with the Dash robot, a further fictional experience was created when preschoolers started to record their own voice and phrases for the toy, which then played them back.

The *functional dimensions* provide *double-wow* experiences, or experiences of discovering unexpected affordances in the toys for their players, meaning that 'hidden affordances' emerge in play situations and the players experience a secondary moment of "wow"—for example, the toys reveal new possibilities for play through sound, light and movement. When playing with the Dash robot, functional experiences manifested when children came up with their own games with Dash and then played these games together in a group. The functional dimension of play occured while the preschoolers created their own game of tag with Dash in which some players tried to catch Dash and some tried to run away as fast as they can. When the preschooler, who was responsible for coding the movement for Dash was touched by the other children, s/he became the coding player. The game allowed the children to engage in physical locomotor play and learn collaboratively by programming the toy quicker and smarter. Functional experiences with the Smart Bear happened in association with learning English language by repeating words together and by listening to the toy's stories.

Double-wow experiences happened while playing with both IoToys: In the case of Dash when the preschoolers realised how a programming exercise could be done to make the toy move. Again, the Fisher-Price Smart Toy Bear provided experiences of double-wow when it told a story and preschoolers chose to change the narrative. In this moment, they created a new role-play episode, and consequently received experiences of double-wow. The preschool children enjoyed to collaborative create, challenge, and in this way, continue the story that the Fisher-Price Smart Toy Bear told them during their free time play session. This was demonstrated in the playing of the preschool children, who completely changed the course of the story, for example, by narrating the adventure about hiking on a mountain to an underwater adventure describing the living of fish.

The *affective experience* with Fisher-Price Smart Toy Bear enabled pleasurable experiences for the preschoolers when they nurtured and played house with the toy. Children received affective experiences when Dash made eye contact with them and when the children taught their peers ways to program some new features for the toy. *Glow* experiences happened when the preschool children explained how they learned something new with Dash, and the playing with the toy left a memory trace not only in the player him/herself, but also in the peers.

Conclusion

Connected digital technologies, such as IoToys have made their way to the Early Childhood Education curriculum (e.g. Stephen & Edwards, 2018). Research examining their integration has shown the pedagogical benefits using technologies to enhance the learning environment and to encourage creativity and imagination, (e.g. Fleer, 2018) and social interactions (e.g. Danby, Fleer, Davidson, & Hatzigianni, 2018). In contrast with traditional touch-screen technology, IoToys with their various affordances enable their players to blur physical learning environments with online ones where both digital and non-digital elements exist in a fluid synchronous way (Marsh, 2017).

In our study we examined the forms of play prompted by two IoToys—Wonder Workshop's Dash and Fisher-Price Smart Toy Bear in order to find out how preschool children and their teachers approach and interact with them to engage with toy-based learning. The toy-based learning approach in early education considers the matching of learning content (culture and environmental) and toys (in this case IoToys), the learning principles that toys incorporate, the design of toy-based learning environments, the effectiveness, sources and institutional use of toys (IoToys) and the design of meaningful play in toys. However, these approaches have predominantly focused on primary schools and higher education. This chapter aimed to investigate the forms of play that the two toys, Wonder Workshop's Dash and Fisher-Price Smart Toy Bear, prompted as a part of free time play sessions in the context of early education in Finland—as a part of five–seven year-old preschoolers' daily lives. In order to explore the various affordances of IoToys unlocked by preschoolers in their course of play, we designed and conducted a 6-month long study of preschoolers' toy-based learning experiences. Studying these in the context of early education enabled us to find out which of the toys affordances could potentially be employed as a part of future early education, first by finding out about the forms of play the IoToys support. Through an analysis we were able to detect six forms of play with the toys, namely exploratory, constructive, creative, pretend, fantasy and socio-dramatic play, physical locomotor, and language or wordplay. Our findings demonstrate that in their exploratory play with the IoToys the children used their creativity in coding and making Wonder Workshop's Dash to move in different ways by building trails and by using other materials in the playroom and by creating their own games with Dash. In their constructive play, the children used Dash for recording their own voice and making the robot move and make sounds according to their design. On some occasions, the children built trails of other play materials while one of them coded Dash to move along the trail. Our findings related to playing with the Fisher-Price Smart Toy Bear show that the toy was employed in play as family member: the children nurtured the Smart Toy Bear and listened to the stories told by the toy. These stories were then developed further by the children in their pretend, fantasy, and socio-dramatic play. In physical locomotor play with the Wonder Workshop's Dash the children also created the "Robot tag game", in which one of the children coded Dash to try to chase after and catch the other children. Moreover, the findings illustrate that language or wordplay with the two IoToys under scrutiny may happen spontaneously by recording one's own voice with the toys and by starting to use both of the toys audio-related affordances as 'humour machines' or by coding Dash to move and sound in a way that the children themselves choose to.

Detailed examples of six types of play purpose framed toy-based learning were given based on these categories in the findings section of the chapter. These detected forms of play could be employed by educationalists as a basis for the planning of playful learning and exercises with more structured learning goals. To give examples, play patterns related to *exploratory and creative play* could be used in emotionally-directed education practices, for example, learning about empathy by nurturing an IoToy like the Fisher-Price Smart Toy Bear. The forms of *language and wordplay* can be used in learning language skills, for example by learning how to pronounce numbers, letters, words etc. The forms of *fantasy and socio-dramatic play* with IoToys can be used in role-playing exercises where preschoolers' may, for example, discuss different emotions and care-taking needs through interaction with the toys. With the forms of *constructive play,* children may collaboratively practice new skills such as coding tasks. The forms of *physical locomotor play* can be encouraged by asking children to create new games that employ movement by playing with Wonder Workshop's Dash. The forms of *creative play* can be facilitated through storytelling exercises, in which children continue the stories of the toy like the Fisher-Price Smart Toy Bear.

Moreover, we used our findings of the study to create the *Dimensions of toy-based learning experience* presented in Fig. 15.1, which builds on earlier work of Heljakka (2020) and describes the four key dimensions of (wow, flow, double-wow and glow) experiences with toys; physical, functional, fictional and affective.

The responsibility of parents and educators is to choose and to learn how to use IoToys in the context of both informal and formal, toy-based learning (Ihamäki &

Heljakka, 2018a). We believe that the results of our longitudinal study will provide ideas particularly for teachers working in the context of early education in how to plan and execute playful exercises with IoToys to fulfil more specific learning goals in preschool. Simultaneously, the study presented in the chapter shows, how these toys invite to free-form, creative play, once offered as a part of unstructured play sessions in the daily lives of the preschoolers. This proves that children aged five–seven years (who, in the Finnish educational context represent preschoolers) are innovative in the ways they incorporate IoToys in their play, which does not limit itself to the employment of the toys' technologically-enabled (or, hybrid) affordances only, but which extend to the realm of more traditional forms of play, previously associated with non-technological toys.

The design and pedagogy related to toy-based learning for early childhood presents unique challenges not relevant to other sectors. The two frameworks introduced, Forms of play with IoToys (Table 15.1) and the Dimensions of the toy-based learning experiences (Fig. 15.1) may provide assistance in future research interested in the many dimensions of current smart, connected toys. The future stages of our research will aim at continuing investigating the long-term capacities and use of technologically-enhanced toys in the context of early education, such as the IoToys presented in this chapter.

References

Cagiltay, K., Kara, N., & Aydin, C. C. (2014). Smart toy based learning. In J. M. Spector, M. D. Merril, J. Elen, & M. J. Bishop (Eds.), *Handbook of research on educational communication and technology* (pp. 703–711). New York, NY: Springer.

Danby, S., Fleer, M., Davidson, C., & Hatzigianni, M. (2018). *Digital childhoods: Technologies and children's everyday lives (part of the international perspectives on early childhood education and development books series)* (Vol. 22). Singapore: Springer.

Fleer, M. (2018). Digital animation: New conditions for children's development in play-based setting. *British Journal of Educational Technology, 49*(5), 943–958. https://doi.org/10.1111/bjet.12637.

Heljakka, K. (2020). Pandemic toy play against social distancing: Teddy bears, window-screens and playing for the common good in times of self-isolation. *WiderScreen, 11*(5), 2020.

Heljakka, K. & Ihamäki, P. (2018). Preschoolers learning with the Internet of Toys: From toy-based edutainment to transmedia literacy. In *Seminar.net–International Journal of Media, Technology, and Lifelong Learning, 14*(1), 86–102.

Heljakka, K., & Ihamäki, P. (2019a). Persuasive toy friends and preschoolers: Playtesting IoToys. In G. Mascheroni & D. Holloway (Eds.), *Internet of toys practices: Affordances and the political economy of children's smart play* (pp. 159–178). Cham: Palgrave Macmillan.

Heljakka, K., & Ihamäki, P. (2019b). *Ready, steady, move! Coding toys, preschoolers, and mobile playful learning*. Paper presented at the International Conference on Human-Computer Interaction 2019 (pp. 68–79). Cham: Springer.

Heljakka, K., & Ihamäki, P. (2019c). Toys That Mobilize: Past, Present and Future of Phygital Playful Technology. In *Proceedings of the Future Technologies Conference 2019* (pp. 625–640). Cham: Springer.

Heljakka, K., Ihamäki, P., Tuomi, P. & Saarikoski, P. (2019). *Gamified coding: Toy robots and playful learning in early education*. Paper presented at the International Conference on Computational Science and Computational Intelligence (CSCI) (pp. 800–805). IEEE.

Hinske, S., Langheinrich, M., & Lampe, M. (2008, February). Towards guidelines for designing augmented toy environments. In *Proceedings of the 7th ACM conference on designing interactive systems* (pp. 78–87).

Holloway, D., & Green, L. (2016). The internet of toys. *Communication Research and Practice, 2* (4), 506–519. Taylor & Francis.

Hughes, B. (2002). *A playworker's taxonomy of play types* (2nd ed.). London: PlayLink.

Hutt, S. J., Tyler, S., Hutt, C., & Christopherson, H. (1989). *Play, exploration and learning: A natural history of the pre-school*. London and New York: Routledge.

Ihamäki, P. & Heljakka, K. (2018a) *Smart toys for game-based and toy-based learning–A study of toy marketers', preschool teachers' and parents' perspectives on play*. In CENTRIC 2018: The Eleventh International Conference on Advances in Human-oriented and Personalized Mechanisms, Technologies, and Services, At: CENTRIC 2018, 14–18. October, Nice, France.

Ihamäki, P., & Heljakka, K. (2018b). The internet of toys, connectedness and character-based play in early education. In K. Arai, R. Bhatia, & S. Kapoor (Eds.), *Proceedings of the future technologies conference 2018* (pp. 1079–1096). Cham: Springer. https://doi.org/10.1007/978-3-030-02686-8_80.

Ihamäki, P. & Heljakka, K. (2018c). Smart, skilled and connected in the 21st century: Educational promises of the Internet of Toys (IoToys). In *Proceedings of the 2018 Hawaii University International Conferences, Arts, Humanities, Social Sciences & Education, Prince Waikiki Hotel, Honolulu, Hawaii.*

Johnson, J. E., & Christie, J. F. (2009). Play and digital media. *Computers in the Schools, 26*(4), 284–289.

Kara, N., Aydin, C. C., & Cagiltay, K. (2013). Investigating the activities of children toward a smart storytelling toy. *Journal of Educational Technology & Society, 16*(1), 28–43.

Lampe, M., & Hinske, S. (2007). Integrating interactive learning experiences into augmented toy environments. In *Pervasive Learning Workshop at the Pervasive Conference* (pp. 13–16).

Magid, L. (2019). *Children's rights in the digital environment*. Accessed from: https://smarterinternet.org/comment-on-the-concept-note-for-a-general-comment-on-childrens-rights-in-relation-to-the-digital-environment/

Marsh, J. (2017). The internet of toys: A posthuman and multimodal analayse to connected play. *Teachers College Record, 119*(120305), 1–32.

Marsh, J., Plowman, L., Yamada-Rice, D., Bishop, J., & Scott, F. (2016). Digital play: A new classification. *Early Years, 36*(3), 242–253.

Monighan Nourot, P. (2006). Sociodramatic play pretending together. In D. P. Fromberg & D. Bergen (Eds.), *Play from birth to twelve: Contexts perspectives, and meanings* (2nd ed.). New York: Routledge.

Piaget, J. (1951). *Play, dreams, and imitation in childhood*. New York: W.W. Norton & Company.

Plowman, L., & Stephen, C. (2005). Children, play, and computers in pre-school education. *British Journal of Educational Technology, 36*(2), 145–157.

Rose, K., Eldridge, S. & Chapin, L. (2015). *The Internet of Things: An overview*, [online]. Accessed from: http://www.bu.edu.eg/portal/uploads/Engineering,%20Shoubra/Electrical%20Engineering/2460/crs-13718/Files/Internet%20of%20Things%20-%20Overview.pdf; http://www.bu.edu.eg/portal/uploads/Engineering,%20Shoubra/Electrical%20Engineering/2460/crs-13718/Files/Internet%20of%20Things%20-%20Overview.pdf

Smart toy bear, Fisher-Price, Mattel. (2020). Accessed from: https://m.service.mattel.com/us/Technical/productDetail?prodno=DNV31&siteid=27

Stephen, C., & Edwards, S. (2018). *Young children playing and learning in a digital age: A cultural and critical perspective*. London: Routledge.

Vygotsky, L. S. (1978). *Mind in society: The development of higher psychological processes*. London: Harvard University Press.

White, J., O'Malley, A., Toso, M., Rockel, J., Stover, S., & Ellis, F. (2007). A contemporary glimpse of play and learning in Aotearoa New Zealand. *International Journal of Early Childhood, 39*(1), 93–105.

Wonder Workshop website, Curriculum. (n.d.). [Online]. Accessed from: https://education.makewonder.com/curriculum/

Wonder Workshop website, Education. (2018). [Online]. Accessed from: https://www.makewonder.com/education/

Wood, L. (2010). Developing integrated pedagogical approaches to play and learning. In P. Broadhead, J. Howard, & E. Wood (Eds.), *Play and learning in the early years* (pp. 9–26). London, UK: Sage.

Yelland, N. (1999). Technology as play. *Early Childhood Education Journal, 26*(4), 217–220.

Pirita Ihamäki (PhD) is an international business developer and researcher. She currently works as the leader of the Gamecoast Network (Robocoast DIH Consortium, Finland). She is developing new gamified services to extend technology innovations. Her research interests include social robots, gamified service design and co-creation.

Katriina Heljakka (Doctor of Arts) leads the Pori Laboratory of Play research group at University of Turku (Digital Culture Studies). She is a toy researcher, visual artist and designer of playful interventions. Her research interests include connected toys, playful environments, tools and techniques in the workspace, and the visual, material, digital and social cultures of play.

Chapter 16
Digital Play Objects as Part of Preschool Children's Imaginative Play

Kristín Dýrfjörð and Anna Elísa Hreiðarsdóttir

Introduction

Play and the objects children play with are not the same as even a few decades ago. In a way, the world has shrunken, and play and play objects along with it. How children from different cultures and groups play is possibly more alike now than in any time before in history. What has not changed, however, is that children use play to make sense of and understand their world and to deliberate spaces and relationships (Grindheim, 2017). It should be remembered that according to the United Nations Convention on the right of the Child (UN General Assembly., 1989) children have right to play. Children's experiences are part of their play, and they use imagination and creativity to support and enrich their play. According to ideas from new materialism, the playthings and objects afforded to children affect how children play (Marsh, 2017; Sellers, 2013). Hence, the play environment and objects to which children have access and the support or non-support they get from adults' matter. In this chapter we look at research on play, imagination and digital objects used as part of play. We then demonstrate through our case study how children incorporate in their play; digital play objects they have not encountered before in an environment they know with support from well-known materials. We examine the connection between children's imaginative play and coding devices, a little beetle (Blu-Bot) and coding blocks (Cubelets). Our aim is to investigate children's relationships with both each other and devices and to explore the creativity, emotions and problem solving that appear during play. We explore how the culture of children's

This research was carried out with a grant from the University of Akureyri Research Fund.

K. Dýrfjörð (✉) · A. E. Hreiðarsdóttir
University of Akureyri, Akureyri, Iceland
e-mail: dyr@unak.is; annaelisa@unak.is

D. Holloway et al. (eds.), *Young Children's Rights in a Digital World*, Children's Well-Being: Indicators and Research 23,
https://doi.org/10.1007/978-3-030-65916-5_16

local community permeates and becomes a significant part of their play. This chapter contributes to the discourse that children's play can be imaginative and full of learning opportunities even when they play with materials that, if used in pre-scripted ways, can transform and control play and play behaviours. In other words, if children get a chance to play with these materials as with any other play objects (toys), they do so.

Play, Creativity and Learning Environments

Theorists and researchers from different fields in child development have reached the conclusion that there is a link between imaginative play and creativity (Russ & Wallace, 2013). Imaginative play gives children "opportunities to express many different processes—cognitive, affective, and interpersonal—important for creativity" (Russ & Wallace, 2013, p. 136). Play is a creative act, and one can observe creativity as it occurs, in the moment it happens. To be able to play imaginatively, children need teachers who see play, along with time and a supportive environment, as important for children. Fleer (2018) argued that the place of imagination in play needs to be reclaimed. Fleer used Vygotsky's concept of leading activity as the gateway to her argument. Fleer summarised four important factors that teachers should keep in mind to facilitate play: play is a leading activity, play is dramatic, play is a source of child development, and play matures through the dynamic between its reproductive and productive natures. To be able to support play, teachers must know how to create learning environments. Davies et al. (2013) conducted metanalyses on creative learning environments in education. According to the findings, the pedagogical environment needs to be flexible, and children need to have control and ownership and to be able to work on their own pace without pressure. Davies et al. (2013) also pointed out that:

> An important feature of the pedagogic environment that can promote creativity is the nature of the relationship between teachers and learners, including high expectations, mutual respect, modelling of creative attitudes, flexibility and dialogue. There is strong evidence that pupil creativity is closely related to opportunities for working collaboratively with their peers, which can productively extend to peer and self-assessment. (Davies et al., 2013, p. 89).

These findings are important when thinking about and planning projects that include both play and creativity. According to Beghetto and Kaufman (2014), it is not enough to know that the environment influences the development of potential creativity; it is also necessary to understand what it takes to develop a supportive, creative learning environment.

Russ and Wallace (2013) explained that imaginative play is a tool, children carry with them regardless of their circumstances. Enhancing the ability to play should give children an advantage in creative problem-solving and expression as they grow older. Russ and Wallace pointed out that the challenge for the early childhood field is to prove this to be true, they point out: "For children, creativity in daily life often takes the form of pretend play. Pretend play becomes, then, a child's creative product. Thus, researchers might study pretend play as either a measure of creativity or as an outlet for it" (Russ & Wallace, 2013, p. 138). Davies et al. (2013) provided reasonable evidence suggesting that creative learning environments can enhance students' creative thinking, leading to greater originality in open tasks, particularly when students are given incubation or cultivation time. Davies et al. point out that it has been suggested that children's creativity is associated with a shift towards 'less prescriptive' (p. 88) lesson planning.

Shifflet, Mattoon, and Toledo (2012) concluded that children use their imagination and participate in an array of hands-on experiences, and through active play, they construct new ideas and understandings of their world. Shifflet et al. (2012) say that it is an assumption that when technology is added to the early learning environment, "the interface may be different, but the principles [of play and learning] remain the same" (p. 37).

Digital Play

Digital media and objects of play have stirred various emotions among both educators and parents (Cowan, 2019). Some fear that they will damage play and childhood, whereas others see opportunities and new ways of learning and connecting (Marsh, 2019). It can be argued that part of fulfilling children's right to education the acknowledgement of digital technologies as an integral part of children live must be made. To explain these new opportunities mentioned above, Marsh (2017) used the notion of connected play that takes place among children and play objects. Marsh described connected play as also taking place along a physical/virtual continuum, enabling smart toys to cross virtual and physical world boundaries. However, digital play can never occur in a vacuum. Edwards (2013) emphases that technologies are enabled by the social settings in which young children encounter them. Edwards argued that from the perspective of digital play, context matters, and it is important to understand the digital in relationship to its contexts, cultural meaning-making and participation in social relationships.

The digital in children's world has changed from being mostly inside computers and tablets to being included in the toys with which children play (Marsh, 2017, 2019). Among these objects, different types of robots and coding devices have become part of the affordances of play in both the home and preschool. Some of these play-objects even have human aspects or are designed to connect directly with children's empathy and humanity. They are intended to allow children (and adults) to embody them in play and communicate with them as living things (Marsh, 2017, 2019). Consequently, for many, the question remains: is the digital device a Trojan horse or an authentic play object? Does the material form the child, or the child the material? The types of materials chosen and environment designed, along with the teacher's behaviour, probably depend on individual teachers and their own theoretical frameworks and professional ideologies. In this paper, the perspective held is that an object in play can be part of child-centred, imaginative play, and therefore, digital play objects can be part of empowered play among peers.

Fleer (2016) analysed and identified the main characteristics of digital play from a cultural-historical perspective, finding five categories: (1) technical behaviours, in which children focus on clicking, swiping and other technical dimensions; (2) imaginary digital situations, in which digital devices create a context for reproducing play in a digital form and act as the context for imaginary play itself; (3) digital talk, in which imaginary digital situations give a new sense to digital objects and actions; (4) the making, naming and modifying of icons and text symbols to create imaginary situations, giving a new sense to the digital situation; and (5) porous boundaries between digital play and social pretend play situations. A digital imaginary situation has rules of engagement and roles for avatars and characters taken from children's real lived worlds. Fleer also pointed out that children draw on their digital play experiences to enhance their play in physical social-material contexts.

These categories can be helpful when looking at digital play. They can be viewed as the stages children passes through while playing with digital objects, they can show how to help children develop play experiences, and they can be used to observe the uncanniness of digital play. Fleer's (2016) categories of connected play, along with Marsh's (2017) types, can be applied to help teachers select digital play objects that satisfy their pedagogical aims and methods. Having helping tool is especially important in light of teachers' insecurities and hesitations concerning the pedagogy of digital apps and play. According to Cowan (2019), teachers are concerned that most digital tools for young children in the market are closed, procedural apps, leaving teachers on their own to select digital play objects that are open ended and support creativity and collaboration. Those findings align with Johnson and Christie's (2009) conclusion that software and toys designed to be open ended and to facilitate problem solving can foster creativity and social learning. However, Johnson and Christie (2009) also claim that if the focus is on drilling programs, it can put limitations on inventiveness, and there is no guarantee of positive developmental influences.

Adults' Roles in Play

Ebbeck and Kam (2016) described how teachers can support children's learning and creativity thorough well-planned, short-term projects based on inquiry and creativity. Henricks (2016) pointed out that adults may be useful for play and can set rewarding challenges based on rules that function both to limit and guide. Henricks claimed that rules can teach children about the reality of otherness. Craft, McConnon, and Matthews (2012) pointed out the benefits of teachers being alongside children during play: teachers have opportunities to become collaborative critics and provide valuable dynamics in learning.

It is worth mentioning that researchers have warned against instrumentalising play, for example, by smuggling into it control and teachers' own ideas about what children should be learning (Grindheim, 2017; Öksnes, 2010). Teachers might make play so occupied with societies' agendas such as twenty-first century skills that they become the main aims of play, not play itself. The danger is that play becomes too controlled, agenda-oriented and, in some cases, joyless. Playing with digital devices can be risky, so how digital play objects can be used to pave the road for instrumentalising play in early childhood settings should be addressed and taken into account. Doing so is especially important in preschools where teachers select and decide everything from learning materials to preferred behaviours, in contrast to preschools that highlight children's free play as children's right, a philosophy and a main learning method.

The Study

This case study was conducted with participation of nine five-year-old children (four girls and five boys), their teacher and a master student in early childhood education in a rural area of Iceland. The study was part of the European Union project; Makerspaces in the Early Years: Enhancing Digital Literacy and Creativity. The project component presented here was Makerspaces in Icelandic Preschool. The preschool children learned about and with digital technology in a creative, playful manner. The children attended six hour-and-half workshops in their preschool twice a week for three weeks. The children were selected by the teacher with the intention to reflect the diversity of the class and the preschool in sex, social status and learning dispositions. Informed consent was gathered from all concerned authorities, teachers, parents and children. During each workshop, informed consent was discussed with children (Hreiðarsdóttir & Dýrfjörð, 2019).

The preschool teacher and the master student implemented activities by following an open curriculum plan or a frame proposed by the researchers. According to the plan, the children started the workshop by discussing drawing and making their own characters they did not get any instructions and made characters based on their own ideas and imaginations (see Fig. 16.1). They were then introduced to coding, by their teacher who showed them the buttons on top of the Blue-bot, she involved in discussions on how the buttons worked and encouraged them to try and to make stories involving both their characters and the Blue-bot and the Cubelets. The aim was to observe how the children interacted and creatively engaged with robots added to their familiar play world, how the children's coding skills developed through play and how they collaborated among themselves. Data were gathered through different methods, including recordings by video cameras on tripods, hand-held cameras, a GoPro camera (fastened on one child's chest) and an iPad controlled by the children. Field notes and research diaries were written, and interviews were conducted.

The project sessions started with the children gathering in a circle. A researcher described the research procedures and told the children that the research-team would be videotaping and photographing the sessions. After the introduction, the classroom teacher took over and led a conversation with the children about stories and how stories and characters are created. She asked the children to describe and imagine all kind of characters. Afterwards, the children created their own characters that they

Fig. 16.1 Children's characters

used throughout the project (Fig. 16.1). To protect the characters and prolong their lifetime, the researcher laminated them.

The children created their own play stages on big paper that could fit under the plastic Blue-bot mat (see the example in Fig. 16.2). Over the next three weeks, the play evolved, and children familiarised themselves with the coding devices and learned how to code, they used the play scenes as a base to let the robots travel, they used tic-tac to fasten the characters on them, so the devises became part of the play. The children were generally happy and showed excitement during the temporary makerspace project. It is important to also mention that after the project ended, the preschool received all the materials for a few weeks. The aim was to give the other children in the classroom who had not taken part in the research the opportunity to play with them.

During the fifth workshop, the researcher asked the children how they would like to plan the last day. The children answered that they wanted to use all the materials but on their own during free play. In the next section, we describe and analyse the data from a free play session during the last workshop.

Fig. 16.2 Children's play stages

We decided to analyse three short videos recorded by the GoPro camera. It was not attached to child's chest; instead, the children either held it, or it was on the floor. We also looked at the videos from the main camera that was on a tripod little bit outside the main playing area, to get better visual information about where the children were situated in the room and how they used their bodies during play. We have put the data into tables with three columns, the first describing the children's actions, the second reporting the verbal dialogues that took place and the third presenting our comments and thoughts. Sellers (2013) outlined how children perform and become the curriculum through their bodies, how they embody during their experiences and ideas through play. In the next section, we look at closely at the three connected episodes from the sixth workshop (GoPro nr. 5 0.00–4.23, with a total of 9.27 minutes of play).

Children's Imaginative Play

Four children, one girl Anna and three boys, Jon, Gunnar and Hilmar (pseudonym names) play on a floormat with one of the play stages the children designed (see Fig. 16.3). They use robots they have made of Cubelets and Blue-bot and the drawings they have made of different characters. They have fastened the characters on the robots. At the beginning of the recording, all the children have characters/ robots in their hands and they have built obstacles with Lego for their robots to travel through. The camera is on the floor beside Anna.

Episode 1

Hilmar silently makes an airplane-like vehicle and uses putty to attach his character to it. He moves it around both on the stage and in the air, and he seems to be watching and listening to the others. Gunnar moves his Lego character over the stage. Meanwhile, Anna sings in a low voice and moves her body with the song, Jon is engaged with the Blue-bot (Table 16.1).

Drawing on Fleer's (2016) definition of digital play and Marsh's (2017) notion of connected play, the children in this play episode cross boundaries and make connections, using the digital play objects as any other playthings. During this scene, the children play together and with the objects. They also connect their play to their cultural experiences. It is worth mentioning that in Iceland, there is a strong culture surrounding public swimming pools, and for many children, going to the swimming pool is a favourite family pastime. Most pools have water slides and sunbathing benches. If we look at Fleer's (2016) categories, we can say that these children employ the digital play objects according to her third and fourth categories and partly the fifth category as they use avatars they have made as part of digital play. During

Fig. 16.3 Robots and Legos on a play stage

Table 16.1 Play including a swimming pool and pairing

Action	Dialogue	Analysis
Jon codes the blue-bot with his character on it, stands up and addresses Gunnar:	*I am going to your home, Gunnar.*	The play stages give Jon the opportunity to move around when he needs it. Gunnar can work alone and simultaneously be in the other children's play and create play worlds with them.
Gunnar answers:	*Are you going to sunbath . . . or to the swimming pool?*	
Jon moves around the room, comes back and says:	*I can't miss my sunbath.*	Jon and Gunnar's chat indicates that they are both in shared play world.
Hilmar is building an airplane with Legos and watches Anna's actions.	[overlaps with the following dialog with Anna]	The play includes interactions, pretending and making stories.
Anna puts two characters (drawings) in front of the camera and says:	*They should be attached together*	The GoPro camera becomes part of the play when Anna uses it to document her characters
She finds two other characters and continues:	*. . . and they should be attached together*	
Anna puts them with the other characters.		She pairs the characters together and then the stack reminds her of a sandwich.
She takes the characters and says, laughing:	*Hey, this is like a sandwich*	
She flips the stack of characters and keeps on singing.		
Gunnar is moving robots around. Jon joins in, and they talk and play.		
Gunnar talks but is not otherwise active.	Sounds muffled	Jon and Hilmar are in their own play world, whispering
Hilmar watches the others and reconstructs his airplane.		
Jon starts the blue-bot character, stands up and addresses Gunnar:	*I am going to your home.*	Jon does not watch the robot move. He just code it and does not seem to care where it goes.
Gunnar says:	*Are you going to sunbath . . . or to the swimming pool?*	Jon takes a walk, comes back and is still involved in the same play.
Jon moves around the room, comes back and says:	*I can't miss my sunbath.*	
Gunnar says:	*I am going to the swimming pool.*	Gunnar and Jon interact and let their characters play in the swimming pool.
Gunnar moves his character. Jon takes one character in his hand and walks around the room.		
Hilmar says:	*I want to go first.*	

the next episode, we follow the play and look at the children's collaboration and empathy towards each other.

Episode 2

Hilmar and Anna play all by themselves on their own while Jon and Gunnar interact. This example shows that the children are all aware of each other and their actions and dialogues, even though it might look as if they are in their own individual play worlds (Table 16.2).

The connectedness (Marsh, 2017) in the children's play occurs among themselves. The children in episode 2 seem to be relaxed, but at the same time, they are aware of Hilmar who is quiet and not participating. They give him emotional support and try to draw him into their play by moving the scene to his house. The children also make serval cultural connections in both the song Jon sings and their conversations about children being allowed to travel between houses. In Iceland, especially in smaller towns, young children are often allowed to play outside their houses alone and are commonly allowed to visit friends who live close by—if, for example, their parents can see where they are going from the window.

Episode 3

The children play seemingly without interacting with each other. Anna and Gunnar keep attempting to involve Hilmar in play, and although they do not seem to draw him into play, he is aware of his surroundings and watches what they are doing (Table 16.3).

The play continues until Jon takes the camera and records the children's play and laughter. He finds Gunnar with the camera, and they play hide-and-seek, starting another type of play.

The connectedness between the children becomes more noticeable. In this episode, the digital play objects are in the background, and the characters and the play dialogue are in the foreground. During this episode, the children show respect and high levels of emotional awareness towards each other. The other children can at last draw the shy boy, Hilmar, closer to the play. He involves himself on the outskirts of their play space.

Discussions and Conclusions

The children in this study participated in a short-term project (Ebbeck & Kam, 2016) in a play world structured and framed by the teachers and researchers. The frame, however, was open, and within it, the children and teachers negotiated how the space

Table 16.2 Playing with airplanes and visiting friends' homes

Action	Dialogue	Analysis
Anna sings in a low voice and moves her body with the song while simultaneously playing with her character. **Anna** chat with Jon, in a singing voice:	*I want to fly in an airplane. . . . can I fly in an airplane?*	Anna plays with her character, on her own in the play-world. She uses her imagination in a creative way and makes up a story in which her character wants to fly.
Gunnar:	*Yes, . . . you can use mine if you want to.*	Gunnar and Anna interact in play. Gunnar allows Anna to use his creation and helps her develop her play.
Anna answers in a questioning tone:	*Really?*	
Gunnar:	*Yes.*	
Anna points to a Lego airplane and asks:	*This one?*	
Gunnar:	*No.*	
Anna points again:	*This one?*	
Gunnar:	*Yes, this is mine.*	
Anna's character takes the airplane and returns it to the same spot.		
Anna, **Jon** and **Gunnar** move their characters around the scene, and **Hilmar** watches the other children.	Play noises but no words.	The children play without words and move things around, but they are each in their own world, fixing their planes and characters. Hilmar makes up dialog between his characters.
Jon:	*I am going to slide in the swimming pool.*	
Jon: Jon's characters talk to each other.	*Hello, hello, friend.*	Gunnar takes a break from play. The other children do not respond.
Gunnar crawls around, sticks his head between the feet of a table and says:	*Hey, my head fits here.*	
Jon moves his characters around and sings:	*These grownups are so strange. They are always scolding us. Even when we do nothing, they are still scolding us.*	The boy sings an Icelandic children's song about all the thing children cannot do because of the grownup's rules
Gunnar plays with his airplane, moving it around and between other characters and vehicles.		Anna plays in Jon's personal space, and although it could interfere with his play, it does not seem to disturb him.
Anna flies her airplane around, between Jon's arms and over his head. Jon keeps on playing, singing and moving characters around		

(continued)

Table 16.2 (continued)

Action	Dialogue	Analysis
Anna:	*It is sad that nobody is in Hilmar's place.*	
Hilmar looks at what she is doing and says:	*You can go there because we see you, because we are close to the house. We see you through the window*	Anna and Gunnar engage in quiet, passive play. Gunnar moves the play scene to his place in the scene.
Anna and **Gunnar** arrange things so that they all has better access to Hilmar's spot. He watches.		

Table 16.3 Interactions of all the children in a play scene

Action	Dialogue	Analysis
Gunnar says [in questioning tone]:	*I am going out. I am going home. Hi, belle, or . . .?*	Grimm [Disney] fairy-tale characters become part of the play. Anna helps Gunnar remember the name of her character without belittling him or drawing attention to him in any way.
Gunnar looks towards Anna inquiringly.		
Anna whispers back:	*Cinderella.*	
Gunnar [loudly and clearly]:	*Hi, Cinderella. Hi, Cinderella. Hi, Cinderella. Would you like to take a sunbath or go to the swimming pool?*	
Jon smiles and watches with Hilmar.		
Anna:	*Yes, yes.*	The children help each other keep the play going.
The play continues, and **Anna** holds out each character as she speaks for them in a playful tone:	*We wanted to ask. ... Can we . . . can we go on the slide with you?*	The positive interactions and helpfulness shown here are typical of the children's communication and play. Here. Gunnar's tone (truism) is that it goes without saying.
Gunnar answers **Anna**:	*Yes, . . . you don't have to ask. . . . you don't have to do that.*	
Hilmar moves, so he is near the play. He watches and after a little while, he lifts his airplane to show Anna. Then he stands up and flies the plane around the room.	The dialogue is wordless.	Hilmar plays although he does not communicate verbally. He moves his character around the others, and he is involved although he might seem a little detached.

and play developed. The children decided which characters they wanted to create and made play stages based on their own experiences, preferences and imaginations. The stages became part of the children's play and narratives as they both framed and directed the play taking place. During the last workshop, the children used both the digital play objects and their own creations in any way they desired. What is clear from analysing these play episodes is that even though the interface (Shifflet et al., 2012) and affordances of the play (Marsh, 2019) were different from what the children were used to, the play itself was contextualised within the children's environment. In this way, it can be said that they used the play stages to understand and contextualise the digital play objects by framing and moving the play into their own play orbit (Grindheim, 2017). The children displayed pleasure and joy in play, as do children who have opportunities to play, according to Beresin and Sutton-Smith (2010). The children also showed that they were aware of their playmates' feelings and actively tried to draw their shy playmate into the play. They first were aware of him not participating, and then they moved the play objects closer to him and said aloud that they felt bad that he was not collaborating with them. When the children got the opportunity to play freely with the digital objects and their own play stages and characters, they collaborated, used their imagination and showed empathy towards each other. The digital play objects became natural parts of play.

According to Davies et al. (2013), it is important for children's creativity to develop to give them incubation time. The play analysed was from the last of six workshops over three weeks during which the children gradually but increasingly took over how they used the space and the objects in it. From the beginning, they had some freedom and space to explore and back out but also clear tasks they had to complete (e.g. making characters and learning to code). It can be said that the children did not fully control the physical learning space, but how they used the materials supplied and played with objects became within their power, and they used these opportunities to create both their own narratives and play. Looking at the data, the workshops gradually shifted from being planned and even somewhat prescriptive towards being open ended and driven by the children. When that happened, beautiful, connected play occurred. The importance of an open, supportive, creative learning environment for developing imaginative play was evident.

References

Beghetto, R. A., & Kaufman, J. C. (2014). Classroom contexts for creativity. *High Ability Studies, 25*(1), 53–69. https://doi.org/10.1080/13598139.2014.905247.

Beresin, A. R., & Sutton-Smith, B. (2010). *Recess battles*. Jackson, MS: University Press of Mississippi. https://doi.org/10.14325/mississippi/9781604737394.001.0001.

Cowan, K. (2019). Digital meaning making: Reggio Emilia-inspired practice in Swedish preschools. *Media Education Research Journal, 8*(2), 11–29.

Craft, A., McConnon, L., & Matthews, L. (2012). Child-initiated play and professional creativity: Enabling four-year-olds' possibility thinking. *Thinking Skills and Creativity, 7*(1), 48–61. https://doi.org/10.1016/j.tsc.2011.11.005.

Davies, D., Jindal-Snape, D., Collier, C., Digby, R., Hay, R., & Howe, A. (2013). Creative learning environments in education—A systematic literature review. *Thinking Skills and Creativity, 91*, 80–91. https://doi.org/10.1016/j.tsc.2012.07.004.

Ebbeck, M., & Kam, S. (2016). Play and the inquiring mind. In M. Ebbeck & M. Waniganayake (Eds.), *Play in early childhood education: Learning in diverse contexts* (2nd. ed., pp. 147–160). Melbourne, Australia: Oxford University Press.

Edwards, S. (2013). Digital play in the early years: A contextual response to the problem of integrating technologies and play-based pedagogies in the early childhood curriculum. *European Early Childhood Education Research Journal, 21*(2), 199–212. https://doi.org/10.1080/1350293X.2013.789190.

Fleer, M. (2016). Theorising digital play: A cultural-historical conceptualisation of children's engagement in imaginary digital situation. *International Research in Early Childhood Education, 7*(2), 75–90.

Fleer, M. (2018). Conceptual playworlds: The role of imagination in play and learning. In *Early years*. https://doi.org/10.1080/09575146.2018.1549024.

Grindheim, L. T. (2017). Children as playing citizens. *European Early Childhood Education Research Journal, 25*(4), 624–636. https://doi.org/10.1080/1350293X.2017.1331076.

Henricks, T. S. (2016). Reason and rationalization: A theory of modern play. *American Journal of Play, 8*(3), 287–324.

Hreiðarsdóttir, A. E., & Dýrfjörð, K. (2019). *Mat leikskólabarna á þátttöku í tilviksrannsókn. Netla–Veftímarit um uppeldi og menntun.* Íslands: Menntavísindasvið Háskóla. https://doi.org/10.24270/serritnetla.2019.34.

Johnson, J. E., & Christie, J. F. (2009). Play and digital media. *Computers in the Schools, 26*(4), 284–289. https://doi.org/10.1080/07380560903360202.

Marsh, J. (2017). The internet of toys: A posthuman and multimodal analysis of connected play. *Teachers College Record, 119*(December), 1–32.

Marsh, J. (2019). The uncanny valley revisited: Play with the internet of toys. In G. Mascheroni & D. Holloway (Eds.), *The internet of toys: Practices, affordances and the political economy of children's smart play*. London, UK: Palgrave Macmillan. https://doi.org/10.1007/978-3-030-10898-4_3.

Öksnes, M. (2010). *Legens flertydighed: börns leg i en institutionaliseret barndom*. Copenhagen, Denmark: Hans Reitzels Forlag.

Russ, S. W., & Wallace, C. E. (2013). Pretend play and creative processes. *American Journal of Play, 6*(1), 136–148.

Sellers, M. (2013). *Young children becoming curriculum: Deleuze, Te Whariki and curricular understanding*. London, UK: Routledge.

Shifflet, R., Mattoon, C., & Toledo, C. (2012). Touch tablet surprises: A preschool teacher's story. *YC Young Children, 67*(3), 36–41.

UN General Assembly. (1989). *Convention on the Rights of the Child*. Treaty Series; New York: United Nations, Volume 1577, p. 3. Accessed from https://www.refworld.org/docid/3ae6b38f0.html

Kristín Dýrfjörð is an Associate Professor at the University of Akureyri. She has experience as a preschool principal, and has taken part in the development of the national curriculum with the Ministry of Education. Her research interests include policy studies and the connection between science and creativity in early childhood.

Anna Elísa Hreiðarsdóttir is an Assistant Professor at the Faculty of Education, University of Akureyri, Iceland. She has experience as a preschool teacher and a preschool principal. Her research interests are play, creativity and computers and technology and gender in preschool settings.

Chapter 17
Co-Creating Hybrid Toys as an Approach to Understand Children's Needs in Play Experience

Tamara Pinos Cisneros, Felipe Escobar Vega, Ben Kröse, Ben Schouten, and Geke Ludden

Introduction

Physical and occupational therapy can make important contributions to improving people's wellbeing, however despite efforts to make therapies engaging (Aarts et al., 2012), patients often find them challenging and uninviting due to their repetitiveness. The rise of emerging technologies, including hybrid toys, gives therapists the opportunity to explore other tools that can be used in assessments and intervention sessions. Eventually, the use of such tools could benefit patients because they could allow for more personalised therapies and improve patient engagement. Throughout this chapter, the terms 'hybrid toy' or 'hybrid play' will refer to the combination of physical and digital elements in toys or play experience. This combination provides new opportunities for play, by adding feedback, automatisation or data collection. Extensive research has shown the importance of play in the development of children (Piaget, 1945; Vygotsky, 1967) and user-centred design's relevance in creating better products (Sanders & Stappers, 2008). Although co-creation, as an approach of user-centred design, has been widely studied, and hybrid toys have been studied to some extent (Tyni et al., 2016), few studies have investigated co-creation of hybrid toys as a way to understand the user's needs.

T. Pinos Cisneros (✉)
University of Twente, Enschede, Netherlands and Amsterdam University of Applied Sciences, Amsterdam, Netherlands
e-mail: t.v.pinos.cisneros@hva.nl

F. Escobar Vega · B. Kröse · B. Schouten
Amsterdam University of Applied Sciences, Amsterdam, Netherlands
e-mail: f.escobar.vega@hva.nl; b.j.a.krose@hva.nl; b.a.m.schouten@hva.nl

G. Ludden
University of Twente, Enschede, Netherlands
e-mail: g.d.s.ludden@utwente.nl

D. Holloway et al. (eds.), *Young Children's Rights in a Digital World*, Children's Well-Being: Indicators and Research 23,
https://doi.org/10.1007/978-3-030-65916-5_17

The study presented in this chapter analyses the impact of prototyping hybrid toys during a co-creation workshop, intended for children 7–12 years old. The research questions addressed are (1) which needs do children have in their hybrid play experience in terms of interactions, topics and type of play and (2) can we better understand those needs by co-creating hybrid toys with them. To answer our first question, data was gathered and analysed via concept mapping, a method that has previously been used to determine users' requirements (Ogden, Barr, & Greenfield, 2017). To gain deeper insight into the results of concept mapping we used network analysis. Furthermore, to answer the second research question we used a questionnaire, as it is standard in collecting participants' feedback (Alreck & Settle, 1994). Understanding children's needs in their hybrid play experience will help designers create new toys with applications in fields such as health, while empowering children in their use of technologies. In this study we have identified different aspects of children's play experience with hybrid toys such as: psychological needs, practical functionality, opinions and requirements.

Hybrid Toys and Understanding the User

Play Objects and Technology

Play is a fundamental part of children's development as it helps them discover the world while developing skills, identity and self-esteem. Despite there being different definitions of play, for the remainder of this chapter, we will refer to 'play' as "the quality of mind during enjoyable, captivating, intrinsically motivated and process focused activities." (Kudrowitz & Wallace, 2009). Current generations of children are growing in a world where technology is ubiquitous. From mobile phones, smart televisions and robot vacuum cleaners to home assistants, children today are surrounded by a diversity of devices. It is only natural that their play experience is becoming more linked to these new technologies. Recently, the diversity and sophistication of toys that include electronics has increased, to those we refer as hybrid toys. These are understood as play objects that have a mixture of physical and digital components (Tyni et al., 2016), using software, sensors, and actuators is common and they may or may not connect to the internet or other devices (Mascheroni & Holloway, 2017). Related terms are: smart-toys, connected toys, internet of toys and augmented toys. These toys have been gaining popularity as younger generations find the borders between the physical and digital less obvious and their familiarity with technology is more natural. They can provide immediate and diverse feedback, data collection, interactivity and personalized behaviour. Examples of hybrid toys are: care toys such as Tamagotchi and Hatchimals, educational toys that help children learn to program like Doc Clementoni and GoTo drawing machine, and advance action figures that expand the video game experience like Amiibo.

Besides being used for recreation, hybrid toys are starting to be used in education and health. In their review of "Smart Toy based learning" (2014), Cagiltay, Kara & Aydin present an analysis of hybrid toys in education. According to these authors, hybrid toys become cognitive tools by providing interactive learning experiences. They are used to teach mathematics, coding, languages, cultural values (Al-Khalifa et al., 2018) music (Luo, Wang, Xiong, Shan, & Zhou, 2018), communication skills, creativity and social interaction (Ihamäki & Heljakka, 2018) to name a few fields. In the health domain, adding sensors in toys, and the possibility to record and store data have likewise opened new opportunities. Smart toys could be (and to some extent already are) used for therapies, sports and assessment of cognitive and motor skills. In this domain, researchers have for example developed toys that encourage social interaction for children with autism (Farr, Yuill, & Hinske, 2012), toys that detect delays in motor skill development (Mironcika et al., 2018; Sander et al., 2017) and toys and games that promote movement (Levac et al., 2010). Although potential uses of these toys have been described in literature, there are limited studies with emphasis on children's perception and preferences in hybrid toys and how technology is affecting their play experience.

Understanding the User

Co-creation is a design practice where the user is considered the expert and is asked to take an active participation in the design process (Sanders & Stappers, 2008). It is used by designers as a methodology to better understand users and their needs. The importance of involving users in earlier levels of this process relates to the needs of relevance, perspective and ownership (Plattner, Meinel, & Leifer, 2012) towards the object or system to be designed. Previous studies have used this approach to design character toys (Ihamäki & Heljakka, 2017) and learning environments (Kangas, 2010). Other examples are Game Jams, which are events that provide educational value for participants, while generating new knowledge by creating prototypes of games or video games in a short amount of time (Deen et al., 2014). With this in mind, we designed a co-creation session where children could build hybrid toys in a space open for experimentation and discussions.

Methodology

Co-Creation

The co-creation workshop "Digital & Physical Play—Digital Camp" took place at the Amsterdam University of Applied Sciences and at the Public Library of Amsterdam, on the 16th and 17th of November 2018. Ten adults and nine children participated. We asked the participating children to reflect on what they like most about play, which themes or topics they are interested in, what type of features they like in toys and how they would use technology. This was done by building a prototype and discussing the rationale behind their actions following a design thinking process (Plattner et al., 2012). This process was structured in five main steps:

1. Empathise—Understand the design challenges. This involves empathising with the user, understanding the needs and the ecosystem around the activity. Participants were asked to create a Persona of their intended user as a way to understand and empathise with the target audience.
2. Define—Summarise requirements of the design, what does it need to have/do to be able to suit the user(s). Participants were asked to reflect on the needs and desires of their persona as a way to extract insights and design criteria.
3. Ideate—Use the defined criteria to come up with ideas and concepts that fit the users. Participants were asked to do brainstorming using lotus blossom (Vangundy, 2004), and pick the concept that best fit their design criteria.
4. Prototype—Build a prototype of the most promising design concept(s). Participants were asked to build high-fidelity prototypes of their concepts.
5. Test—Test the prototype with users. Participants tested their high-fidelity prototypes with their target audience.

Data was collected from the session and explored via concept mapping and network analysis. Finally, we used a questionnaire to analyse the acceptance of the co-creation process (Fig. 17.1).

Concept Mapping

Concept mapping is a structured process that is focused on a topic or problem that needs to be solved (Fig. 17.2). It involves input from several parties, which produces a graphical view (concept map) of their ideas and concepts. It allows participants to easily visualize how these ideas (statements) are interrelated and how they can be clustered (Trochim, 1989). There are six main steps that Trochim defines as part of this process:

1. Prepare project, by choosing a focus, recruiting participants and scheduling the mapping.
2. Generate ideas, where participants will formulate statements that answer the seeding question.
3. Structure ideas, where the previously mentioned statements or ideas will be organised by the participants in clusters.
4. Compute maps, statistical analysis done on the clusters of ideas generated. For this study we used the open source tool R-CMap (Bar & Mentch, 2017) to aid us generate this analysis.

Fig. 17.1 Co-creation cycle used in the workshop

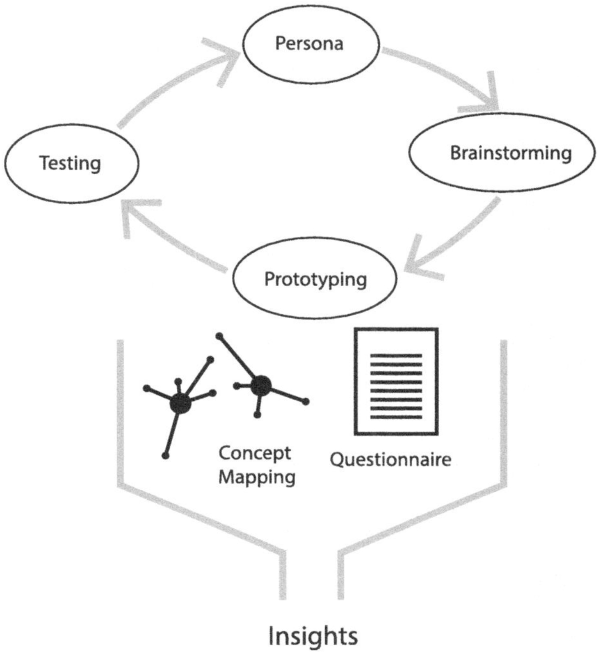

5. Interpret maps: where stakeholders and researchers analyse and interpret the maps.
6. Utilise maps: bring the results into practice, which in our case will be developing new hybrid toys.

We used concept mapping to get an overview of the main clusters that were identified within the statements. Additionally, we wanted to have an overview of the entire network and see how the different statements related to each other; for that we used network analysis.

Network Analysis

Network analysis is a method that allows examining the relationships between nodes (statements) and edges (relationships that connect them). This uncovers the relationship, closeness and clusters of similarity that nodes have within networks (Borgatti, Mehra, Brass, & Labianca, 2019), and an overview of the connections between the different nodes. Network analysis was used to help identify the similarity and links between the network of statements made by the participants of the session.

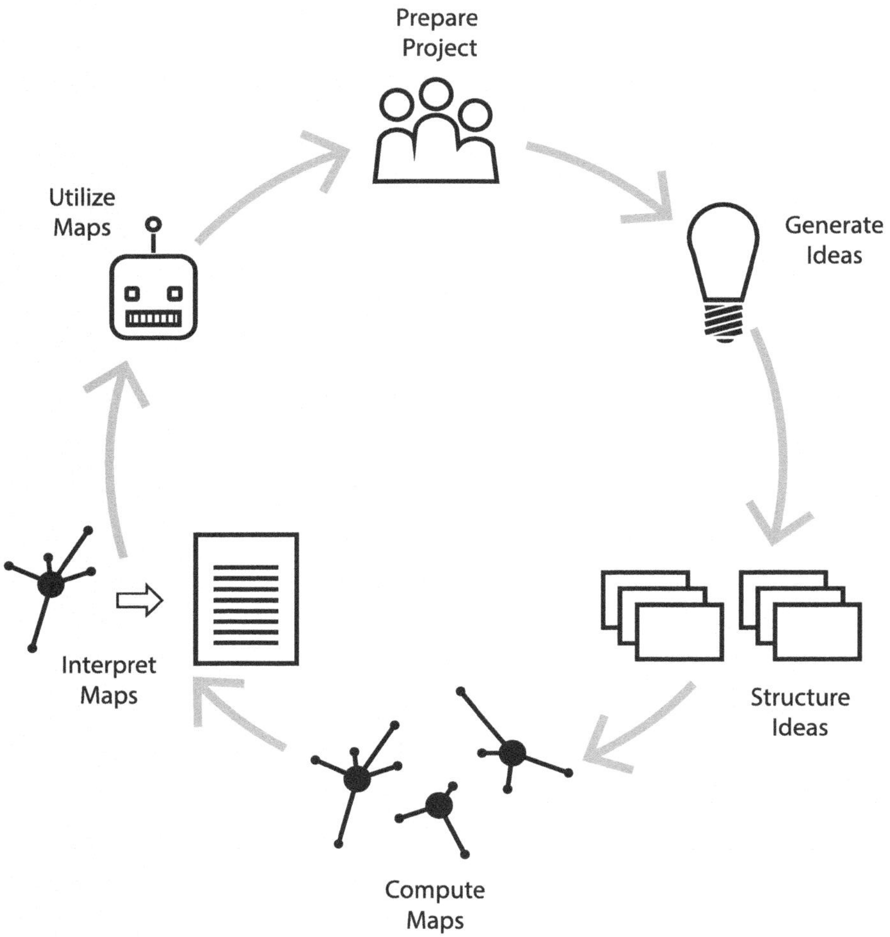

Fig. 17.2 Concept mapping structure

Questionnaire

A questionnaire was filled in by nine of the ten children that participated once the co-creation session ended (one of the participants had to leave early). The questionnaire consisted of open questions and Likert scale questions asking about the children's perception of the co-creation session, their use of technologies, their creativity and further interest in the topic of the workshop.

Study

We designed and conducted an exploratory workshop of co-creating hybrid toys, where children worked together with adult participants in creating a prototype. In this section we will describe the materials, participants and structure of the workshop.

Materials: Participants could use a variety of electronic tools: Nintendo Switch with their programmable interface provided by Nintendo Labo, LittleBits a collection of electronic building blocks, Makey Makey an interface that replaces the keyboard with the use of conductive material, and the Touch Board a microcontroller that can play sounds via the use of electrodes. Participants also had access to Legos, craft material and a laser cutter.

Day 1: Nine design students and design/game professionals (seven females, two males; age-range: 24-42y) participated in a workshop. After reviewing theory about the importance of play and different types and characteristics of hybrid toys (Tyni et al., 2016), participants worked in teams of three to create a prototype of a toy with physical and digital elements. They started by defining a persona, then participants used Lotus blossom for ideation, the result of this process was three different prototypes of hybrid toys. These hybrid toys were used in day two as thought-provoking tools.

Day 2: Ten children (seven boys, three girls; age-range 7-12y) were invited to participate as 'experts of play'. With consent from the children's parents, video and photographs were taken. The children first tested the prototypes created by the observers during the first day of the workshop in order to get them into a playful mood, get comfortable as a group and collect feedback from their play experience. They were divided into teams (one of four and two of three participants) supported by observers (adult participants and coaches) to create their own prototype of a hybrid toy. The role of the observers included taking note of behaviours and comments of the children. Coaches had experience with the tools provided and helped troubleshoot issues. A brainstorming session (Fig. 17.3) helped children identify themes and objects that they found interesting and decide which play object they were going to build. During the prototyping session they experimented with the tools provided, thought about which type of actions they wanted their toys to perform, and made a sketch of what they wanted to build. As result of this session each team built a hybrid toy: 'Monster car', 'Horror House' and 'The roller coaster' (Fig. 17.4). At the end there was a round of testing and feedback. All of their remarks during this process were recorded by the observers as statements.

In Concept Mapping ideas are generated based on a seeding question, for this research the seeding question was: *"Which characteristics of hybrid toys and play experience motivate children to play?"*. During the workshop observers wrote down statements that answered the question and during a closing session they compiled a final list of statements, based on what the children said. This list was later verified with video and audio recordings from the day. To generate a concept mapping analysis, observers clustered statements based on similarity and rated their

Fig. 17.3 Some ideas from the brainstorming session

Fig. 17.4 Prototypes day 2: Monster car, Horror house and Roller coaster

importance (based on children's behaviour and comments) on a Likert scale (one to five) by using a digital tool (16.5) (Brons, 2018) (Fig. 17.5).

Results

The observers (n = 13; eight students/design professionals and five coaches) identified 81 statements, six of them clustered the statements answering the seeding question. By using concept mapping, we clustered those 81 statements into eight clusters (Table 17.1).

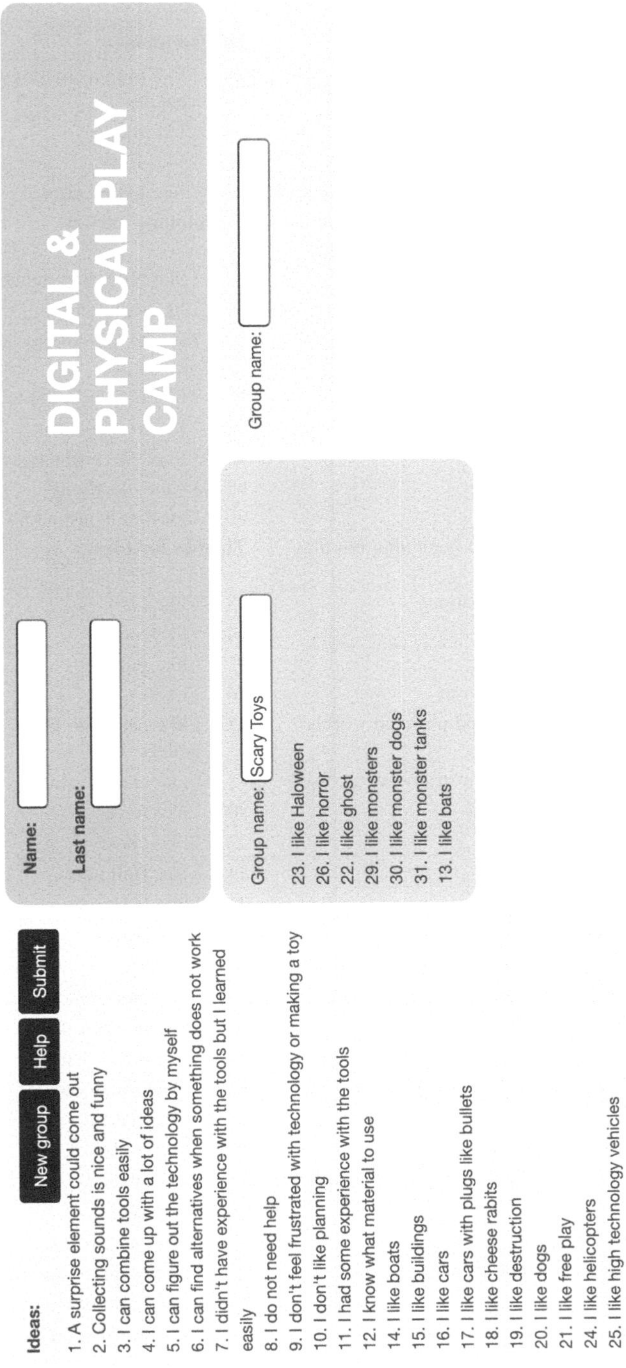

Fig. 17.5 Clustering tool

Table 17.1 Clustered statements

Curiosity and freedom		**DIY technology**	
3	I can combine tools easily	25	I like high technology vehicles
4	I can come up with a lot of ideas	34	I like robots
5	I can figure out the technology by myself	35	I like scary
6	I can find alternatives when something does not work	43	I like to do crafts
7	I didn't have experience with the tools but I learned easily	**Technology variety**	
8	I do not need help	32	I like Nintendo switch
9	I don't feel frustrated with technology or making a toy	39	I like technology
12	I know what material to use	50	I like to see a personalised message
21	I like free play	51	I like to use digital tools
41	I like to be part of participatory design	55	I like videogames
42	I like to be relaxed	57	I liked the birthday cake toy
44	I like to explore	63	I want electricity
45	I like to feel curious	76	LitteBits is just another block
46	I like to have instructions at the beginning but play freely after.	**Themes for toys**	
47	I like to make different combinations	13	I like bats
52	I like to use my imagination	14	I like boats
Hybrid functionality		15	I like buildings
1	A surprise element could come out	16	I like cars
65	I want to combine the digital and physical worlds	17	I like cars with plugs like bullets
69	I want to know which musical instruments are used	18	I like cheese rabits
74	It is not necessary to add digital	20	I like dogs
How to play		22	I like ghost
10	I don't like planning	23	I like Halloween
19	I like destruction	24	I like helicopters
40	I like to be competitive	26	I like horror
48	I like to play with lego freely, without instructions.	28	I like light
56	I like when a game is exciting	29	I like monsters
73	It is nice to play together	30	I like monster dogs
77	Making a toy is playing	31	I like monster tanks
80	The game could have an specific environment	33	I like remote control cars
Game structure		36	I like snakes
11	I had some experience with the tools	37	I like spiders
27	I like lego	38	I like tanks
49	I like to play outside	53	I like trucks
64	I want to build a tomato charger to a phone	54	I like vampires
68	I want to know which action triggers a sound	59	I want a house that walks
72	It is funny to be barefoot	60	I want a robot house
79	The game could be played in the dark	61	I want a roller coaster

(continued)

Table 17.1 (continued)

Curiosity and freedom		**DIY technology**	
Feedback signal		62	I want animals
2	Collecting sounds is nice and funny		
58	I missed the use of sounds		
66	I want to have variety of sounds		
67	I want to hear other sounds		
70	I want to see more emojis		
71	It could use a secret language		
75	It is super cool to send and receive a signal		
78	More messages could be sent		
81	The use of remote control is fun		

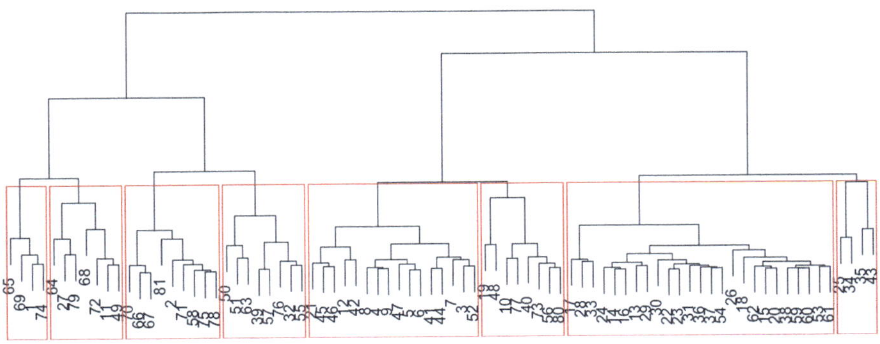

Fig. 17.6 Dendogram plot

The dendogram plot (Fig. 17.6) and the ray cluster map (Fig. 17.7) show the clustered statements. Table 17.2 shows that the cluster "Curiosity and freedom" scored highest on importance. This cluster shows that children enjoyed combining tools easily, explore, feel curious and have free play as a component of their play activity. This gives indications of the importance of open play as a format for toys. In the second highest scoring cluster: "Hybrid functionality", the contradictory statements "I want to combine the digital and physical world" and "it is not necessary to add digital", show that the addition of digital elements needs to be meaningful. As Goldstein (2012 p. 29) has stated "If a toy is no fun to play with, no amount of technology will increase its desirability as a play object". The "Feedback signal" cluster highlights the importance of sound as a part of the toy. The "Technology variety" cluster shows the acceptance of the tools used in the workshop. In the "Themes for toys" cluster we see a list of topics that children enjoy in their play experience. The "How to play" and the "DIY technology" clusters provide diverse views of elements that can be part of the toys. Furthermore, what is noticed is that the statements identified relate to different needs: psychological needs, practical functionality, opinions and requirements, which encompass a wide range of users' needs.

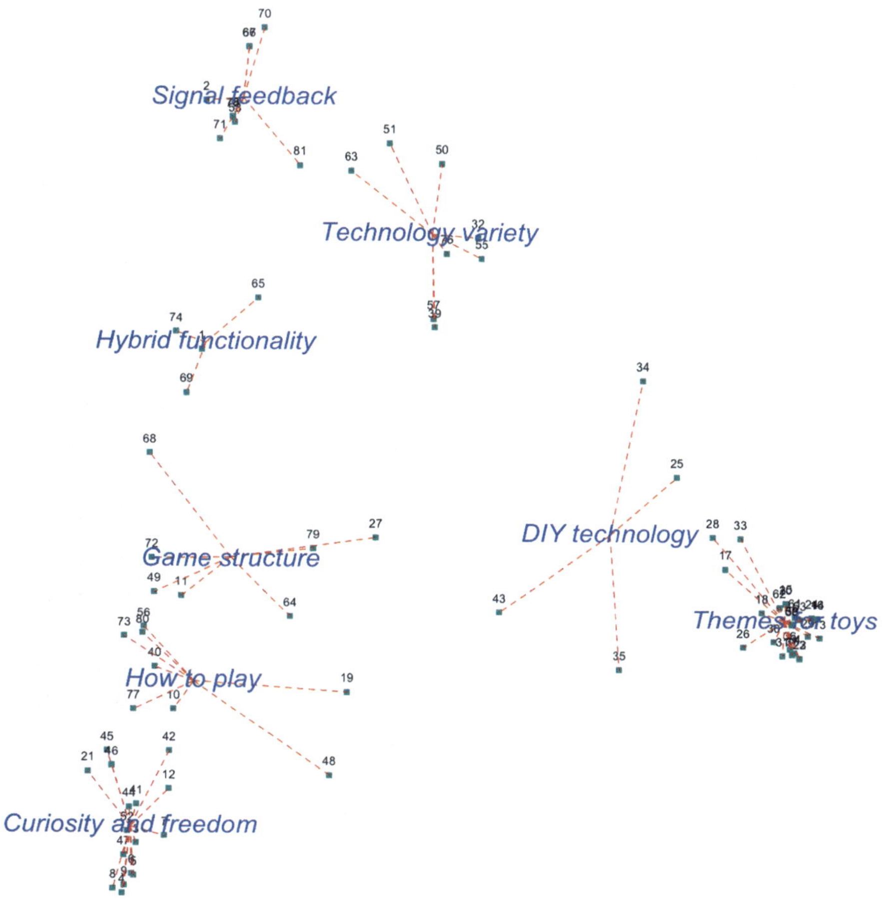

Fig. 17.7 Ray cluster map

Table 17.2 Summary of clusters

Cluster ID	N	Mean	S D
Curiosity and freedom	16	3.92	0.47
Hybrid functionality	4	3.75	0.29
How to play	8	3.35	0.59
Game structure	7	2.90	0.79
Feedback signal	9	2.87	0.69
DIY technology	4	2.67	0.76
Technology variety	8	2.46	0.82
Themes for toys	25	1.73	0.31

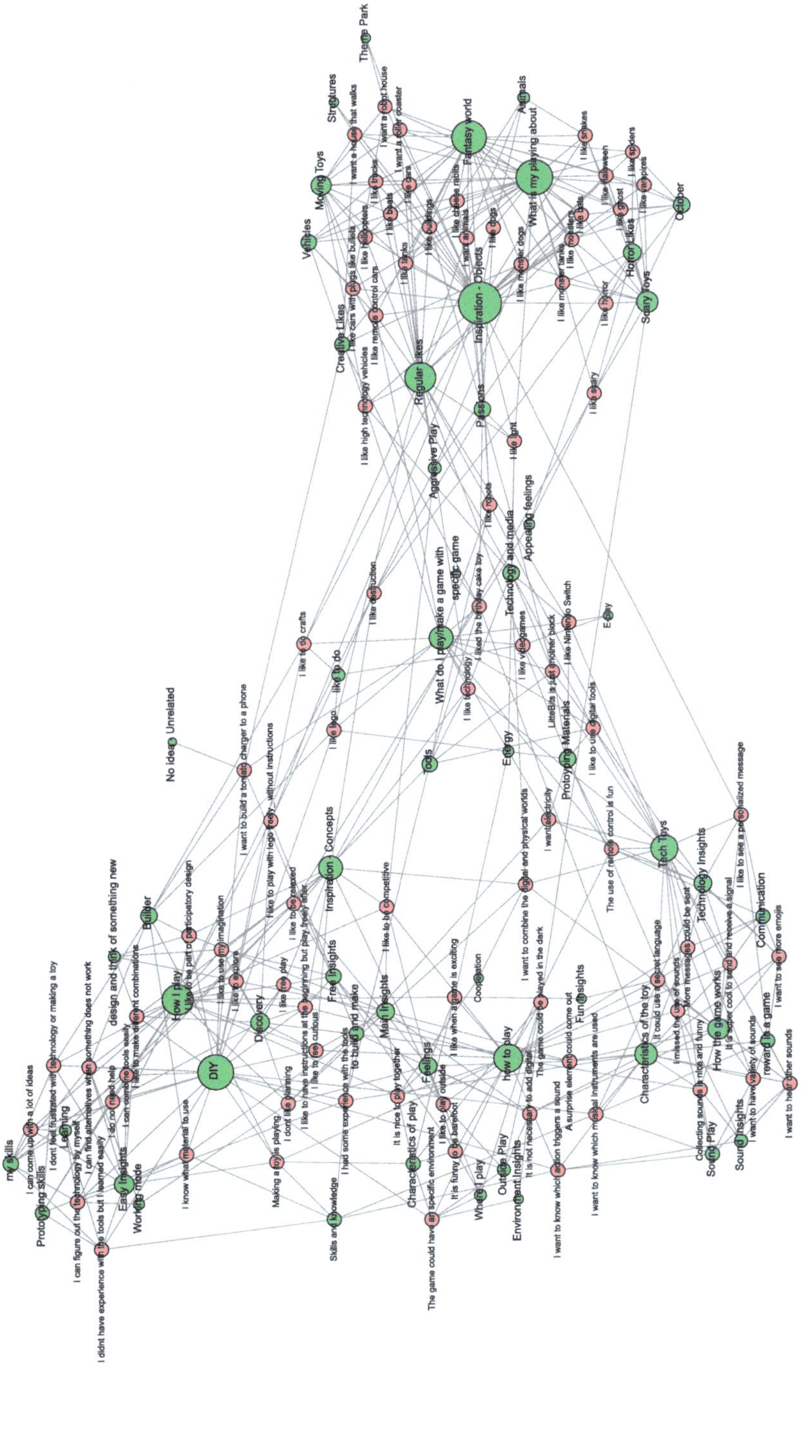

Fig. 17.8 Network analysis

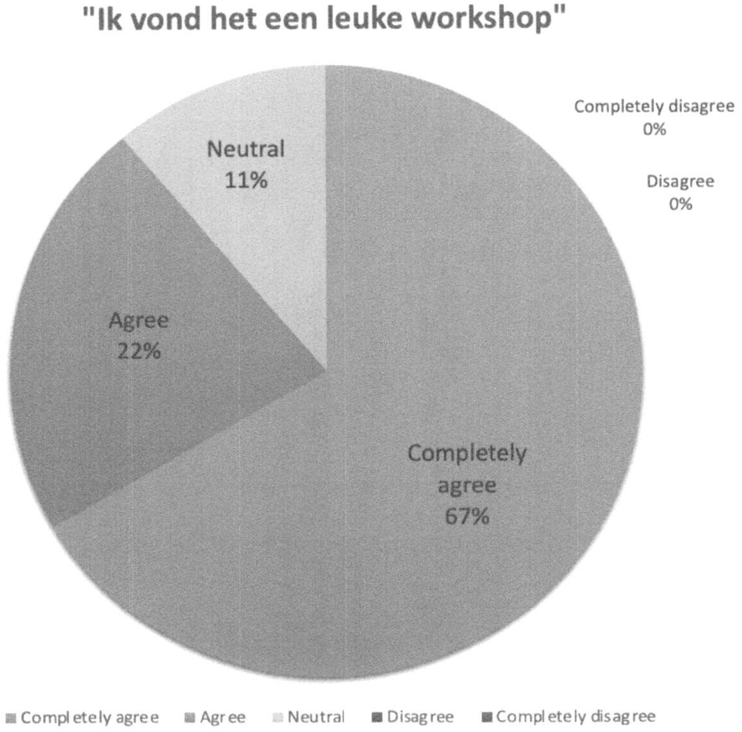

Fig. 17.9 Question on acceptance: "I thought it was a nice workshop"

The network analysis of the statements (Fig. 17.8) shows that there are two main clusters of information. The cluster on the right revolves around conceptual themes that the children would like their toys to have. The second cluster revolves around the activities that they find interesting/fun when playing. This second cluster is divided into two sub clusters that centre around expectations and desires that the children have about the physical characteristics of toys and the interaction expectations and desires about play itself. In the middle of both clusters is a sub-cluster that revolves around what the children find important when making their own toys.

The cluster related to themes is tightly packed and the cluster around activities is spread out, therefore, we can identify themes that are around shared common interests. Moreover, there is a lot of diversity surrounding the concepts that deal with these themes. The statements indicate that the children made no clear distinction between the characteristics of the play experience and the toy (or toy system).

The results of the questionnaire used to assess the children's acceptance of the workshop show their enjoyment (Fig. 17.9) and what they learned (Table 17.3). They also expressed their interest in learning more about robots, programming, making toys and the technical tools used during the workshop. In a short amount of time (five hours), they were able to build the prototypes, learn about the

Table 17.3 Some statements of what children learned

I have learned...
How you can make robots
I have learned that everyone can figure things out
That we can use other things as a controller
How to make things with the switch
You can do more things if you know little bits

technology and discuss what they would like to see in their hybrid toys. As one of the children mentioned, "[these tools] add some magic".

Discussion

Prior studies have noted the value of co-creation in understanding users. The generated concept map further supports the idea of using co-creation of hybrid toys as a playful and educational experience to better understand user's needs. However, our findings may be limited by the number of participants, the lack of diversity among participants and cases of miscommunication with the observers. Although we believe that co-creation workshops can lead to a better understanding of a target group, further research should develop workshops with more diverse groups of children. Moreover, this approach should also include therapists, children and their families to identify what are the specific requirements of hybrid toys that can facilitate therapies.

Although some of the coaches acted as translators, one of the main issues encountered was the language barrier between observers and children. This study was conducted in The Netherlands where most of the participating children only spoke Dutch, while some of the observers did not speak Dutch. This will be taken into consideration for future studies.

Another limitation of the current study is that rating the importance was made by the observers, as an attempt to identify what they have learned from the users. It would be interesting to conduct a future study with bigger and more diverse groups of children and see how they would rate the statements themselves.

Conclusions

The aim of the present study was to understand the needs of children in their hybrid play experience and to examine co-creation workshops as a means to understand the user. This study has compiled a list of insights that can be a starting point for designing new prototypes of hybrid toys for use in health, and in other fields. These insights refer to different aspects: psychological needs, practical functionality, opinions, and requirements. They provide a rich overview of children's needs and thoughts about these types of toys. From a design perspective these insights help designers make informed decisions. This study has also found that empowering children with technology makes ideas more tangible and allows for richer design criteria to develop hybrid toys. The innovative aspect of this workshop is that it was not just about co-creation, but also tinkering with technology. While ideation and concepting provides useful information, prototyping with technology can trigger conversations about the needs of the user in hybrid play experiences. In this setting, children have the opportunity to experiment with these tools, have a more active role in the design of their toys and think about their potential.

The combination of co-creation and concept mapping can lead to valuable results when designing solutions by triggering and documenting ideas. The children who participated in the workshop expressed their interest in learning more about robots, programming, making toys and the tools used (Little Bits, Makey Makey and Nintendo Switch). From this we can conclude that it was a pleasurable experience for the participants and that this format could continue to be used to understand better the needs of the user. Despite the exploratory nature of the study, the results presented show that we, as designers, can identify the users' needs from working and playing alongside with them. In future studies we will analyse which specific elements, like sound for example, can be used to improve physical and occupational therapies.

References

Aarts, P. B., van Hartingsveldt, M., Anderson, P. G., van den Tillaar, I., van der Burg, J., & Geurts, A. C. (2012). The pirate group intervention protocol: Description and a case report of a modified constraint-induced movement therapy combined with bimanual training for young children with unilateral spastic cerebral palsy. *Occupational Therapy International, 19*(2), 76–87. https://doi.org/10.1002/oti.321.

Al-Khalifa, H. S., Faisal, H. R., Rushdi, S. M., AlNawwar, G. M., Al-Gumaei, G. N., & Alabduljabbar, A. A. (2018). Basma: An interactive IoT-based plush toy for Arabic-speaking children. *Journal of Computer Science, 14*(11), 1440–1453. https://doi.org/10.3844/jcssp.2018.1440.1453.

Alreck, P. L., & Settle, R. B. (1994). *The survey research handbook*. New York, NY: McGraw-Hill.

Bar, H., & Mentch, L. (2017). R-CMap—An open-source software for concept mapping. *Evaluation and Program Planning, 60*, 284–292. https://doi.org/10.1016/j.evalprogplan.2016.08.018.

Borgatti, S. P., Mehra, A., Brass, D. J., & Labianca, G. (2019). Network Analysis in the Social Sciences. *Science, 323*(5916), 892–895.

Brons, A. (2018). *Concept Mapping–sorting tool* (Version 1.0) [Computer Software] Amsterdam: HvA.

Cagiltay, K., Kara, N., & Aydin, C. C. (2014). Smart toy based learning. In *Handbook of research on educational communications and technology* (pp. 703–711). New York, NY: Springer. https://doi.org/10.1007/978-1-4614-3185-5_56.

Deen, M., Cercos, R., Chatman, A., Naseem, A., Bernhaupt, R., Fowler, A., ... Mueller, F. (2014). *Game jam*, 25–28. https://doi.org/10.1145/2559206.2559225.

Farr, W., Yuill, N., & Hinske, S. (2012). An augmented toy and social interaction in children with autism. *International Journal of Arts and Technology, 5*(2–4), 104–125.

Goldstein, J. (2012). *Play in children's development, health and Well-being*. Brussels: Toy Industries of Europe.
Ihamäki, P., & Heljakka, K. (2017). *Workshop on the internet of toys*, 251–254. https://doi.org/10.1145/3131085.3131114.
Ihamäki, P., & Heljakka, K. (2018). Smart, skilled and connected in the 21st century: Educational promises of the internet of toys (IoToys). In *Proceedings of Arts, Humanities, Social Science Education Conference* (pp. 1–24). Prince Waikiki Hotel, Honolulu, Hawaii.
Kangas, M. (2010). Creative and playful learning: Learning through game co-creation and games in a playful learning environment. *Thinking Skills and Creativity, 5*(1), 1–15. https://doi.org/10.1016/j.tsc.2009.11.001.
Kudrowitz, B. M., & Wallace, D. R. (2009). The play pyramid: A play classification and ideation tool for toy design. *International Journal of Arts and Technology, 3*(1), 36. https://doi.org/10.1504/ijart.2010.030492.
Levac, D., Pierrynowski, M. R., Canestraro, M., Gurr, L., Leonard, L., & Neeley, C. (2010). Exploring children's movement characteristics during virtual reality video game play. *Human Movement Science, 29*(6), 1023–1038. https://doi.org/10.1016/j.humov.2010.06.006.
Luo, S., Wang, Y., Xiong, N., Shan, P., & Zhou, Y. (2018). An Interactive Smart Music Toy Design for Children. In *Lecture Notes in Computer Science (including subseries Lecture Notes in Artificial Intelligence and Lecture Notes in Bioinformatics)* (Vol. 10921 LNCS, pp. 372–390). Design Industrial Innovation Center, China Academy of Art, Hangzhou, 310024, China. https://doi.org/10.1007/978-3-319-91125-0_31.
Mascheroni, G., & Holloway, D. (2017). *The Internet of Toys: A report on media and social discourses around young children and IoToys*. DigiLitEY. Accessed from http://digilitey.eu/wp-content/uploads/2017/01/IoToys-June-2017-reduced.pdf
Mironcika, S., de Schipper, A., Brons, A., Toussaint, H., Kröse, B., & Schouten, B. (2018). Smart toys design opportunities for measuring children's fine motor skills development. In *Proceedings of the Twelfth International Conference on Tangible, Embedded, and Embodied Interaction–TEI '18* (pp. 349–356). New York, New York, USA: ACM Press. https://doi.org/10.1145/3173225.3173256.
Ogden, K., Barr, J., & Greenfield, D. (2017). Determining requirements for patient-centred care: A participatory concept mapping study. *BMC Health Services Research, 17*(1), 1–11. https://doi.org/10.1186/s12913-017-2741-y.
Piaget, J. (1945). *Play, dreams and imitation in childhood*. London: Heinemann.
Plattner, H., Meinel, C., & Leifer, L. (2012). *Design thinking research: Studying co-creation in practice*. New York, NY: Springer.
Sander, J., de Schipper, A., Brons, A., Mironcika, S., Toussaint, H., Schouten, B., & Kröse, B. (2017). Detecting delays in motor skill development of children through data analysis of a smart play device. In *Proceedings of the 11th EAI International Conference on Pervasive Computing Technologies for Healthcare–PervasiveHealth '17* (pp. 88–91). ACM Press. https://doi.org/10.1145/3154862.3154867.
Sanders, E. B.-N., & Stappers, P. J. (2008). Co-creation and the new landscapes of design. *CoDesign, 4*(1), 5–18. https://doi.org/10.1080/15710880701875068.
Trochim, W. M. K. (1989). An introduction to concept mapping for planning and evaluation. *Evaluation and Program Planning, 12*(1), 1–16. https://doi.org/10.1016/0149-7189(89)90016-5.
Tyni, H., Kultima, A., Nummenmaa, T., Alha, K., Kankainen, V., & Mäyrä, F. (2016). *Hybrid Playful Experiences–Playing between Material and Digital, Final Report*. Accessed from https://trepo.tuni.fi/handle/10024/98900
Vangundy. (2004). *101 activities for teaching creativity and problem solving. Work*. New Jersey: Wiley.
Vygotsky, L. S. (1967). Play and its role in the mental development of the child. *Soviet Psychology, 5*(3), 6–18. https://doi.org/10.2753/rpo1061-040505036.

Tamara Pinos Cisneros is a PhD candidate in the 'Digital Society School', 'Play & Civic Media' and 'Digital Life' research groups at the Amsterdam University of Applied Sciences and the University of Twente. Her research focuses on the use of smart-toys to motivate children with disabilities to do physical therapy.

Felipe Escobar Vega is an UX/UI Designer with a MSc in Design for Interaction from TU Delft. His background is on multimedia creation and industrial design. His interests focus on bridging the gap between digital and physical products by designing meaningful experiences and interactions.

Ben Kröse is Emeritus Professor in 'Ambient Robotics' at the University of Amsterdam and 'Digital Life' at the Amsterdam University of Applied Science. His research focuses on robotics and interactive smart devices for health, safety and wellbeing. He is involved in steering boards on IT and Health, and the platform for practice-oriented IT research.

Ben Schouten is a full Professor in Playful Interactions in Smart Environments at Eindhoven University of Technology and Lecturer of Play & Civic Media Research at Amsterdam University of Applied Sciences. He is an advisor for the European Commission on the 'Internet of Things' and the Dutch Cultural Media Fund.

Geke Ludden is Associate Professor and Head of the Chair of Interaction Design, Department of Design, Production and Management and a research fellow of the DesignLab at the University of Twente. Geke studies the design of products and services that support healthy behaviour and that engage people in therapy.

Chapter 18
Assessing Developmental Difficulties in Children Through Connected Smart Toys

Diego Rivera, Maria Luisa Martin-Ruiz, Luis Cruz-Piris, Kevin van der Meulen, Antonio García, Cristina Serrano García, Susel Fernández, Bernardo Alarcos, and Juan R. Velasco

Introduction

Early detection of developmental disabilities is a fundamental step for the success of intervention programs, which have been proven to have a positive impact on the general quality of life of children (Alcantud & Alonso, 2016). Therefore, many tools and methodologies have been proposed, studied, and validated, which aim to improve the development assessment process (Goodman, Ford, Richards, Gatward, & Meltzer, 2000).

Over the years, these methods have evolved, incorporating new tools and measurements which have improved the assessment outcomes (Miroshnikov, Nasledov, & Zashchirinskaia, 2016). However, the dominant methodology has been essentially unaltered and continues to be a manual observation-based process. Many of these psychology-oriented scales and tools (Roid & Sampers, 2004) are based on the evaluation of certain activities performed by children. For instance, children are asked to perform a specific movement while being observed by an expert who

D. Rivera (✉) · M. L. Martin-Ruiz
Universidad Politécnica de Madrid, Madrid, Spain
e-mail: diego.rivera@upm.es; marialuisa.martinr@upm.es

L. Cruz-Piris · A. García · S. Fernández · B. Alarcos · J. R. Velasco
Universidad de Alcalá, Madrid, Spain
e-mail: luis.cruz@uah.es; antonio.garciah@uah.es; susel.fernandez@uah.es; bernardo.alarcos@uah.es; juanramon.velasco@uah.es

K. van der Meulen
Universidad Autónoma de Madrid, Madrid, Spain
e-mail: kevin.vandermeulen@uam.es

C. S. García
Centro Universitario Cardenal Cisneros, Madrid, Spain
e-mail: cristina.serrano@cardenalcisneros.es

D. Holloway et al. (eds.), *Young Children's Rights in a Digital World*, Children's Well-Being: Indicators and Research 23,
https://doi.org/10.1007/978-3-030-65916-5_18

evaluates the performance using a pre-determined scoring system. The manual nature of the process limits the possible resolution of the measurements and relies heavily on observer expertise in the matter. Moreover, the assessment must be performed in front of the expert and requires the planning of specific test environments and activities.

Although the observation-based methods have been proven to provide excellent results in the early detection of psychomotor delays (Cools, De Martelaer, Samaey, & Andries, 2009), the technological advances of the last decades can contribute to the improvement of the assessment process (Albinali, Goodwin, & Intille, 2012; Taffoni, Focaroli, Keller, & Iverson, 2014), as it has been the case with many other aspects of the healthcare field (Kulkarni & Sathe, 2014). Specifically, the use of sensors to monitor and assess health-related indicators has been a widely studied topic in research (Andreu-Perez, Leff, Ip, & Yang, 2015). It has improved diagnosis and processes in different areas of health (Nguyen, Mirza, Naeem, & Nguyen, 2017). In general, sensors are devices that convert physical quantities into signals which can be read by electronic instruments. The technologies used to perform the measurement and conversion tasks establish the cost and accuracy of this process. The improvements in sensor-based technology, along with the rising of internet-based systems and the growth of internet-based communications, have led to the proposal of platforms and methods which take advantage of both domains to provide new and more efficient health monitoring processes (Yin & Jha, 2017).

The use of sensors in everyday objects, combined with internet-based communications and the capacity to perform some actions autonomously as 'intelligent devices', has materialised in the rise of the Internet of Things (IoT) paradigm. Improvements in sensor and actuation technologies have enabled their use in automating daily tasks. The IoT-based systems have already been used in different and very heterogeneous scenarios such as Smart Cities, Smart Buildings, and of course, health-related scenarios (Samie, Bauer, & Henkel, 2016). For example, a remote medical parameter monitoring system can help in the prevention of diseases, ad hoc diagnosis, or detect an accident that requires medical attention (Bandyopadhyay & Sen, 2011). Some examples of IoT applications in smart building scenarios are the use of motion sensors to improve security, automatically turn on the air conditioning under specific temperature and humidity values, and lighting management (Pallavi & Smruti, 2017). In smart cities, IoT can be applied in the fields of intelligent transport (intelligent transportation systems to minimise traffic congestion, smart parking management, and smart traffic lights), smart water systems, or smart environments and agriculture (automated irrigation systems according to weather conditions and the sensors values) (Lin et al., 2017).

Although connected smart toys have been already described for educational (Cagiltay, Kara, & Aydin, 2014), entertainment (Magerkurth, Cheok, Mandryk, & Nilsen, 2005), and health-related purposes (Mironcika et al., 2018), the use of IoT paradigm in this field has specific challenges, concerns, and promises. Due to the relevance in this area of features such as security and privacy (McReynolds et al., 2017) has arisen a new sub scenario of the IoT, and it is usually called the Internet of Toys (IoToys) (Holloway & Green, 2016). Given that in many of the scales and assessment methodologies, the proposed activities rely on the use of toys and other everyday objects, it would be possible to add technology to such purposes following the IoToys paradigm. In this way, the data acquisition and classification process would be more efficient than applying manual methods and improve the assessment process by providing new information to experts.

In this paper, we present the architectural design requirements, the methodology used, and the necessary elements to build a smart toy-based platform, which aims to provide child development professionals with new tools for increasing their insight and accuracy of performance assessments conducted with children. To evaluate this proposed platform, we have designed and developed a prototype version of the designed system, and we have built smart toys to test them in real scenarios. In the next section, we define the task requirements and the actual design of the proposed platform, which following a reference model and a view-based design approach. We then provide a summary of the current prototype designs and the tests carried out with them. Finally, we discuss the results obtained using the proposed architecture and possible future work lines.

The Smart Toys Platform Architecture

The current methodologies for child development assessment through playing activities are mainly based on scales designed by psychologists. During the development of each activity, experts must collect concrete measures to compare these values with the scales later. In this context, the accuracy of the measurements depends directly on the expertise of the person collecting the data. Besides, the only data obtained in this process are those included in the known scales. We have designed a specific platform that could ease the task of visualising, analysing, and monitoring the data gathered while a child performs activities using the developed smart toys. This platform provides the essential elements to automate the process of collecting, storing and processing the information generated during each play activity. The integration and automation of tasks improve the work of experts, as they do not need to take notes, store manually, and analyse the data obtained.

This platform would act as a secure data storage environment, service provider for data analysis, alerts, etc., and as a centralised communications environment for the IoT devices. These features are important in the platform, as secure management of health-related data is a crucial part of this kind of system. Moreover, our proposal must comply with data protection regulations, which are principally governed by the EU General Data Protection Regulation (GDPR) (Council of the European Union, 2016). It is common to design IoT platforms using this approach (Mohammed et al., 2014), where there are numerous sensor-devices and a centralised element to gather and store the information. Usually, the devices deployed in these scenarios are only dedicated to obtaining information from the environment and, in some cases, interact with it. The huge amount of data generated by the devices must be stored in centralised environments such as gateway devices and cloud servers (Kovatsch,

Mayer, & Ostermaier, 2012). In our case, the nature of the information requires special security measures, as it has been stated in the IoToys related literature after analysing the current status of security and privacy in such devices (Chu, Apthorpe, & Feamster, 2018). The privacy of the data shared through this platform must be protected by determining the threats of our specific scenario and then designing the proper mechanisms to avoid them. We have developed specific secure servers and devices, and we have defined communications to be as anonymised as possible (Rivera et al., 2019). In the next sections, we explain with more detail some of the design principles of our proposal.

The platform has been based on a custom architecture designed specifically for the children assessment platform, and which has been defined using a series of requirements collected from the experts' current necessities and the available technological environment. The creation of specific software and hardware architecture allowed us to comply with the requirements without relying on generic architectures. The requirements have been gathered by interviewing experts and by the observation of their usual activities with children. Moreover, in our research group, we count with some children assessment experts that gave us advice on the design process.

In the next sections, we identify the most important requirements and the high-level design for the platform. Finally, we focus on the main physical entities of the IoT platform.

Design Requirements

The design of this system was heavily based on specific design requirements. We obtained these requirements in coordination with experts in child development, the available literature in this topic, and the technological limitations and availability in the Internet of Things platforms and the security and privacy issues derived from the platform utilisation.

First, we have read about the current assessment through psychology scales and the studies performed with them (Bayley, 2006; Roid & Sampers, 2004), which gave us an overview on the basic functionalities of our system: type of toys, activities to be performed, information used in the evaluation of children, etc. In (Forti et al., 2011), there is, for instance, an analysis of the kinematic development in children, which can be used as a base to determine abnormal behaviour during playing activities.

Other studies can be found in (Cools et al., 2009), where they assess movement skills in preschool children, and (Mari, Castiello, Marks, Marraffa, & Prior, 2003), where they specifically focus on the grasping movement in children with autism spectrum disorder (ASD). Then we have studied similar works in the development of toys and other devices for the assessment of children. For instance, in (Marschollek et al., 2012), the authors analyse wearable sensors to gather health information, and in (Focaroli, Taffoni, Parsons, Keller, & Iverson, 2016), they determine the motor sequences for children with risk of ASD. In (Cordella et al., 2016), the authors develop a device to monitor the grasping movements in these children.

Another approach can be found in (Vega-Barbas, Pau, Ferreira, Lebis, & Seoane, 2015), for the development of toys that can aid in the development assessment. These works have been used as a base for our design both in terms of the hardware and software requirements. Additionally, the multidisciplinary nature of our project has allowed us to count on the child development experts both for the initial requirement gathering and for the actual design of the platform.

Child Development Requirements

These requirements are based on the necessities obtained in consultation with experts and the current methodologies for development assessment through playing activities. In our multidisciplinary research group, we count on children assessment experts (psychologists, physiotherapists) which were available to determine their necessities. Moreover, we were able to attend some of the assessing activities and take notes about how they are usually performed. We obtained a signed consent from children's parents to do this.

For instance, the platform must be able to offer additional data from the current playing activities, which are activities such as playing with balls, stack wooden toys, etc. (Bayley, 2006; Roid & Sampers, 2004). These activities are defined in the psychology scales, and therefore, the platform must be designed to allow further analysis of such information. The information usually gathered by psychology scales is related to how good is the child performance in the activity. This means that, for instance, if the activity consists on throwing a ball as far as possible, the information obtained would be related to how far the ball did go and what that means in the development scales.

On the other hand, the system must be easy to use given that the main users will be young children and people who might not be familiar with the use of technologies in this kind of environments.

More specifically, maintaining the current assessment methodology implies that the developed system must be as transparent as possible to the users. For instance, the battery life in the IoT devices must not be an obstacle to finishing the playing activities, and the size, weight, and form of these devices must be as close as possible to the current tools used in the assessment tasks.

Technological and Environmental Requirements

From a technological point of view, there are some limitations and requirements that should be considered before designing the platform. Some of them are related to the requirements derived from the child development assessment. These requirements are important because any change in the tools used for the assessment might invalidate the already tested scales and would not allow to provide reliable data.

For instance, if we develop stackable toys twice the size of those used in the scales, we don't know if the activity results using our toys would be extrapolated to those obtained with the original activity. These requirements are mainly related with the size, weight, colour, and other physical features of the devices. Given that we are developing electronic devices, energy autonomy is also a key requirement, as the toys should be able to be used as much as those without batteries.

Besides, we have identified other technological requirements derived from the available technologies and the current IoT environments. For instance, we should consider the platform flexibility and extensibility, allowing the future integration of new devices, services, and technologies, and the interoperability of this platform with other data sources and health-related services. In our current platform, we do not expect to interoperate with other data sources, but the design must allow obtaining information from, for instance, clinical data from children, to improve the assessment. The energy consumption of the devices, the communications models, and the data storage and analysis mechanisms must also be considered. The interconnection of devices and the availability of information through internet are also critical requirements of the system.

Security and Privacy Requirements

Security and privacy are crucial aspects of any IoT platform, and they are especially important when dealing with children health-related data. This means that the design must enforce the confidentiality of the information shared within it. Moreover, the platform must consider various user roles when accessing data and, therefore, allow granular access control to the information. Not only the data and the internet-based services in the platform must be protected, but the IoT devices themselves, avoiding any unauthorised utilisation or access.

View-Based Design

Following the requirements identified in the study shown in previous sections, we have designed the main high-level aspects of the platform's architecture. Given that some of the most important requirements for this kind of platform are related to interoperability and interconnection of heterogeneous technologies, we have followed a design methodology proposed by the IoT-A project (Iot-a, 2010) which aims to standardise the design process of custom IoT architectures, maintaining a high degree of compatibility across the platforms. The reference architecture proposed by IoT-A is based on the separate design of architecture views (functional, information, physical entities, context, operational and deployment views), and we have followed it to define our architectural design. IoT-A project only describes the methods for the design of the software and hardware system in the IoT environment. Therefore, we have used it exclusively to determine the way we are building our

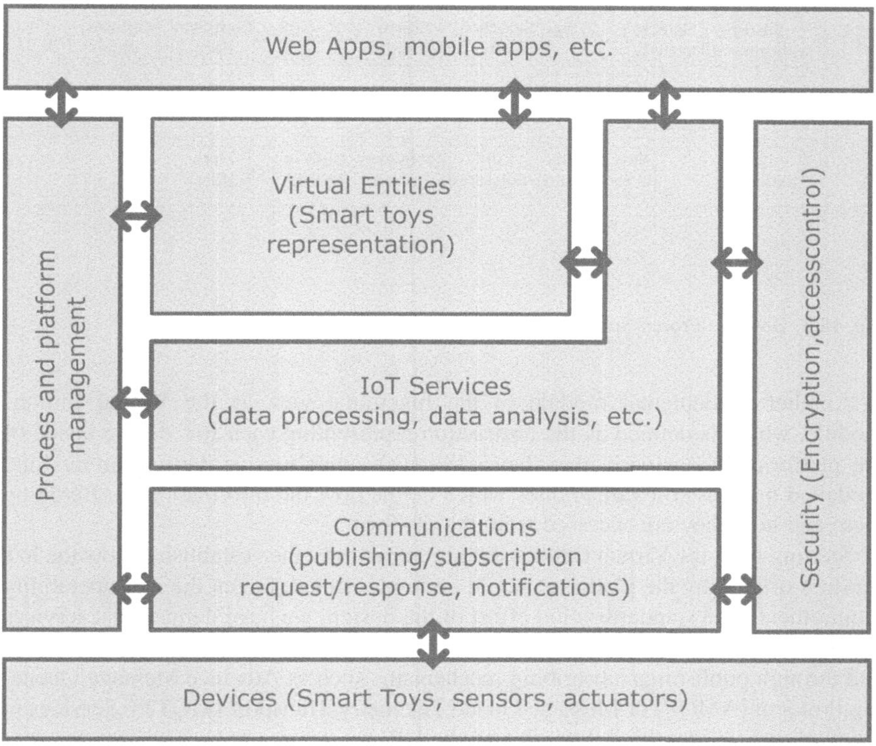

Fig. 18.1 Functional groups of the architecture design, based on the IoT-A reference model (Rivera, 2019)

platform, but not the studies to carry out with it. In the following sections, we explain how we have designed the system according to each architecture view.

From the functional view, the reference architecture proposes the creation of a series of functional modules which group the main functionality of the system (Bauer et al., 2013). Figure 18.1 demonstrates our design for these functional modules, which is essentially a simplified version of the reference model, which considers the specific requirements of our proposal.

The process management module includes the definition of the business processes related to the platform. Using this methodology, we have defined the main playing activities and the other interactions between the platform functionalities. The design has been based on the Business Process Model and Notation (BPMN) (Meyer, Ruppen, & Hilty, 2015), which provides a standardised method for defining this kind of functionalities. In Fig. 18.2, we can see a BPMN diagram showing the interaction between different elements in the platform to complete a playing activity and sensor data gathering process effectively.

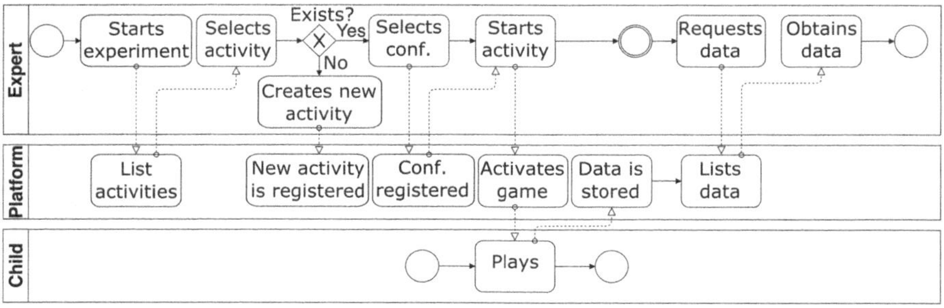

Fig. 18.2 Business Process modelling of activities in the platform (Rivera, 2019)

Another fundamental module of the functional view is the Virtual Entities module, which is defined as the abstraction representing each IoT device inside of the platform. In our case, they have designed smart toys as virtual entities and modelled them as software entities which define how the information is offered by them and how they are accessed from the platform.

Starting from the Virtual Entities definition, we have then established how the IoT services offered by the platform should be implemented. Given the interoperability requirements and standardisation effort of the design, we have defined that services will be available using both through REST (Representational State Transfer) APIs and through publishing/subscribing mechanisms such as Advance Message Queueing Protocol (AMQP) or Message Queue Telemetry Transport (MQTT). Services in the platform are identified through specific URLs.

The security issues have been considered in different ways. HTTPS protocol and data encryption have been enforced in all communications. In cases where no standard communications are used (i.e., when the IoT devices use non-standardised radiofrequency wireless communications) we have designed custom authentication and confidentiality methods, as shown in (Rivera et al., 2019). We have also incorporated into the design the access control methods described in (Cruz-Piris, Rivera, Marsa-Maestre, De La Hoz, & Velasco, 2018).

The information view has allowed us to identify the Virtual Entities models using Relation-entity diagrams such as the one shown in Fig. 18.3, which is the definition of the smart toys and the current specific implementations of each one. In this view, the relationship between the functional modules and the information generated in the devices, from its gathering to its consumption through the appropriate services.

In this view, we also have designed the information lifecycle, which can be shown summarised in Fig. 18.4. All data generated by the IoT devices can be processed and then stored, analysed, composed, and then consumed by the services. Eventually, the stored information will be deleted from the platform.

As for the context view, we have defined the relationships between the platform and the external actors, which can be both users (experts, children, parents, teachers, etc.) or other technological platforms (health-related databases). These relationships are shown in Fig. 18.5.

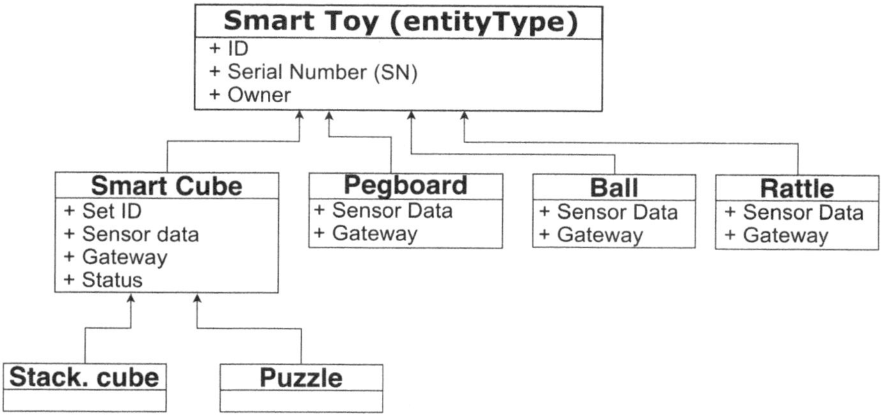

Fig. 18.3 Entity-relation diagram for the design of virtual entities in the platform (Rivera, 2019)

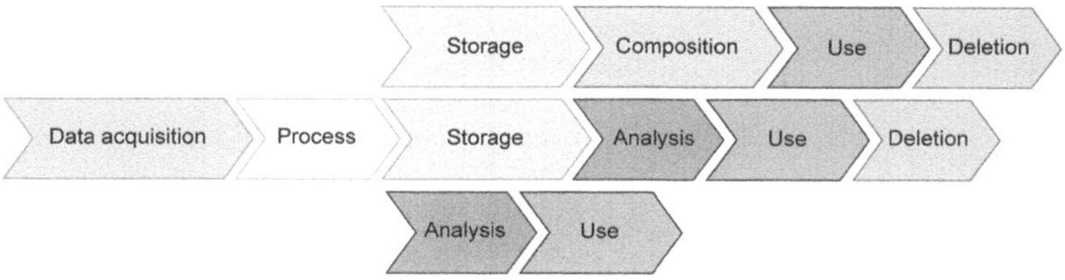

Fig. 18.4 Information lifecycle on the platform. Each line represents a possible flow for the information generated and contained in the system (Rivera, 2019)

Physical Entities

The Physical Entities are related to the actual hardware implementations composing the platform. They are related to the already designed Virtual Entities, but in this case, we have defined the actual hardware devices and their relationships. There are four main physical entities in the platform, as shown in Fig. 18.6. The main physical entity corresponds with the main virtual entity and is the smart toys. Although their physical design can vary from one to another, they all share common design specifications and use the same schemes for access and data retrieval. Their communications are based on a centralised element in each local sensor network, which is called 'collector' and is the second main physical entity in the platform. This entity is designed as a small box which should be placed near the toys and provides certain functionalities and information about the system (for instance, allowing powering up or down the system and providing the current status through a small screen). All the functionalities regarding data communication and collector management are automated and transparent to end-users. This means that the toys can be used as regular

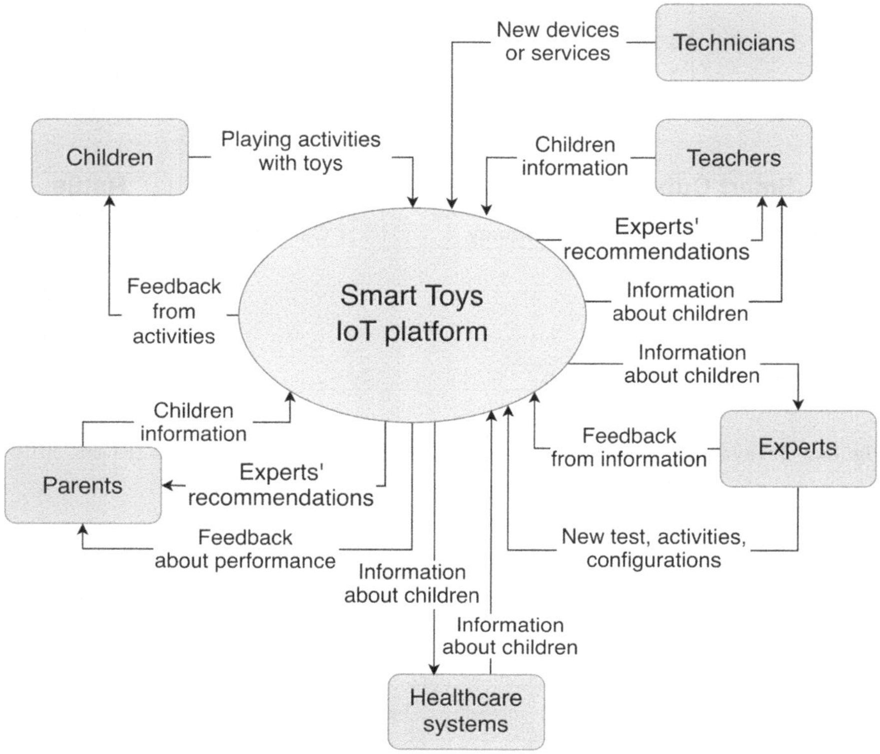

Fig. 18.5 Context view of the platform: Relationships with external actors and systems (Rivera, 2019)

toys and the experts do not need to interact at all with the system, except for turning on and off the devices and start or stop the activities. Once the data is stored, it can be analysed using web-based applications in the internet servers.

The collector can communicate smart toys and the rest of the internet-based platform, that is, it performs a similar function as a gateway in a communication network. The collector making point to point communications with each device on the platform (smart toys, tablets, and servers) and adapt the format of the messages for each of them. This device also guarantees the security of communications by using the encryption mechanisms defined for each information flow. It must be equipped with more resources than the IoT devices and is connected to the internet wirelessly. They are designed to, in absence to reliable internet connections, store securely and temporarily the data obtained from the smart toys.

The activities to perform using smart toys must be monitored and authorised by expert users, just as regular assessing activities. For that, they use a client interface, which will communicate with the toys via the collector. These interfaces, designed to be installed in PCs or mobile devices (smartphones, tablets, etc.), allow the users to

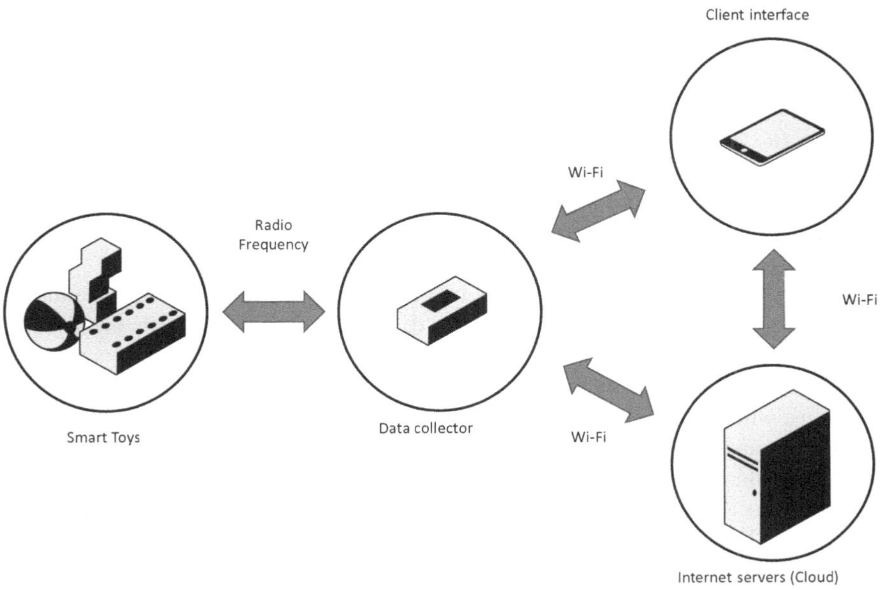

Fig. 18.6 Physical entities of the platform

control the activities and visualise the data. They are designed to be able to connect themselves to the storage and analysis servers to obtain information and display it. The user interfaces are designed to be as simple and easy to use as possible, so they do not become an obstacle to the assessment task.

Finally, the servers are designed to store the data and offer the IoT services (via REST APIs and subscription/publishing methods) to provide useful information to the users. These servers can be accessed through internet using the specific clients or other interfaces using the deployed APIs. Moreover, the servers can execute data analysis and other techniques to obtain useful information from the raw data generated by sensors.

Prototypes and Tests

A prototype version of the platform has been developed and tested. This prototype version is composed of sets of specially designed smart toys, a collector, a mobile-based client interface, and a proof-of-concept server deployed and accessible through the internet.

The smart toys designed for the prototype have been selected from some of the most used toys in the development assessment scales: Sets of stackable cubes and a pegboard. The specification of the most advanced design, the cubes, has been published in (Rivera et al., 2016).

The stackable cubes are based on a wooden set of cubes, which are usually used for the assessment of motor development by asking children to perform certain activities like building a tower or other figures. We have designed 3d-printed cubes with a similar size but including a microcontroller, a battery, and sensors inside. The sensors (a 9-axis accelerometer/gyroscope/magnetometer and various light-dependent resistors (LDRs)), allow us to obtain data about the movements performed with each cube during the activity. Each cube is also equipped with a radio-frequency wireless communication system to receive commands and send data to the collector. In Fig. 18.7, there is a picture of the cube design.

The pegboard, on the other hand, is based on the boards used to measure motor development and manual dexterity (Poole et al., 2005). These boards are usually composed by a series of holes located in one or more rows and a set of pegs that can be inserted into the holes. We have included photo-interrupter sensors in each hole to determine when each peg is inserted or removed, and it is also equipped with a radio frequency transceiver to allow wireless communication with the rest of the platform. These modifications are shown in Fig. 18.8.

These toys are managed through a prototype collector who has been developed in a Raspberry Pi Board. It includes two wireless interfaces to provide Wi-Fi and Radio Frequency wireless connections with the toys and with the rest of the platform. This device can send commands to the toys, receive data from them, store it temporarily, and send it to the platform servers. It is also the element to which the user interfaces are connected. Two user interfaces have also been developed, based on mobile environments (Android) and a Web application.

Prototypes have been subject to tests in experimental and real-life scenarios. These tests have been used to validate the designed devices and platform, and as a base to determine which data we should prioritise for its transformation in useful information and, later, in recommendations, alerts, etc.

For the tests, we have used the sets of smart cubes described earlier, and we have selected a set of variables which can be measured or derived from the sensors embedded in the toys. Among these variables, we have chosen various measurements related to accelerations, movement speed, shakings (in this variable, we aim to measure the doubt or decision of a movement with a cube), time, etc.

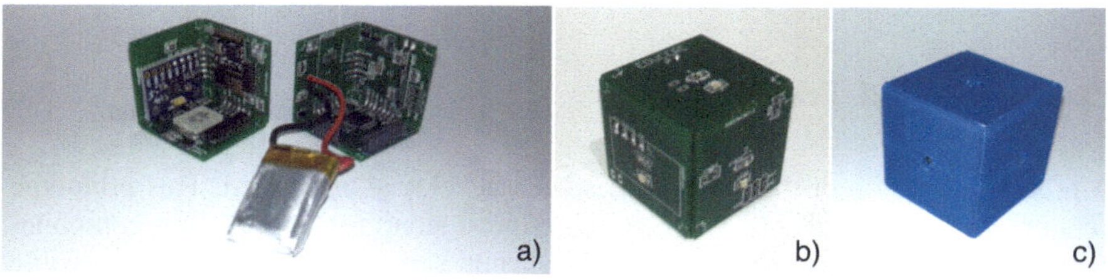

Fig. 18.7 Stackable 'Smart Cube' design: Cube PCB interior and sensors (**a**, **b**), Cube PCB (**b**) and 3D-printed case (**c**)

Fig. 18.8 Internal view of the board showing the sensors, LEDs, microcontrollers, and battery added

More than 60 children from different schools were asked to perform an activity in the presence of a development expert. The selected activity was to build a tower using the smart cubes. Stacking cubes can be a complicated activity for children between the ages of two and three. The difficulty of this play can be increased by reducing the size of the cubes and increasing the height of the tower. This activity was selected because it is an activity already used by experts, and that is part of the assessment scales (Roid & Sampers, 2004). When a 24-month-old child fails to complete a tower of five cubes, different items must be analysed to determine whether he may have a psychomotor or language development problem (e.g., due to not understanding the instructions). The tests were video-recorded, and we studied the relationship between the expert's evaluation through the videos and the data obtained from the sensors. In (Gutiérrez García, Martín-Ruiz, Rivera, Vadillo, & Valero Duboy, 2017) we published a preliminary study of the results obtained in these experiments, which showed a high correlation between experts' scores and the sensor data (around a 96 per cent for the experts' scores taken individually and 98 per cent when comparing the average rating with the sensors' data).

Discussion and Future Work

The platform proposed in this work has been designed to generate a future ecosystem of smart toys and other connected devices which can provide sensor information to an internet-based system. The rationale behind the architectonic design is that the platform should be as extensible and interoperable as possible. With that goal in mind, we have followed specific design reference models, and we have incorporated the functional, context, and information designs that should be considered to accomplish the requirements of the platform. Moreover, we have designed the platform's basic virtual entities, and then we have built physical entities in the form of prototype IoT devices.

The proposed reference framework that includes the elements that make up the platform, communication scheme, the definition of the messages exchanged, and the data structures where the results are stored, makes it easier to design new activities and extend them to new toys in a simple way. These new elements will be added depending on the experts' needs to add new toys or explore new features in activities included in the platform. The validation tests have shown that there exists a high correlation between sensor data and the visual evaluation carried out by experts. This result can be used as a base for the future design of services, which ultimately can provide the experts with useful information about child development. We have already begun to analyse the information provided by the smart toys to automate the detection of the movements which compose the playing activities (Rivera et al., 2018) using artificial intelligence and other related techniques.

Acknowledgments This work has been partially supported by the Spanish Ministry of Science, Innovation, and Universities through the TRASGO project (RTI2018-101962-B-I00 (MCIU/AEI/FEDER, UE)), by the Spanish Ministry of Economy and Competitiveness through the EDUCERE project (TIN2013-47803-C2-1-R and TIN2013-47803-C2-2-R (MINECO/BIE/FEDER, UE), and by the University of Alcalá through the SMARTY project (UAH-GP2018-5).

References

Albinali, F., Goodwin, M. S., & Intille, S. (2012). Detecting stereotypical motor movements in the classroom using accelerometry and pattern recognition algorithms. *Pervasive and Mobile Computing, 8*(1), 103–114.

Alcantud, F., & Alonso, Y. (2016). Predictive value of the Merrill-palmer-R scale applied during the first year of live. *Psicología Educativa, 22*(2), 87–92.

Andreu-Perez, J., Leff, D. R., Ip, H. M., & Yang, G. Z. (2015). From wearable sensors to smart implants–toward pervasive and personalized healthcare. *IEEE Transactions on Biomedical Engineering, 62*(12), 2750–2762.

Bandyopadhyay, D., & Sen, J. (2011). Internet of things: Applications and challenges in technology and standardization. *Wireless Personal Communications, 58*(1), 49–69.

Bauer, M., Boussard, M., Bui, N., De Loof, J., Magerkurth, C., Meissner, S., et al. (2013). IoT reference architecture. In *Enabling things to talk* (pp. 163–211). Berlin, Heidelberg: Springer.

Bayley, N. (2006). *Bayley scales of infant and toddler development* (3rd ed.). London: Pearson.

Cagiltay, K., Kara, N., & Aydin, C. C. (2014). Smart toy based learning. In *Handbook of research on educational communications and technology* (pp. 703–711). New York, NY: Springer.

Chu, G., Apthorpe, N., & Feamster, N. (2018). Security and privacy analyses of internet of things children's toys. *IEEE Internet of Things Journal, 6*(1), 978–985.

Cools, W., De Martelaer, K., Samaey, C., & Andries, C. (2009). Movement skill assessment of typically developing preschool children: A review of seven movement skill assessment tools. *Journal of sports science & medicine, 8*(2), 154.

Cordella, F., Taffoni, F., Raiano, L., Carpino, G., Pantoni, M., Zollo, L., & Formica, D. (2016, August). Design and development of a sensorized cylindrical object for grasping assessment. In *Proceedings of the 38th Annual International Conference of the IEEE Engineering in Medicine and Biology Society (EMBC)* (pp. 3366–3369). IEEE.

Council of the European Union, European Parliament. (2016). *Regulation (EU) N° 2016/679 of the European Parliament and of the Council of 27 April 2016 on the protection of natural persons with regard to the processing of personal data and on the free movement of such data, and repealing Directive 95/46/EC (General Data Protection Regulation).*

Cruz-Piris, L., Rivera, D., Marsa-Maestre, I., De La Hoz, E., & Velasco, J. (2018). Access control mechanism for IoT environments based on modelling communication procedures as resources. *Sensors, 18*(3), 917.

Focaroli, V., Taffoni, F., Parsons, S. M., Keller, F., & Iverson, J. M. (2016). Performance of motor sequences in children at heightened vs. low risk for ASD: A longitudinal study from 18 to 36 months of age. *Frontiers in Psychology, 7*, 724.

Forti, S., Valli, A., Perego, P., Nobile, M., Crippa, A., & Molteni, M. (2011). Motor planning and control in autism. A kinematic analysis of preschool children. *Research in Autism Spectrum Disorders, 5*(2), 834–842.

Goodman, R., Ford, T., Richards, H., Gatward, R., & Meltzer, H. (2000). The development and Well-being assessment: Description and initial validation of an integrated assessment of child and adolescent psychopathology. *Journal of Child Psychology and Psychiatry, 41*(5), 645–655.

Gutiérrez García, M. A., Martín-Ruiz, M. L., Rivera, D., Vadillo, L., & Valero Duboy, M. A. (2017). A smart toy to enhance the decision-making process at children's psychomotor delay screenings: A pilot study. *Journal of Medical Internet Research, 19*(5), e171.

Holloway, D., & Green, L. (2016). The internet of toys. *Communication Research and Practice, 2* (4), 506–519.

Iot-A-Internet of Things Architecture fp7 project (2010). Online. Accessed from https://cordis.europa.eu/project/rcn/95713/factsheet/es

Kovatsch, M., Mayer, S., & Ostermaier, B. (2012, July). Moving application logic from the firmware to the cloud: Towards the thin server architecture for the internet of things. In *Proceedings of the 2012 Sixth International Conference on Innovative Mobile and Internet Services in Ubiquitous Computing* (pp. 751–756). IEEE.

Kulkarni, A., & Sathe, S. (2014). Healthcare applications of the internet of things: A review. *International Journal of Computer Science and Information Technologies, 5*(5), 6229–6232.

Lin, J., Yu, W., Zhang, N., Yang, X., Zhang, H., & Zhao, W. (2017). A survey on internet of things: Architecture, enabling technologies, security and privacy, and applications. *IEEE Internet of Things Journal, 4*(5), 1125–1142.

Magerkurth, C., Cheok, A. D., Mandryk, R. L., & Nilsen, T. (2005). Pervasive games: Bringing computer entertainment back to the real world. *Computers in Entertainment (CIE), 3*(3), 4–4.

Mari, M., Castiello, U., Marks, D., Marraffa, C., & Prior, M. (2003). The reach–to–grasp movement in children with autism spectrum disorder. Philosophical transactions of the Royal Society of London. *Series B: Biological Sciences, 358*(1430), 393–403.

Marschollek, M., Gietzelt, M., Schulze, M., Kohlmann, M., Song, B., & Wolf, K. H. (2012). Wearable sensors in healthcare and sensor-enhanced health information systems: All our tomorrows? *Healthcare Informatics Research, 18*(2), 97–104.

McReynolds, E., Hubbard, S., Lau, T., Saraf, A., Cakmak, M., & Roesner, F. (2017, May). Toys that listen: A study of parents, children, and internet-connected toys. In *Proceedings of the 2017 CHI Conference on Human Factors in Computing Systems* (pp. 5197–5207). ACM.

Meyer, S., Ruppen, A., & Hilty, L. (2015, June). The things of the internet of things in BPMN. In *International conference on advanced information systems engineering* (pp. 285–297). Springer, Cham.

Mironcika, S., de Schipper, A., Brons, A., Toussaint, H., Kröse, B., & Schouten, B. (2018). Smart toys design opportunities for measuring children's fine motor skills development. In

Proceedings of the Twelfth International Conference on Tangible, Embedded, and Embodied Interaction (pp. 349–356). ACM.
Miroshnikov, S. A., Nasledov, A. D., & Zashchirinskaia, O. V. (2016). Criterion validation of the scale of psychomotor development (SPMD) in developmental delay study of pre-school children. *Procedia-Social and Behavioral Sciences, 233*, 220–224.
Mohammed, J., Lung, C. H., Ocneanu, A., Thakral, A., Jones, C., & Adler, A. (2014, September). Internet of Things: Remote patient monitoring using web services and cloud computing. In Proceedings of the *2014 IEEE International Conference on Internet of Things (iThings), and IEEE Green Computing and Communications (GreenCom) and IEEE Cyber, Physical and Social Computing (CPSCom)* (pp. 256–263). IEEE.
Nguyen, H. H., Mirza, F., Naeem, M. A., & Nguyen, M. (2017, April). A review on iot healthcare monitoring applications and a vision for transforming sensor data into real-time clinical feedback. In *Proceedings of the 2017 IEEE 21st International Conference on Computer Supported Cooperative Work in Design (CSCWD)* (pp. 257–262). IEEE.
Pallavi, S., & Smruti, R. S. (2017). Internet of things: Architectures protocols and applications. *Journal of Electrical and Computer Engineering, 2017*, 1–0147.
Poole, J. L., Burtner, P. A., Torres, T. A., McMullen, C. K., Markham, A., Marcum, M. L., et al. (2005). Measuring dexterity in children using the nine-hole peg test. *Journal of Hand Therapy, 18*(3), 348–351.
Rivera, D. (2019). *Contributions to the data acquisition technologies and the methods for automatic detection and classification of activities in a smart toys environment* (Doctoral Disertation). Accessed from https://ebuah.uah.es/dspace/handle/10017/42706
Rivera, D., Cruz-Piris, L., Fernandez, S., Alarcos, B., García, A., & Velasco, J. R. (2018). A novel method for automatic detection and classification of movement patterns in short duration playing activities. *IEEE Access, 6*, 53409–53425.
Rivera, D., García, A., Alarcos, B., Velasco, J., Ortega, J., & Martínez-Yelmo, I. (2016). Smart toys designed for detecting developmental delays. *Sensors, 16*(11), 1953.
Rivera, D., García, A., Martín-Ruíz, M. L., Alarcos, B., Velasco, J. R., & Oliva, A. G. (2019). Secure communications and protected data for a internet of things smart toy platform. *IEEE Internet of Things Journal, 6*(2), 3785–3795.
Roid, G. H., & Sampers, J. L. (2004). Merrill-palmer-revised: Scales of development.. Stoelting.
Samie, F., Bauer, L., & Henkel, J. (2016). IoT technologies for embedded computing: A survey. In *Proceedings of the Eleventh IEEE/ACM/IFIP International Conference on Hardware/Software Codesign and System Synthesis* (p. 8). ACM.
Taffoni, F., Focaroli, V., Keller, F., & Iverson, J. M. (2014). A technological approach to studying motor planning ability in children at high risk for ASD. In *Proceedings of the 2014 36th Annual International Conference of the IEEE Engineering in Medicine and Biology Society* (pp. 3638–3641). IEEE.
Vega-Barbas, M., Pau, I., Ferreira, J., Lebis, E., & Seoane, F. (2015). Utilizing smart textiles-enabled sensorized toy and playful interactions for assessment of psychomotor development on children. *Journal of Sensors*, 2015.
Yin, H., & Jha, N. K. (2017). A health decision support system for disease diagnosis based on wearable medical sensors and machine learning ensembles. *IEEE Transactions on Multi-Scale Computing Systems, 3*(4), 228–241.

Diego Rivera is an Assistant Professor at Universidad Politécnica de Madrid. He was awarded a Bachelor Degree in Computer Science, Master's Degree and his PhD. Degree (2019) from the Universidad de Alcalá His research interests include Internet of Things (IoT) architectures, artificial intelligence, and computer networks.

María Luisa Martín-Ruiz is an Associate Professor at Universidad Politécnica de Madrid. She received the Laurea (Graduate Diploma) Degree in Informatics Engineering from Universidad Carlos III (Spain), and her PhD Degree from the Universidad Politécnica de Madrid (2014). Her research interests include intelligent systems for e-health.

Luis Cruz-Piris is an Assistant Professor at Universidad de Alcalá (Spain). He was awarded a Degree in Telematics Engineering, the MS Degree in Telecommunications Engineering and his PhD was from the Universidad de Alcalá (Spain, 2019). His research interests include computer science intelligent agents, Internet of Things (IoT) architectures, and intelligent transportation systems.

Kevin van der Meulen is an Associate Professor at Universidad Autónoma de Madrid. He received the MS Degree in Psychology from the Radboud University in Nijmegen (Netherlands), and the PhD (2003) Degree from the Universidad Autónoma de Madrid (Spain). His research interests are in the field of psychological development in childhood.

Antonio García is an Associate Professor in Universidad de Alcalá. He received a Degree in Statistical and Operative Research from the Universidad Complutense de Madrid (Spain) in 1993, and Degrees in Telematics and Electronic Engineering from the Universidad de Alcalá (Spain) in 1997 and 2002. His research interests include Internet of Things (IoT) and networks.

Cristina Serrano Garcia is an Associate Professor at Cardenal Cisneros University. She received a Degree in Speech Therapy, an MS Degree in Neuropsychology and a PhD Degree in Psychopedagogy. Her research interests include topics related to attention to diversity and classroom environment in the educational field.

Susel Fernández is an Assistant Professor at Universidad de Alcalá. She received the MS Degree in Computer Science from the Universidad de Oriente (Cuba) in 2003 and the PhD Degree from the Universidad de Alcalá (Spain), in 2013. Her main research interests are artificial intelligence, multiagent, and knowledge-based systems.

Bernardo Alarcos is an Associate Professor at Universidad de Alcalá. He was awarded Degrees in Electronic Equipment and in Telecommunications from the Universidad Politécnica de Madrid (Spain), and his PhD degree from the Universidad de Alcalá (Spain) in 2004. His research interests include Internet of Things (IoT) and cybersecurity.

Juan R. Velasco is a Professor and Vice-President for Strategy and Planning at Universidad de Alcalá. He received the MS Degree in Telecommunications Engineering and the PhD Degree from the Universidad Politécnica de Madrid (Spain) in 1987 and 1991. His research interests include service personalization, Internet of Things (IoT), and ontology mapping.

Chapter 19
Young Children Learning to Code: A Digital Technologies Framework for the Early Years

Karen Murcia

Introduction

Young children today are growing up in a digital era. The evolving changes made by digital technologies to childhood and children's life experiences are undeniable and enduring. Children, like their parents and families, engage with digital technologies for different purposes; such as entertainment, learning, creating and communicating. Digital technologies are different from other types of technology as they incorporate microprocessors that convert information into numbers or digits. These electronic devices and systems can generate, store and process information. People use many different types of digital technologies, such as computers, iPads, smart televisions, smartphones, and smart watches. Digital networks enable people to access, create, store or communicate information daily. These digital devices are now almost seamlessly integrated into life. For example, the National Association for the Education of Young Children (NAEYC) and Fred Rogers Center position statement on *Technology and Interactive Media as Tools in Early Childhood Programs Serving Children from Birth through Age 8* (2012), states that "technology is here to stay and in essence that these electronic tools have already changed the lives of many in how we communicate with and reach others, how we teach, and how professional development is delivered" (p. 18). Digital experiences are embedded into daily life, and as such, we are seeing young children both exposed to and actively engaging with digital technologies. Consequently, young children require guidance and support in their use of such technologies, to ensure they develop important foundation digital literacy skills and positive and productive usage behaviours.

K. Murcia (✉)
Curtin University, Perth, WA, Australia
e-mail: Karen.murcia@curtin.edu.au

D. Holloway et al. (eds.), *Young Children's Rights in a Digital World*, Children's Well-Being: Indicators and Research 23,
https://doi.org/10.1007/978-3-030-65916-5_19

Educators are increasingly faced with both challenges and opportunities as to how the affordances of digital technologies can be integrated meaningfully and safely into early years learning environments. Children require access to age appropriate digital technologies in order to develop critical twenty first century learning skills and capabilities. The NAEYC (2012) stated,

> there has been much debate in the early childhood education field with regard to the use of technology in the curriculum, but there is now general consensus that children need some access in order to become familiar with a range of hardware and software. (p. 3)

However, a significant range of challenges must be addressed to enable educators to effectively and safely integrate digital technologies into the early years learning environment. Researchers Marsh, Kontovourki, Tafa and Salomaa's (2017) review of digital technologies in childhood, identified barriers to the use of technology in early years settings. These included educator beliefs and attitudes, lack of knowledge and skills, lack of equipment/resources, lack of training, classroom condition constraints, educators' lack of confidence, lack of appropriate educational software, lack of support, IT technical problems, lack of funding, lack of time, EC curriculum and guidelines and the nature of the EC sector itself. The Early Childhood Australia's (ECA, 2018) *Statement on young children and digital technologies,* produced by researchers Edwards, Straker and Oakey, was developed in response to such barriers and the identified need of early childhood educators and professionals for guidance. The guidelines offer support to decision-making about the role and optimal use of digital technologies in early childhood education and care. Researchers Zabatiero, Straker, Mantilla, Edwards, and Danby (2018) highlight that early year's education practices with digital technologies should be based on evidence and aim to provide learning experiences that promote young children's cognitive, and socio-emotional development.

The purpose of the current study was to document, categorise and generalise early childhood educators' practice and observations of young children's developing digital literacy, as they playfully learned with tangible coding technologies or 'robots'. In this chapter, we begin by reviewing the literature to establish our conceptual framework, which included types of digital technologies, play-based learning with digital technologies and educators' pedagogy. The action research approach used in the study is detailed and the grounded analysis of the data is explained. We draw from this empirical evidence and the literature to propose the early childhood DigiTech Framework that was a significant outcome of the research.

Making Choices: Touchscreens and Tangible Coding Technologies (TCTs)

Touchscreen devices such as iPads/tablets and smartphones are among the most commonly used digital technologies by children. They are watching their parents and other significant adults' use the devices and using the devices themselves for different purposes. Marsh, Plowman, Yamada-Rice, Bishop and Scott (2016) described young children's use of touchscreen devices as playful, collaborative and interactive. The activities they described ranged from viewing digital media to creating new images through digital painting, photography and video. They also documented digital devices being a connector that enabled relationships through the sharing of digital media. It is evident that not all screen time is passive for children (Arnott, 2016). There is potential for active and creative use of digital technologies and young children themselves are reported to use terms such as make, build, paint and learn when describing how they use digital touchscreen devices (Mertala, 2016).

We know young children are also using computers, printers, online games and toys, game consoles, digital toys, coding toys, robots, and voice-activated technologies. The active positioning of children with these technologies suggests that digital play in early childhood can provide opportunities for exploration and experimentation (ECA, 2018). Positioned in this way, the question should be, what are children actively doing with digital technologies rather than simply, how much time with a digital device is appropriate for young children? For example, tangible coding technologies (TCTs) or robots are a hands-on digital technology experience which is recognised as "providing young children with opportunities to learn to code in a playful situation, while supporting collaborative learning and social interaction" (Murcia, Campbell, & Aranda, 2018, p. 246). Children can touch and physically manipulate a TCT as they program it for action. These types of devices have tangible controllers such as coding blocks and mechanical push buttons so a robot can be programmed without a computer screen.

There are TCTs on the education resource market, which are finding their way into early years environments. For example, *Cubetto* is a tactile wooden box on wheels with a separate control board that is driven by a tangible coding language (see https://www.primotoys.com). Children place a sequence of colourful shaped pieces onto the control board in patterns that command *Cubetto* to move and explore environments represented by the accompanying range of floor grid maps. As children play with *Cubetto*, they are using and developing computational thinking skills such as breaking a task down into logical steps to reach a destination, then recognising cause and effect in coded patterns and sequences in relation to the device's movements. The concrete and enduring presence of the coded sequence of coloured shapes on the control board allows children to try out sequences and then correct or de-bug if errors occur. Another example of TCT historically used more widely in early years education is *Bee-Bot*. This device also provokes and develops young children's computational thinking and problem-solving skills as they play with the colourful bee-like character (see https://www.terrapinlogo.com). This device has a mechanical push button coding interphase on the top of the *Bee-Bot*. Directional arrows are pushed in sequence to program movement and a pathway forward or backwards.

Children are drawing on a range of semiotic resources as they playfully learn and develop foundation coding capabilities with these TCT's. Language is not limited to words and it includes a range of representational modes such as mathematical symbols, images, gestures and physical objects (Kress & van Leeuwen, 2001).

Multimodal representations of direction, locations and actions are evident in the design of both the *Cubetto* and *Bee-Bot* coding devices and these are crucial to the way young children learn and make meaning (Murcia, 2018). Bruner (1990) argued that learning often occurs through three different stages of thinking with the use of representations. These stages are enactive (concrete, hands-on, materials based), iconic (representing images, modelling, 'mind's eye') and moving to increasing abstraction with symbolic representations (words, symbols, social conventions). Understanding the design features of a digital device and the opportunities for active hands-on play coupled with the multi-modal affordances offered by the coding interphase should assist early childhood educators in making informed choices about the types of digital technologies introduced to children's play-based learning environment.

Play Based Learning with Digital Technologies

When age appropriate digital technologies are intentionally introduced by an educator into an early years learning environment, children will have the opportunity to learn and develop digital literacies in the same way they learn in other curriculum areas such as language, science and mathematics (ECA, 2018). Play-based learning with digital devices such as coded 'robots' can be child initiated when they are made available and accessible to the children. For example, empirical research conducted in the U.S. and Australia documented young children's self-initiated play with a tangible coding device or 'robot' and then subsequently, with the guidance of educators, the device became a tool that provoked inquiry and enabled exploratory play (Berson, Murcia, Berson, Damijanovic, & McSporran, 2019). Research led out of the U.K. also identified children's exploratory play with digital technologies and how social interactions with co-playing peers and adults helped build children's knowledge and technical expertise (Marsh, Hannon, Lewis, & Ritchie, 2017). In addition, Australian research has identified and documented how play-based learning opportunities with digital coding devices were extended as educators guided the play, asked open questions and held intentional learning conversations with the children (Murcia, 2018). These examples of international research provide empirical evidence of young children being introduced to important digital literacies through play.

Play is well-recognised and promoted as a critical component in quality early childhood education. We know that discovery and imaginative play generates opportunities for young children to explore, problem solve and create while engaging with others (Pyle, DeLuca, & Danniels, 2017). It is well recognised that young children are curious and capable learners who develop cognitive and emotional capabilities in positive social environments (DEEWR, 2009). Playing with others provides social engagement for young children and opportunities for building collaboration and communication skills. Social engagement interconnected with playful learning not only builds cognitive capabilities but also fosters strong relationships between children and with significant adults.

The Western Australian Department of Education (WADoE, 2018) actively supports and promotes the significance of play and has released guidelines titled "Importance of play-based learning" (2018). The WADoE has identified and defined three types of play-based learning in these guidelines. These are:

1. Child-initiated: Freely chosen by the child with little direct adult involvement or interaction. It is spontaneous in nature.
2. Guided: Initiated by the child or adult, with adults joining in to extend learning through questioning and demonstrating.
3. Adult-led: Organised and directed by an adult and may include instructions but remains open ended. Intentions are clear, specific and promote high level thinking skills.

When educators guide or lead children's play, they make pedagogical decisions based on children's interests, current capabilities and developmental needs. The important role of the educator in children's play has been recognised and promoted in the WADoE (2018) guidelines. They state, "play-based learning involves careful decisions and support from teachers, so your child grows to accept responsibility for their learning while receiving purposeful guidance and feedback. It promotes important opportunities for your child to understand how to learn, develop critical thinking skills, adapt to change, and work independently as well as with others" (p. 3). In this the digital era, educators are now making pedagogical decisions about the place and nature of children's play with digital technologies.

Current research findings, coupled with the WADoE (2018) play-based learning guidelines and the ECA statement on young children and digital technologies could help address and overcome some reported barriers in early childhood education by developing new ideas about children's digital play that helps educators recognise children's activities with technologies in a play-based way (Bers, 2012). The integral nature of pedagogy to the successful implementation of digital technologies into early childhood education has been consistently highlighted. For example, Early Childhood Australia's Guidelines for Digital Technologies (2018) states "play and pedagogy involve children using a range of digital devices for exploration, meaning-making, collaboration and problem-solving" (p. 18). Educators make informed decisions based on the context, learning needs and interests of the children about the use (or non-use) of digital technologies for learning (Bird & Edwards, 2015). In this decision-making process, educators should be critiquing the design elements and interfaces of digital technologies and considering their suitability to the developmental needs of young children.

International research has highlighted that many early years educators lack the confidence, technical knowledge and appropriate integration pedagogies required to effectively introduce digital technologies into children's learning environment. It is apparent that in some situations, educators may not have the digital or ICT skills needed to provoke, guide and lead young children's play with tangible coding technologies (Murcia et al., 2018). Yet we understand that educators' technological pedagogical content knowledge is key to the quality of learning experiences with digital technologies for young children (Bers, Seddighin, & Sullivan, 2013). These concerns identified in the research literature and observed in practice were a major driver for the research reported in this chapter. Specifically, we questioned and sought to understand; how do early childhood educators see and understand learning opportunities for young children as they play with tangible coding technologies?

Our Action Research Project

We use action research methods and grounded theory to explore educators' integration of tangible coding technologies (*Cubetto* & *Bee-Bots*) into an early childhood learning environment. This project was a collaboration involving researchers and educators from a University's Early Childhood Centre located on their metropolitan campus. The Early Years Centre provided long daycare and education services to the children of University staff and students. There were four educators participating, who worked as a pair in each of the Centre's two kindergarten rooms. For the purpose of the research, the educators selected two focus groups of children (ages three & four) from their kindergarten program, based on parents return of a signed ethics consent form and children's interest and engagement with the TCT's during the 6 months research period.

Action research methods informed the design and protocols for working with the educators and positioned them as practitioner researchers. Action research was established by Kurt Lewin (1946) as a term describing the integration of action, implementing a plan, with research that is developing an understanding of the effectiveness of this implementation (Murcia, 2005). This approach valued the classroom expertise of the educators as they partnered with the researchers in understanding the impact of TCT's and pedagogy on children's learning and development. The educators' planning, acting, observing and critically reflecting on children's learning, as elaborated by Murcia and Tang (2019) occurred in two cycles of action research. Importantly, learnings from each cycle informed the planning of the next.

The TCTs, *Cubetto* and *BeeBot* were introduced alternatively across kindergarten rooms and used as a vehicle for the children's exploration of the world around them. In the first cycle of action, educators planned and presented intentional experiences with the tangible coding technology to spark play-based learning experiences. In the second cycle, educators swapped the TCT between the kindergarten rooms. Critical reflection on the outcomes of cycle one informed the educators' second cycle planning. In both cycles, the educators were identifying and designing inquiry opportunities for children to develop coding and computational thinking skills while playing with TCT's. When designing and guiding learning experiences educators made pedagogical decisions aiming to ensure activities were developmentally appropriate and meeting the play and learning needs of the children. An assertion of the research team was that TCTs should sit integrated across the learning areas and

Table 19.1 Questions guiding critical reflection

Stage of critical reflection	Guiding questions
	What did you observe in your classroom as a result of your implementation? What did the educator do? What did the children do? What were the children saying?
So what (I think)	What questions come to mind during the implementation? Why did this happen? Was this what was expected? Have you observed this before?
Now what (I wonder)	What did you learn from the implementation? What worked and what didn't? What would you do differently next time? What will you plan for the child's learning going forward?

create provocation for children's inquiry, development of computational thinking and potentially creativity. The educators based their planning on the children's interests and the curriculum design emerged from their play.

Critical reflective practice was used throughout the action research cycles. The model of reflection used was based on the Harvard Visible Thinking strategy; I see, I think, I wonder (Ritchhart, Church, & Morrison, 2011). This model underpinned the researcher's field notes and the educators' digital research journal. It also provided a structure and expected depth in the educators' critical reflection during team research meetings where they debriefed their actions, observations and learnings. Importantly, these collegial meetings provided time for the team to share practice and to build common language and understanding about children's digital literacy. Digital photographs were also taken by the educators of the children playing and learning with the tangible coding technologies. These photos were deidentified and used as further evidence to inform the observational notes and checklists made by educators during the action research cycles. A semi-structured de-briefing interview with the educators was also conducted at the conclusion of the project.

In addition, educators wrote reflective learning stories about the children's engagement with the TCTs. Learning stories were a normal part of practice in this Centre and required educators to focus on an individual child's experience. In the learning story, educators described 'what' happened in the learning experience, followed by 'so what' was significant in this observation, and then 'now what' in terms of how learning from the experience could be used to inform activities going forward. This was framed to align with the Harvard visible thinking strategy, and as shown in Table 19.1, questions were used to guide the educators' reflection and writing.

Grounded Theory Analysis and Emerging Themes

Grounded analysis was used to identify emerging issues and key themes in the data collected. Initially, critical episodes were identified in the data sets and these were interrogated in detail for evidence of computational thinking, creativity, technology skills and children's social and emotional capabilities. This grounded approach, as described by Glaser and Strauss (1967) allowed us to generate new understanding of young children's digital literacy by examining the empirical data collected first rather than imposing a pre-determined curriculum focus.

Multiple illustrative learning stories were generated by the participating educators. In brief and as an example, children's play with the code-able robot Cubetto led to an inquiry project where children explored outer space and the planets. Children debated if Cubetto could live on the various planets and created their own exploration map and stories about the robots adventures. By including the robot into the children's play experience, they were problem solving, collaborating and communicating ideas (Berson et al., 2019). The children were observed developing and using foundation skills such as matching, sequencing and decomposing a problem situation. These capabilities were integral to children's computational thinking and evident in their coding of Cubetto. The tangible coding interface of Cubetto also scaffolded children's learning of symbolic representations and contributed to the development of multimodal digital literacies (Murcia & Tang, 2019).

Children also displayed computational thinking, mathematical number sense and reasoning while learning to code with the *BeeBot* robot, which they named 'Willbee' after a character in their favourite storybook. As the play with Willbee evolved, the children were being increasingly creative with new stories, coded sequences and constructed environments. It was evident in these learning episodes that the children were following the creative planning of the educator and also trying out creative ideas of their own (Murcia, Pepper, & Joubert, 2020).

The following capability categories were being demonstrated by focus children in their play with the tangible coding technologies.

- Sequencing, identifying patterns, number awareness and computational thinking.
- Counting with numbers, using one to one correspondence (pointing).
- Predicting and hypothesising.
- Location and orientation, using directional language and non-standard units of measurement.
- Children showing a sense of ownership about where the play would lead and their investigations.
- Technical knowledge about digital hardware.
- Social and emotional capabilities, including creating solutions, collaborating and communicating.

It became evident through field observations and debriefing activities carried out in each action research cycle that a scaffold would assist educators to 'see' and support children's developing digital literacies. In response, an observational checklist was created for use by the educators while working with children playfully learning with digital technologies. To produce the checklist, the identified capability categories were integrated with relevant aspects of existing documents including the Australian Curriculum: Digital Technologies—pre-primary to year 10 (SCSA, 2018) and guidelines for early childhood education such as the Early Years Learning Framework (DEEWR, 2009), National Quality Framework (ACECQA, 2017) and Western Australian Kindergarten Guidelines (SCSA, 2016). The resulting checklist was used by the educators to monitor the children's development of digital literacy capabilities.

The educators' use of the checklist evolved and it was used to inform planning. Intentional planning and guided play opportunities were a response to the children's interests and driven by play-based learning with the tangible coding technologies. Over time, educators were increasingly aware of young children's ability to develop the fundamentals of computer science such as patterns, sequences and symbolic representations. The concepts and skills initially included in the checklist were revised both through use and critical reflection amongst the research team. In this process, some concepts and skills were removed or integrated into other areas or categories. The revised checklist was further refined and then generalised as the theoretical framework.

An Early Years DigiTech Framework: Coding and Creating

An outcome of this research was an innovative early year's digital technologies (DigiTech) framework, which aims to assist early childhood educators to see and respond to learning opportunities for young children as they playfully learn to code with digital technologies. This DigiTech framework was based on the empirical evidence obtained through the research and drew on concepts and processes such as number sense, algorithms (making steps and rules), mathematical reasoning (patterns and generalisations), data and digital knowledge and creating solutions (Fig. 19.1).

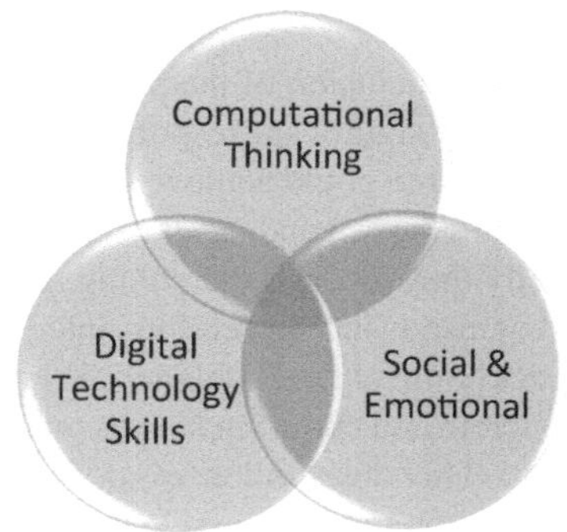

Fig. 19.1 Early Years DigiTech framework: coding and creating

Table 19.2 Digital literacy domains and children's capabilities

Digital literacy domain	Capabilities
Computational thinking: Number sense	Uses the words more and less appropriately Anticipates a change in quantity as bigger than, smaller than or same Distinguishes spoken numbers from other spoken words Distinguishes numerals from other written symbols Says numbers once and in the conventional order Solves small number stories (e.g. one more; $2 + 1 = 3$)
Computational thinking: Mathematical reasoning	Patterns & sequences; spotting and using similarities Decomposition; breaking a problem down into its parts Removing; taking away unnecessary detail Predicting; cause and effect Evaluating; what worked and why?
Digital technology skills	Awareness of digital software and hardware Operates simple digital hardware Represents 'data' with pictures and symbols Uses and create a 'data' sequence to complete a simple task or problem
Social and emotional capabilities	Creativity Reflective thinking Reasoning and decision-making Collaboration Perseverance Self-motivated and directed activity (agency)

The Early Years DigiTech Framework positions digital literacy as a multi-faceted construct that sits at the intersection of young children's computational thinking (number sense and mathematical reasoning), digital technology skills and social and emotional capabilities. To elaborate, based on the empirical evidence obtained during the action research project, the three domains of young children's digital literacy are summarised in Table 19.2.

Discussion and Conclusion

At the conclusion of the project, when de-briefing and reflecting with the educators, the most significant response from the team was, how surprised they were by what the children could achieve with a tangible coding technology. For example, "It's just amazing how 3 and 4-year-olds can actually use these pieces of technology and how easily they grasped the coding. It was amazing to see their progress". The educators recognised that the coding interphase was an important factor impacting on how quickly the children learned. When comparing the technologies, they noted that the concrete and enduring shapes on the control board of *Cubetto* assisted the children to code. This observation would be consistent with Bruner's (1990) stages of learning and moving from concrete hands-on representations (enactive) before moving to the more abstract (symbolic) representations used in coding. One of the educators commented that "The children learn differently. The visual learners were helped with the colours, with Cubetto's red for right and yellow for left. By using different strategies, the children picked it up easier".

It was also evident that the collegial nature of the action research approach and protocols for critical reflection enhanced both the educators' confidence and understanding of how children code and develop digital literacy. An educator explained, "we do reflect on the curriculum, but we are analysing each individual's experience, we're doing it in a lot more depth, so you are thinking about it a lot more". A significant feature of the project identified by the team of educators, that impacted on their confidence and ability to plan and guide children's learning, was the digital literacy checklist. For example,

> you're thinking about these things in the checklist when you're guiding the children, or observing them, and then you're looking back at them and seeing how their actions might relate to it, or how you could plan for the skills that might come next.

To conclude, early childhood educators were empowered through this action research project to both build their own digital literacy and to guide children's learning and development. The value in the project and the importance of the DigiTech framework is captured in this final educator quote, "when you're learning, you're challenging yourself; it's positive because you learn as an educator and think about things you wouldn't normally see in children's learning."

References

Arnott, L. (2016). An ecological exploration of young children's digital play: Framing children's social experiences with technologies in early childhood. *Early Years, 36*(3), 271–288.

Australian Children's Education and Care Quality Authority (ACECQA). (2017). *Guide to the National Quality Framework* (February 2018). Sydney, NSW: ACECQA. Retrieved from www.acecqa.gov.au/sites/default/files/2018-03/Guide-to-the-NQF_0.pdf

Bers, M. (2012). *Designing digital experiences for positive youth development: From playpen to playground.* New York, NY: Oxford University Press.

Bers, M., Seddighin, S., & Sullivan, A. (2013). Ready for robotics: Bringing together the T and E of STEM in early childhood teacher education. *Journal of Technology and Teacher Education, 21* (3), 355–377.

Berson, I., Murcia, K., Berson, M., Damijanovic, V., & McSporran, V. (2019). Tangible digital play in Australian and U.S. Preschool. *Kappa Delta Pi Record, 55*(2), 78–84.

Bird, J., & Edwards, S. (2015). Children learning to use technologies through play: A digital play framework. *British Journal of Educational Technology, 46*(6), 1149–1160.

Bruner, J. (1990). *Acts of meaning.* Cambridge, MA: Harvard University Press.

Department of Education, Employment and Workplace Relations (DEEWR). (2009). *Belonging, being & becoming: The early years learning framework for Australia.* Canberra, ACT: Commonwealth of Australia.

Early Childhood Australia (ECA). (2018). *Statement on young children and digital technologies.* Canberra, ACT: ECA. https://doi.org/10.23965/ECA.001.

Glaser, B., & Strauss, A. (1967). *The discovery of grounded theory: Strategies for qualitative research.* Chicago: Aldine Publishing Company.

Government of Western Australia School Curriculum and Standards Authority. (2016). *Kindergarten curriculum guidelines.* Retrieved from https://k10outline.scsa.wa.edu.au/media/documents/outline_downloads/Western-Australian-Kindergarten-Curriculum-Guidelines-pdf.pdf

Government of Western Australia School Curriculum and Standards Authority. (2018). Western Australian curriculum: Digital technologies—pre-primary to year 10. Retrieved from https://k10outline.scsa.wa.edu.au/__data/assets/pdf_file/0005/364550/Digital-Technologies-Curriculum-Pre-primary-to-Year-10.PDF

Kress, G., & van Leeuwen, T. (2001). *Multimodal discourse: The modes and media of contemporary communication*. London: Arnold.

Lewin, K. (1946). Action research and minority problems. *Journal of Social Issues, 2*(4), 34–46.

Marsh, J., Hannon, P., Lewis, M., & Ritchie, L. (2017). Young children's initiation into family literacy practices in the digital age. *Journal of Early Childhood Research, 15*(1), 47–60.

Marsh, J., Kontovourki, S., Tafa, E., & Salomaa, S. (2017). *Developing digital literacy in early years settings: Professional development needs for practitioners*. A White Paper for COST Action IS1410. http://digilitey.eu

Marsh, J., Plowman, L., Yamada-Rice, D., Bishop, J., & Scott, F. (2016). Digital play: A new classification. *Early Years, 36*(3), 242–253.

Mertala, P. (2016). Fun and games—Finnish children's ideas for the use of digital media in preschool. *Nordic Journal of Digital Literacy, 2016*(4), 207–226. https://doi.org/10.18261/issn.1891-943x-2016-04-01

Murcia, K. (2005). Using action research principles for professional development: The Waikiki school experience. *Action Learning and Action Research Journal, 10*(1), 3–11.

Murcia, K. (2018). *Creative coding in early childhood*. DigiLitEY. European cooperation in science & technology. 6th Project Meeting. Riga. Retrieved from http://digilitey.eu/sixth-project-meeting-riga-latvia

Murcia, K., Campbell, K., & Aranda, G. (2018). Trends in early childhood education practice and professional learning with digital technologies. *Pedagogika, 68*(3), 235–250.

Murcia, K., & Tang, K. S. (2019). Exploring the multimodality of young children's coding. *Australian Educational Computing, 34*(1), 1–15.

Murcia, K., Pepper, C., & Joubert, M. (2020). Creative coding in early childhood. *Early Years: An International Research Journal* (under review).

NAEYC. (2012). *Technology and interactive media as tools in early childhood programs serving children from birth through age 8*. NAEYC Joint Position Statement with the Fred Rogers Center. Retrieved from http://www.naeyc.org/files/naeyc/PS_technology_WEB.pdf

Pyle, A., DeLuca, C., & Danniels, E. (2017). Context and implications document for a scoping review of research on play-based pedagogies in kindergarten education. *Review of Education, 5* (3), 352–353.

Ritchhart, R., Church, M., & Morrison, K. (2011). *Making thinking visible: How to promote engagement, understanding and independence for all learners*. San Francisco, CA: Jossey-Bass.

Western Australian Department of Education. (2018). *The importance of play-based learning*. Retrieved from www.education.wa.edu.au/play-based-learning

Zabatiero, J., Straker, L., Mantilla, A., Edwards, S., & Danby, S. (2018). Young children and digital technology: Australian early childhood education and care sector adults' perspectives. *Australasian Journal of Early Childhood, 43*(2), 14–22. https://doi.org/10.23965/AJEC.43.2.02.

Karen Murcia is an Associate Professor in the School of Education, Curtin University. Her research interests include scientific literacy for citizenship, creativity with digital technologies and STEM education. She is a researcher with the Australian National Centre of Excellence for the Digital Child, Chair of Child Australia and a Scitech Discovery Centre Board Member.

Part IV
Privacy and Protection

Chapter 20
Researching Representations of Children and Childhood on Instagram: Ethical and Methodological Considerations

Madeleine Dobson and Jenny Jay

Introduction

Children and childhood have become highly visible across social media. The impact of this presence appears to be an underdeveloped and under-theorised aspect of contemporary childhood. Choi and Lewallen (2017, p. 1) note that Instagram has become a "photo album for children" and report that a search on the platform in 2017 for the tag #children revealed 13 million photos. More recently, in 2019, our own search reveals 24.5 million photos of children available via this hashtag—and, evidently, there are many more available beyond that particular hashtag. What appears to have emerged is a curated and perhaps idealised narrative of childhood, where certain trends and tropes come to the fore. As we explored a variety of Instagram posts featuring children, we became curious about the image of the child that exists in this context. Given how pervasive social media has become, we were keen to explore how the image of the child might be swayed or shifted. In particular, we wondered how parents' actions on social media may impact on the identities of children. This curiosity led to the development of a research project focusing on the representation of children on Instagram.

This chapter explores our emerging research project on the image of the child. As educators in Early Childhood Education and Care (ECEC), we hold an appreciative image of children which acknowledges children as significant and capable individuals (Department of Education, Employment, and Workplace Relations [DEEWR], 2009). Our multidimensional image respects the identities, competencies, and contexts of all children. Diversity and difference is honoured, and childhood is privileged as a crucial life stage. We view children as agentic subjects (Robinson

M. Dobson (✉) · J. Jay
Curtin University, Perth, WA, Australia
e-mail: madeleine.dobson@curtin.edu.au; jenny.jay@curtin.edu.au

D. Holloway et al. (eds.), *Young Children's Rights in a Digital World*, Children's Well-Being: Indicators and Research 23,
https://doi.org/10.1007/978-3-030-65916-5_20

& Jones-Diaz, 2016) with rich potential (Robertson, 2006) who ought to be engaged actively in all matters influencing their lives (DEEWR, 2009). They deserve to be viewed and treated as citizens in their own right (Robinson & Jones-Diaz, 2016) who are becoming confident and creative individuals (Ministerial Council on Education, Employment, Training and Youth Affairs, 2008). This image underpins and guides how we work with and advocate for children, and was integral to the development of our project. In this chapter, we explore the existing literature and methodological and ethical considerations that influenced the creation and evolution of our project.

Exploring the Literature

The ECEC sector—which in Australia focusses on children aged birth to eight—holds a clear and well-articulated understanding of young children and their rights to be agentic individuals that are active and capable in their environments (e.g. DEEWR, 2009; Robinson & Jones-Diaz, 2016). The Early Years Learning Framework (DEEWR, 2009, p. 9) states that children should be viewed "as active participants and decision makers" who "recognise their own agency, capacity to initiate and lead… and have rights to participate in decisions that affect them". When children are viewed through this lens at this significant life stage, this positively shapes the way in which they are parented, educated, listened to, and generally permitted to be themselves.

Robinson and Jones-Diaz (2016) explore the construction of childhood and the changing nature of children's worlds, acknowledging that the discourse of childhood innocence can be used to regulate children's lives and limit their power. This belief creates a binary between childhood and adulthood, implying that children are unable to participate in the construction of their own childhood and identity. From this perspective, a power differential is created, thus allowing adults more control (Robinson & Jones-Diaz, 2016).

In exploring the 'image of the child' we must consider the Reggio Emilia philosophy, which has led to the impression of the 'rich child' (Rinaldi, 2004). This philosophy considers all children to be intelligent and capable of making sense of the world. The 'rich child' is constantly involved in constructing knowledge, building an identity, and should be listened to (Rinaldi, 2004).

Bandura (2001) explores the concept of human agency, claiming people are agents of experience who, rather than merely being exposed to life experiences are able to act upon these experiences through active engagement. He states it is "not just exposure to stimulation, but agentic action in exploring, manipulating, and influencing the environment that counts" (p. 4). This theory can be applied to the lives of children as, from an ECEC perspective, children are not passive beings having life experiences 'done to them' but rather active agents of their own lives.

In the twenty-first century, many children's lives have become highly visible and accessible as representations of their everyday experiences are shared through social media in great volume and with great speed by family members. Literature exploring the representation of children on social media has focussed on the rights of the child to expect privacy (Chalklen & Anderson, 2017; Elmer, 2015), the possible long-term ramifications of this lack of privacy on children's lives and identity formation (Choi & Lewallen, 2017; Leaver & Nansen, 2017), the use of children as 'digital labour' to generate family income (Abidin, 2017), how mothers as consumers turned producers or 'prosumers' frame their children online (Archer, 2019), and the sharing of information about family on social media by parents, dubbed 'sharenting', in terms of ethics (Blum-Ross & Livingstone, 2017) and risks (Autenrieth, 2018). Swist and Collin (2017, p. 677) explain that as "technology diversifies and intensifies communication, it is reconfiguring human expression and interaction at the level of self and society". This means that the constructs of 'child', 'children', or 'childhood' are subject to representation on social media in ways that could be changing beliefs about who and what children are at both the level of self and society.

Bandura (2001, p. 22) writes:

> through agentic action, people devise ways of adapting flexibly to remarkably diverse geographic, climatic and social environments; they figure out ways to circumvent physical and environmental constraints, redesign and construct environments to their liking, create styles of behaviour that enable them to realise desired outcomes, and pass on the effective ones to others by social modelling and other experiential modes of influence.

The influence, place, and value of social media has risen in significance as technology becomes more accessible. Bandura's (2001) notion of 'social modelling' and 'modes of influence' can be applied to social media platforms. On these platforms (e.g. Instagram, Twitter, and Facebook) the phenomena of 'influencers' and 'micro-celebrities' have resulted in the regular recording and presentation of family life and, consequently, children's lives as a genre to generate income (Abidin, 2017). In this environment, children of 'family influencers' also become micro-celebrities through the regular, curated recording of their life.

Regarding the phenomena of micro-celebrity and influencer parents on social media, Leaver (2017) notes the emergence of deliberately curated narratives of family life and childhood. As popular parent influencers attract large numbers of followers, the pressure to sustain engagement and generate promotional opportunities will shape the content and composition of blogs, videos, and images. The potential power of these 'digital estates' to raise family revenue via product or service sponsorship will influence the subject and setting of each photo to create an interest in the commodity and generate followers to return to each new post (Abidin, 2017). As for potential influence, Leaver (2017, p. 7) states "micro-celebrity parents can be quite influential in promoting and normalising certain shifts in parenting practices".

As what was previously private becomes increasingly public, family influencers attract followers who view and interact with their content. In all of this, young children have emerged as a group with little say, or, as Leaver (2017, p. 2) states, "no direct self-representational agency, whose online presence is crafted by other people". This can impact on identity formation. On this note, Choi and Lewallen (2017) emphasise that young children shape their own identities by observing other people and media. For many children, this includes images of themselves shared on Instagram by their parents. In this process, decisions will have been made about the child's representation—this may be determined entirely by the parent, thus omitting the child's voice, choice, and agency. Choi and Lewallen (2017) add that parents have the power to represent their children in stereotypical and non-stereotypical ways, thus influencing the way in which children are viewed in regards to critical aspects of identity such as gender and race. Their findings indicated that while Instagram may afford greater visibility to marginalised groups, stereotypes remained unchanged. Choi and Lewallen (2017) contend that further research in this area is imperative as children have become more vulnerable and deserve protection from wrongful representation. Matters of visibility and consent are also examined by Jorge and Marôpo (2017) in their examination of the circumstances and rights of children of celebrities. Furthermore, Locatelli (2017) explores the depiction of children by their mothers in her research about the representation of breastfeeding on Instagram. Locatelli discusses various approaches to the depiction of children, including whether or not parents show their child's face or reveal their name, and marks this as an area for further research.

While previous literature in relation to children and the Internet deals primarily with online safety, there is increasing emphasis on children's rights in online contexts in regards to data gathering and privacy (Livingstone & Third, 2017; Molina, Oltra-Gutierrez, & Sarabdeen, 2010; Swist & Collin, 2017). Children have been positioned in early discussions of Internet governance and rights as simply 'internet users' or a part of the general public with the result of their rights falling into the category with all human rights. Further discussion has focussed on children's safety when using the Internet and digital platforms. Livingstone, Carr and Byrne (2015, p. 5) warn that taking a protectionist view of children's rights, considering mainly protection against child abuse, largely undermined "their freedom of expression or traded children's particular needs off against adult freedoms online". Molina et al. (2010) advise that children should be provided with the tools to help them decide whether their images are uploaded and who would be able to view the images, suggesting, at that time a more restricted framework in regards to children's images should be considered. Livingstone et al. (2015) caution that parents are not always aware of the consequence of online engagement or have the necessary resources to protect and empower their children on-line. They suggest that increasingly some responsibility to uphold children's online rights must fall to internet services and providers through best practice guidance.

An investigation and subsequent discussion about how young children are perceived in the twenty-first century and how social media may influence this perception has led to the development of this project, which brings together the emerging research on the power and significance of social media and the understanding of the community's view of children within the early years of their lives. The following sections detail our methodological and ethical approaches to examining the representation of children on Instagram and how this compares/contrasts to our image of the child.

Methodological Considerations

This section details the design of our project—including conceptual origins, research questions, and the emergence of a multi-phase approach; ethical considerations; the recruitment of participants; and, data collection and analysis. Throughout, we identify complexities that we confronted along the way.

Designing the Project
Our first step in determining the methodology of our project was to consider the image of the child. We discussed the image held in ECEC, where children are honoured as capable, agentic, and multidimensional citizens who have their own voice and views. A central curiosity was whether this image was apparent in the representation of children on Instagram. Stemming from this were questions around what other images might be apparent in this context and what potential influence this might have. As such, the following research question emerged: *What 'image of the child' is apparent in posts shared on Instagram by influencer parents, popular children's brands, and high-profile celebrity parents?*

In the early stages of developing the project, our focus was solely on parents, and in particular, influencer parents. In recognition of their high profiles and their potential sway over followers, we wanted to understand how they were representing their children and what image of the child was inherently apparent in that representation. As the project progressed, our focus expanded and a multi-phase approach emerged:

1. A single case study of an influencer parent's Instagram posts
2. Multiple case studies of popular children's brands Instagram posts
3. Multiple case studies of high-profile celebrity parents' Instagram posts

The case studies integrated various sources of data and involved several qualitative analysis techniques, which are discussed later in this chapter. We selected a case study approach to engage in an exploration of individuals, brands, and the Instagram context, and capture rich data to build a nuanced understanding (Hamilton & Corbett-Whittier, 2013).

At the time of writing, we are contemplating further expansion to include other data sources. Our overall intent is to create a comprehensive dataset which speaks to the multitude of ways in which children are represented across different realms including education and care contexts and different genres of media and types of marketing.

Engaging Ethically

Researching within social media contexts involves complex ethical considerations. As we developed this project, our conversations revolved around the ethics of it all and primarily encompassed themes of access, consent, privacy, and respect.

All of our prospective participants held publicly available Instagram accounts and had anywhere between tens of thousands to over a million followers. We were reluctant to conclude that this implied automatic access to potential data. While Internet sources are easily accessible, there are tensions around what is public versus what is private. We considered provocations around public/private as posited by Eastham (2011) regarding research involving the analysis of publicly available blogs. The view of public and private as existing along a spectrum, rather than in binary opposition to each other, was particularly critical. The provocations around blog attributes and bloggers' privacy choices—which we translated to the context of Instagram—were important in determining project boundaries and ethical procedures. Furthermore, a particular point of resonance was Eastham's (2011) description of "the paradox of personal intimacy in a publicly available location"—this triggered further thinking how we could engage in best ethical practice. The internet-specific ethical questions delineated by Markham and Buchanan (2012) also proved formative in understanding ethics in online contexts, as did guidance from Livingstone and Locatelli (2014) on ethical research in digital and online environments, and provocations from Warfield, Hoholuk, Vincent, and Camargo (2019) regarding ethical considerations for researchers working with images of bodies on social media.

We ultimately embraced a holistic view of ethics which involved moving beyond standard ethical requirements and the immediate context. We considered the short, medium, and long term outcomes for participants and their families. For example—how can privacy be preserved in the context of research dissemination, such as sharing findings at conferences and in publications? Rose (2016, p. 365) points out that while gaining consent is imperative, there are inherent complexities at play. Regarding this, she writes, "Once an image has gone public—and online, in particular—it is very hard to ensure that it will only be seen by particular audiences, or in the context in which you have carefully placed it". This was an important consideration as we developed resources—including information letters and consent forms—to share with prospective participants.

We also decided to approach this project with a mind towards our ethics as ECEC educators and researchers, where a key priority is building and sustaining respectful relationships with families. While we held no pre-existing relationships with the prospective participants, we chose to view them in this same light. The Early Childhood Australia [ECA] Code of Ethics (ECA, 2016) lays out principles for relating to families which include listening to and learning with families, developing respectful relationships with families based on open communication, and respecting families' rights to privacy and confidentiality. We also foregrounded recommendations by Markham and Buchanan (2012) and Rose (2016) in our ethical processes. Stemming from this, we sought comprehensive informed consent from all prospective participants. The information letters and consent forms we provided meticulously detailed the project, what their participation would involve, and foreseeable outcomes. They were assured that if they consented, their posts would be carefully de-identified. This process is explored further in the following section, which details methodological considerations.

By committing to a carefully considered approach to ethics, we were able to plan for and engage in our research with confidence. In particular, we were guided by the following assertion from Ward (2011, p. 1): "Ethics at its best is reflective engagement with urgent problems, in light of where we have been and where we hope to be tomorrow". The ethics of our project hinged on reflective engagement with a mind towards the image of the child: that known in ECEC, what might be apparent on Instagram, and how we can use this knowledge to sustain a strong image of the child and support the rights of children.

Recruitment

The selection and recruitment of participants varied across the three stages (see Table 20.1). Across all phases, we identified key users and immersed ourselves in their online contexts. Information and insights emerged through this process—critically, we were able to identify who they were following and who followed them, which allowed us to identify a range of users for recruitment. We remained engaged in this process until we hit the point of saturation—that is, where few new profiles were appearing and we were looping back through already identified profiles.

With regards to the first phase, challenges in the recruitment process did arise. As our only connection to influencers was through Instagram—or, in some cases, via email addresses listed on their profiles—there were limited ways to follow up. Many messages and emails went unseen or unanswered, and when we did receive a response, this wasn't always fruitful. Some declined, others said they might be willing in future, and a few said they would be willing but never returned the consent forms and didn't respond to follow-up emails. A point of reassurance was the feedback provided by those who did respond. Universally, they voiced that they felt the project was interesting and important.

Table 20.1 Approach to recruitment

Phase	Approach to recruitment
Phase 1	• Popular influencer parents identified • Profiles with 15,000+ followers approached • Informed consent sought for all prospective participants
Phase 2	• Popular children's brands identified • A small sample (n = 3) with 200,000+ followers selected • Data collected under Australia's fair dealing principles
Phase 3	• Popular celebrity parents identified • A small sample (n = 3) with 1,000,000+ followers selected

Data Collection and Analysis

Each phase involved a month of data collection. Two password-protected and securely stored iPads were used to access Instagram on a daily basis and screenshot posts in their entirety: the photo/video/story being posted and any caption, tags, and comments. Capturing stories was an issue due to time constraints—as such, these were collected on a more ad-hoc basis.

All data was anonymised (see Fig. 20.1 for an example). This included:

- Redaction of usernames, location tags, and any other identifying text (e.g. potentially identifying information in the caption/comments/tags).
- Full captions were collected for the purpose of analysis only—for dissemination, only sections of captions were to be used with the phrasing amended to prevent the wording being searchable and risking the re-identification of participants.
- Crystallisation of facial features and any other identifying features (e.g. visible birthmarks on subjects' bodies, parents' tattoos, personal artefacts such as wedding rings, belongings customised with names).

Admittedly, we were concerned about what may be lost in this process—for example, subjects' emotional expressions. To retain any such information, captions were added to the images.

Once anonymised, all posts were uploaded to NVivo 12 for analysis. To inform and guide this process, we developed an analytical matrix to assess the different elements of an Instagram post: the photograph/video, caption, and tags (see Table 20.2). The matrix combined aspects of compositional interpretation to assess the imagery in terms of its production and components (Rose, 2016) and thematic text analysis (Kuckartz, 2014). It was built as a coding tree and supported an initial deductive analysis of the posts, after which inductive analysis revealed key themes. The comment section of each post was also explored and analysed inductively, if there was commentary relating to the child(ren) featured in the post.

We were also interested in exploring the perspectives and experiences of parents participating in Phase 1. Specifically, we wanted to give them an opportunity to share insights about their family, their intent and process when using Instagram, and any parameters to their Instagram presence. As well as representing an important dimension of the overall dataset, which could contribute richness and participant voice, this also reflected our commitment to a respectful research ethic. Seeking the perspectives of parents contributed to how we honoured the following ECA Code of Ethics principle: *Learn about, respect, and respond to the uniqueness of each family, their circumstances, culture, family structure, customs, language, beliefs, and kinship systems.*

Initially, we invited prospective Phase 1 participants to consider taking part in an interview in a context of their preference (face-to-face, over the phone, or via a web-conferencing platform with which they were familiar, e.g. Skype or Zoom) and at a time most convenient for them. This then evolved to include a survey option at the request of our first participant, Alice, who lives in a rural area with limited access

Fig. 20.1 Example of an anonymised post

to a reliable internet connection. This ruled out a web-conference and Alice further expressed she would prefer not to speak over the phone, and offered to answer any questions via email. To support Alice's willingness to contribute, we developed a Qualtrics survey. This included questions exploring her motivations to use

Table 20.2 Overview of the analytical matrix

Element	Aspects	Elaboration
Photograph/ video	Curation	How is the image curated? E.g. evidence of editing, styling
	Action/ activity	What action or activity is depicted? Is the subject still or in motion? Is there evidence of planning or spontaneity?
	Presentation	How is the child presented (clothes, hairstyle, etc.)?
Caption	Language	What kind of language features in the caption?
	Focus	Where is the focus placed in the caption?
	Elaboration	Does the caption provide elaboration about the photo, its subject, context, circumstances, etc.?
Tags	Categorical	Tags that categorise the image, e.g. *#children #play*
	Commentary	Tags that add to the caption with further commentary, e.g. *Look at Jamie play! #hecouldplayallday*
	Location	A tag pinpoints the location
	Brand	A brand or brands are tagged
	Individual	A specific individual is tagged
	Identifiable	Tags that identify the subject or their family, e.g. #annabellacharliedavies

Instagram, its significance in her life, her processes when posting, and if/how her family were involved (see Appendix 1 for the full list).

While a survey limits opportunities for further questioning, we pursued it as an additional option to afford our participants flexible options. All questions required a response and were designed with essay text-boxes to accommodate detailed answers. Our willingness to accommodate Alice's request served our relationship with her well—she thanked us, enthusiastically signed on to participate, and contributed in substantive and meaningful ways.

Our project is currently ongoing. As it continues, we hope to create a rich and diverse dataset that speaks to the multitude of ways in which children may be represented on Instagram, and which will support us in seeking out what image of the child exists in this context.

Conclusion

This chapter has explored our emerging research project about the representation of children on Instagram. The image of the child is a significant construct which influences the ways in which we relate to children and the rights and opportunities afforded to them. It is particularly important for educators, researchers, and families to consider and critically reflect on. Our project, which has recently concluded its first phase and has progressed to its second, explores how the image of the child may change across contexts. In particular, we are seeking to compare and contrast the image held by ECEC educators with the image or images apparent on Instagram. This chapter has explored the origins of this project, our navigation of ethics, and methodological considerations. It may offer insight and guidance to researchers working in this area, where there are a variety of complexities with which to contend. Ongoing research will support our evolving understanding of children's rights in the digital age, an issue well worth centring in dialogue about children and childhood.

Appendix 1: Interview Questions

- Can you tell me about yourself and your family?
- When did you start using Instagram and what motivated you to do so?
- Has your Instagram account changed over time in terms of content, focus, etc.? If so, how so?
- What are the things you like best about using Instagram?
- Do you have any other social media accounts (e.g. Twitter, Facebook, YouTube)? If so, how are they similar or different to your Instagram?
- Are your immediate family aware of your Instagram and what you post there?
- When taking photos to share on Instagram, what is your process? How do you decide what photos to take and share?
- Are your posts spontaneous, planned, or a mix of both?
- On average, when you take photos to share on Instagram, how many versions of that photo do you take?
- What do you look for in the photos you choose to share?
- Do your children help with making decisions about the photos being taken and shared? If so, how so?
- Do you have any personal rules for what you will/won't post? If so, how did you decide on these rules?
- Who would you say has the most creative control over your Instagram?
- Overall, what does your Instagram account mean to you?
- Is there anything else you would like to share? (e.g. further information you would like us to know about, any questions or feedback about this study)

References

Abidin, C. (2017). #familygoals: Family influencers, calibrated amateurism, and justifying young digital labor. *Social Media + Society, 3*(2). https://doi.org/10.1177/2056305117707191

Archer, C. (2019). How influencer 'mumpreneur' bloggers and 'everyday' mums frame presenting their children online. *Media International Australia, 170*(1). https://doi.org/10.1177/1329878X19828365.

Autenrieth, U. (2018). Family photography in a network age: Anti-sharenting as a reaction to risk assessment and behaviour adaptation. In G. Mascheroni, C. Ponte, & A. Jorge (Eds.), *Digital parenting: The challenges for families in the digital age* (pp. 219–231). The International Clearinghouse on CHILDREN, YOUTH & MEDIA: Gothenburg.

Bandura, A. (2001). Social cognitive theory: An agentic perspective. *Annual Review of Psychology, 52*(1), 1–26.
Blum-Ross, A., & Livingstone, S. (2017). "Sharenting," parent blogging, and the boundaries of the digital self. *Popular Communication, 15*(2), 110–125. https://doi.org/10.1080/15405702.2016.1223300.
Chalklen, C., & Anderson, H. (2017). Mothering on Facebook: Exploring the privacy/openness paradox. *Social Media + Society, 3*(2). https://doi.org/10.1177/2056305117707187
Choi, G. Y., & Lewallen, J. (2017). "Say Instagram, Kids!": Examining sharenting and children's digital representations on Instagram. *Howard Journal of Communications, 29*(2), 144–164. https://doi.org/10.1080/10646175.2017.1327380.
Department of Education, Employment, and Workplace Relations [DEEWR]. (2009). *Belonging, being, and becoming: The early years learning framework for Australia.* Retrieved from https://www.education.gov.au/early-years-learning-framework
Early Childhood Australia. (2016). *ECA Code of Ethics – Core principles.* Retrieved from http://www.earlychildhoodaustralia.org.au/our-publications/eca-code-ethics/code-of-ethics-core-principles/
Eastham, L. A. (2011). Research using blogs for data: Public documents or private musings? *Research in Nursing and Health, 34*, 353–361. https://doi.org/10.1002/nur.20443.
Elmer, G. (2015). Going public on social media. *Social Media + Society, 1*(1). https://doi.org/10.1177/2056305115580341
Hamilton, L., & Corbett-Whittier, C. (2013). *Using case study in education research.* London: Sage.
Jorge, A., & Marôpo, L. (2017). Born to be famous? Children of celebrities and their rights in the media. In J. O'Connor & J. Mercer (Eds.), *Childhood and celebrity.* London: Routledge.
Kuckartz, U. (2014). *Qualitative text analysis: A guide to methods, practice & using software.* London: Sage.
Leaver, T. (2017). Intimate surveillance: Normalizing parental monitoring and mediation of infants online. *Social Media + Society, 3*(2).https://doi.org/10.1177/2056305117707192
Leaver, T., & Nansen, B. (2017). Infancy online: An introduction. *Social Media + Society, 3*(2). https://doi.org/10.1177/2056305117707193
Livingstone, S., Carr, J., & Byrne, J. (2015). *One in three: The task for global internet governance in addressing children's rights.* London: Centre for International Governance Innovation and Chatham House.
Livingstone, S., & Locatelli, E. (2014). Ethical dilemmas in qualitative research with youth on/offline. *International Journal of Learning and Media, 4*(2), 67–75. https://doi.org/10.1162/IJLM_a_00096.
Livingstone, S., & Third, A. (2017). Children and young people's rights in the digital age: An emerging agenda. *New Media & Society, 19*(5), 657–670. https://doi.org/10.1177/1461444816686318.
Locatelli, E. (2017). Images of breastfeeding on instagram: Self-representation, publicness, and privacy management. *Social Media + Society, 3*(2). https://doi.org.dbgw.lis.curtin.edu.au/10.1177/2056305117707190
Markham, A., & Buchanan, E. (2012). *Ethical decision-making and internet research (version 2.0): Recommendations from the AoIR Ethics Working Committee.* Retrieved from https://aoir.org/reports/ethics2.pdf
Ministerial Council on Education, Employment, Training and Youth Affairs. (2008). *Melbourne declaration on educational goals for young Australians.* Melbourne, VIC: Curriculum Corporation.
Molina, M., Oltra-Gutierrez, J. V., & Sarabdeen, J. (2010). An exploratory study on the privacy of children's images in Spain's most widely used social network sites (Tuenti and Facebook). *International Review of Law, Computers & Technology, 24*(3), 277–285. https://doi.org/10.1080/13600869.2010.523953.

Rinaldi, C. (2004). *In dialogue with Reggio Emilia: Listening, researching and learning*. Abingdon, Oxon: Routledge.
Robertson, J. (2006). Reconsidering our images of children: What shapes our educational thinking? In A. Fleet, C. Patterson, & J. Robertson (Eds.), *Insights: Behind early childhood pedagogical documentation* (pp. 37–54). Mt Victoria, New South Wales: Pademelon Press.
Robinson, K., & Jones-Diaz, C. (2016). *Diversity and difference in childhood: Issues for theory and practice* (2nd ed.). Berkshire, UK: Open University Press.
Rose, G. (2016). *Visual methodologies: An introduction to researching with visual materials* (4th ed.). London: Sage.
Swist, T., & Collin, P. (2017). Platforms, data and children's rights: Introducing a 'networked capability approach'. *New Media & Society, 19*(5), 671–685. https://doi.org/10.1177/1461444816686319.
Ward, S. J. A. (2011). *Ethics and the media: An introduction*. Cambridge: Cambridge University Press.
Warfield, K., Hoholuk, J., Vincent, B., & Camargo, A. D. (2019). Pics, dicks, tits, and tats: Negotiating ethics working with images of bodies in social media research. *New Media & Society, 21*(9), 2068–2086. https://doi.org/10.1177/1461444819837715.

Madeleine Dobson is a Senior Lecturer and Director of Student Experience and Community Engagement in Curtin University's School of Education. Her research interests include media, digital technologies, social justice, children's rights, and resilience and wellbeing within educational contexts.

Jenny Jay is an Associate Professor in Early Childhood Education at Curtin University. Her passion is ensuring high-quality play-based education for young children, and supporting pre-service teachers to become strong advocates and leaders. Her research interests include play-based curriculum, respect and agency of children in the online environment, and teacher resilience.

Chapter 21
The 'Sharent' Trap: Parenting in the Digital Age and a Child's Right to Privacy

Anna Potter and Renee Barnes

Introduction

Article 16 of the UN Convention on the Rights of the Child asserts a child's right to a privacy that is free from "arbitrary or unlawful interference" (UNESCO, 1989). Technology, and particularly digitisation, have introduced multiple ways of interfering with a child's privacy, in all aspects of their lives, from conception. While many of the threats posed to contemporary children's privacy are from corporations in pursuit of their data for its monetisation, parents' growing embrace of digital technologies also raises new questions about privacy-related practices in relation to their children. The willingness of many parents to share their children's images and information on social media platforms such as Facebook and Instagram, while at the same time playing a critical role in protecting their children's privacy, creates a significant conflict of interest. This conflict of interest, between "a parent's right to narrate their children's stories"—without asking for their children's consent—and their role as "gatekeeper of their children's privacy" has profound implications, for children's rights and wellbeing (Steinberg, 2017).

In this chapter, we consider parental attitudes to children's rights to privacy on social media in the light of this conflict of interest. We use the data we collected from a 2019 online survey of 613 Australian parents to illustrate some of the pressures and paradoxes created by these seemingly incompatible parental roles and responsibilities. We consider the highlighting of these pressure points important for two key reasons. First, because children's status as a vulnerable audience deserving of protection has underpinned Australian screen and media policy and public discourse for decades. Yet societal concerns around the protection of children's right to

A. Potter (✉) · R. Barnes
University of the Sunshine Coast, Sunshine Coast, Australia
e-mail: apotter@usc.edu.au; Renee.barnes@usc.edu.au

D. Holloway et al. (eds.), *Young Children's Rights in a Digital World*, Children's Well-Being: Indicators and Research 23,
https://doi.org/10.1007/978-3-030-65916-5_21

privacy as the object of a meditatised gaze through sharenting—the practice of parents sharing information about themselves and their children online (Blum-Ross & Livingstone, 2017)—remain surprisingly muted in comparison. Second, we suggest that the idea that children are somehow less deserving of privacy when their own parents are breaching it through sharenting is emblematic of a larger cultural shift in the relationship between children, parents and the media, one encouraged by digitisation. Similar breaches of privacy have been obvious for some time in other often highly commercialised content in which parents are gatekeepers of their children's privacy, including reality television series like *Supernanny*, *World's Strictest Parents*, *Toddlers and Tiaras* and *Brat Camp* (Potter & Goldsmith, 2017) and YouTube content including toy unboxing channels (Craig & Cunningham, 2017).

The proliferation in media content through digitisation has eroded children's right to privacy on many fronts. Digitisation amplifies parents' opportunities to expose their children to a public and enduring gaze, with little consideration of whether children will be comfortable with such a level of exposure as they grow up. Improving our understanding of Australian sharenting practices may help mitigate against some of these privacy concerns. To that end, this chapter explores these emerging issues and concerns by drawing on a pilot study of Australian parents and the practice of sharenting on Facebook and Instagram.

Parents, Social Media Use and Issues of Privacy

Understanding the significance of sharenting, is important, because of the ethical and practical concerns the practice generates, and its enormous potential for breaching children's privacy (Blum-Ross & Livingstone, 2017). Concerns about sharenting are at times seen as emblematic of the moral panics that characterise the relationship between children and the media. Online, however, where images are potentially more visible, and shareable to a large and diverse audience (boyd, 2006), the question of what, and how often parents are sharing about their children has a significantly larger impact.

A recent study of UK parents, for example, found 75% of those surveyed shared pictures and videos of their children online (Livingstone, Blum-Ross, & Zhang, 2018). In Australia, parental attitudes to sharing content about their children on social media have been the subject of limited scholarship (see for example, Chalklen & Anderson, 2017), although Australians have high rates of Internet use, with 84% of the population accessing the Internet daily. Social networking forms a significant part of this Internet use, with up to 79% of users owning a social media profile. Among these, Facebook is the most popular social networking site for Australian users (94%), followed by Instagram (46%) (Tolliday, 2017).

Research into Australian parents' use of social media highlights some of the benefits new mothers derive from using Facebook, as well as their concerns about its possible ill effects. The activities of 'mumpreneur' influencers who create mummy blogs featuring their children's information and images have also been examined (Archer, 2019; Archer & Kao, 2018). Facebook has the potential to offer significant support to parents, including the ability to ask for advice, engage with a virtual parenting community, and maintain social contact (Archer & Kao, 2018). Facebook also allows women who are physically separated from their children to keep in touch with them (Lupton, Pedersen, & Thomas, 2016). New mothers remain wary however about its possible negative impacts on their family dynamics, and on the behaviour they are modelling to their children when using it (Archer & Kao, 2018). Motherhood appears to increase women's engagement with Facebook nonetheless, including through posting, with 50% reporting greater engagement after the birth of their first child, and 51% reporting posting 'daily, weekly or every so often' (Archer, 2019). The use of Facebook as a virtual family photo album has also become popular, despite some concerns about the children's digital footprints created in the process (Holloway & Green, 2017). Parents have embraced pregnancy apps as well, with reported low levels of concern about any digital footprints created; a 2015 survey of Australian mothers using these apps found they were not particularly worried about the collection and use of their and their unborn children's data (Lupton & Pedersen, 2016).

Women are also using social media, in particular blogs, to create their own parental identity, in activities which subvert some of the clichés and stereotypes about how mothers should behave. These maternal blogs have gathered considerable followings; many include information and images about the authors' children. Their reach can be amplified considerably if they are picked up by mainstream media, in what are often critical treatments (Archer, 2019). Instagram influencers also regularly share information and images about their children including, of course, high-profile celebrities. Defined as 'micro-microcelebrities', the children of Instagram influencers are frequently and deliberately featured by their parents in overtly commercialised posts, the primary purpose of which is marketing. These make no effort to afford their young subjects any privacy, even before birth (Abidin, 2015).

Parental attitudes to children as the subject of an audience's social media gaze contrast sharply with the societal understanding that children constitute a special audience of media, and one whose immaturity and vulnerability render them deserving of our protection (Buckingham, Davies, Jones, & Kelley, 1999; Lemish, 2007). Such protections are formalised in media policy, television editorial guidelines, and widely used systems of screen content classification (Potter, 2017). Children's use of the internet has also attracted widespread concern as well as regulatory scrutiny, with fears of children using the internet to communicate with strangers, to access pornography and to bully one another characterising its increasing use by children during the 2000s. Parental supervision and monitoring of children's use of screen media are encouraged because children are seen as immature and in need of adult safeguarding from the malign effects of these forms of media, including social media. While constructions of childhood remain contested, and are historically, socially and culturally bound, in most countries the protection of the child as a media user remains a priority (Buckingham, 2000, 2005). In contrast, concerns about children's rights to protection as the object of an audience's gaze on social media, and the lack of privacy the practice of sharenting entails appear much more muted.

Clearly parents enjoy the benefits their use of social media entails and appear comfortable sharing their data—and their children's data—in return. How Australian parents consider the risks involved in the sharing of their children's information and images through their social media activities is less clear, given the dearth of research in this emerging field. Thus, this study aimed to ascertain how Australian parents are sharenting on social media, and to understand both their awareness of risk, and their perceptions of their children's rights to privacy. In doing so the study addressed the following research question:

RQ1: How are Australian parents using Facebook and Instagram to share content about their children?

While children's use of media including social media continues to attract more attention, concerns have also been raised about sharenting, including the extent to which it facilitates children's data mining, facial recognition and Facebook profiling (Holloway, 2019; Webb, 2013). Some also warn that posting images and video online may expose children to sexual harm or online grooming (Tait, 2016). At the same time, mainstream media and some scholars have argued that concerns about sharenting constitute a moral panic (Saner, 2018), a common accompaniment to any technological innovation with which children and young people engage. Parents' perception of these risks is, therefore, a vital component in understanding the nature of, and ramifications of sharenting. Therefore, our second research question was:

RQ2: How risky do Australian parents perceive the sharing of content about children on Facebook and Instagram?

Government efforts to legislate to protect children's privacy tend to focus on data collection by corporations via, for example, Internet-connected toys, children's websites and apps, and baby wearables. These corporations (particularly Google, Facebook and Amazon) treat children as economic objects as well as consumers, and the source of data that can subsequently be monetised, including by its on-selling (Holloway, 2019). Significant scholarship is being undertaken around the commercial and educational mining of children's data and associated privacy concerns (see for example Holloway, 2019; Holloway & Green, 2016; Mascheroni & Holloway, 2019). Less research has been conducted with parents, however, around their sharing of their children's images, information and data online, particularly through social media platforms like Facebook and Instagram. This is significant given the conflict of interest that exists where parents are the primary caretakers of their children's right to privacy and come under scrutiny, particularly in relation to protecting children from practices of sexting and cyberbullying. However, these activities, while subject to well-founded concerns for children's wellbeing, do at least entail a degree of agency for the child (Steinberg, 2017). The practice of sharenting, however, frequently erodes children's autonomy and rights. Additionally, clear disagreement exists between parents and children in terms of permission seeking in a recent US study conducted on this subject. Here children believed their parents should ask permission to post their images more often than their parents thought they should, and parents thought they had asked permission more often than they had. Children were also very averse to parents sharing anything they deemed 'embarrassing', although they were less concerned about positive or complimentary posts (Moser, Chen, & Schoenebeck, 2017). Similarly, Lipu and Siibak found Estonian children were frustrated by their parents' sharing of information about them on Facebook, leading to considerable 'privacy boundary turbulence' as children sought to restrict their parents' sharenting practices (Lipu & Siibak, 2019: 59). A study of Flemish 12–14 year olds found they too were concerned about their parents breaching their privacy, and also posting photos that might undermine their carefully curated public personas (Ouvrein & Verswijvel, 2019). The breakdown of trust and the feelings of betrayal once young people realise what their parents have shared online are indicative of the disconnect between parents, and their children, in terms of their social media behaviours (see for example, Bokhari, 2019). This study therefore sought to examine a third research question:

RQ3: How do Australian parents consider their children's right to privacy?
Significant research has been undertaken on parental strategies that increase digital opportunities for their children, while also reducing risks (see Livingstone et al., 2017). This research has tended to focus however, on parental guidance and management of children's access to digital media, rather than parents' own practices. Within this context research shows that high digital aptitude and skills are decisive in whether people are able to take full advantage of the opportunities online, while avoiding associated harms (Hargittai, 2002; Helsper, 2002; Litt, 2013). Some research has shown strategic parental mediation of online experiences is practiced more by educated and/or digitally skilled parents, as well as parents of younger children (see Clark, 2011; Nikken & Schols, 2015; Pasquier, Simões, & Kredens, 2012). Research into practices of sharenting by British parents found many actively managed privacy by changing their privacy settings, removing people from their contact lists and deciding which information to share online. These digital skills were not evenly distributed however. Younger parents were more able to manage their privacy settings and contact lists, and parents of younger children were better at managing their privacy settings (Livingstone et al., 2018). The final research question, then, for this study was:

RQ4: What measures, if any, do Australian parents take to protect their children's privacy when sharing content about them on Facebook and Instagram?
By addressing the four outlined research questions, this study sought to unpack emerging issues related to the conflict of interest inherent in parents' social media sharing practices and children's right to privacy.

Method

The pilot study focused on Australian parents of children aged 0–13 years. Participants (N = 613) were initially recruited through a news web site *ABC Online* (www.abc.net.au/life) and Facebook and Instagram advertisements and subsequently by snowballing. The news article and advertisements informed potential participants that the research was investigating parents' social media usage and provided a link to a web-based version of the survey. This study used a cross-sectional survey design. The focal independent variables were Facebook and Instagram usage related to their children. The demographic variables age group and age of children were also considered. The dependent variables were views on children's privacy reported by participants and measures used to protect privacy online. Items on the survey were randomised to minimise order effects.

Participants' sharenting activities, defined as sharing photographs, videos, stories and information about their children was assessed using a five-point Likert scale, where 1 = never, 2 = rarely (1–3 times), 3 = sometimes (3–10 times), 4 = often (once a week) and 5 = regularly (every day). Participants were asked to nominate how risky they felt sharing content about their children on Facebook and Instagram on a scale of 0–10 with 0 being none, 1 being low and 10 being high. Participants were also asked to nominate an age at which children had a right to privacy and answer yes or no to questions about asking for children's permission before posting. Finally, participants were asked to nominate which measures, if any, they took to protect children's privacy online.

Results

Of the 613 participants, 11% (n = 67) were male and 89% (n = 546) were female. They were aged between 18–55 years with the majority (51%, n = 313) in the 35–44 range, followed by the 25–34 (36%, n = 223). Participants were also highly educated, with 49% (n = 297) with a Bachelors degree and 22% (n = 136) with a Masters degree.

RQ1 It is significant to note that 93% of respondents have a Facebook account, whereas only 66.07% have an Instagram account. Parents aged 25–34 years are significantly more likely to have an Instagram account than those aged between 35–54 (76% for 25–34; 61% for 35–44; and 55% for 45–54). Of those with a Facebook account, 15% post to Facebook weekly or more often about their children and photos and video constituted 86% of parents Facebook posts about their children. By comparison 13% of respondents post to Instagram weekly or more often (Fig. 21.1).

Only 14% of 25–34 year olds and 19% of 35–44 year olds posted information and stories, rising to 31% of 45–54 year olds doing the same. These differences could

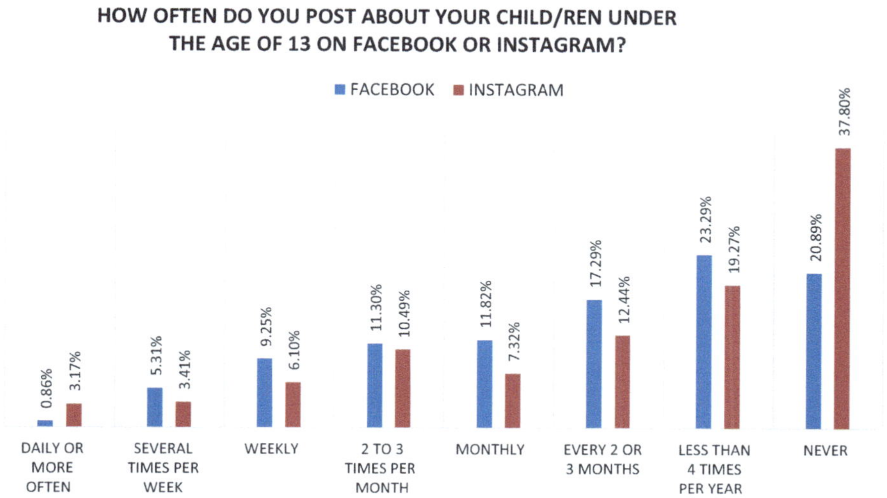

Fig. 21.1 Frequency of posting on Facebook and Instagram by parents about their children

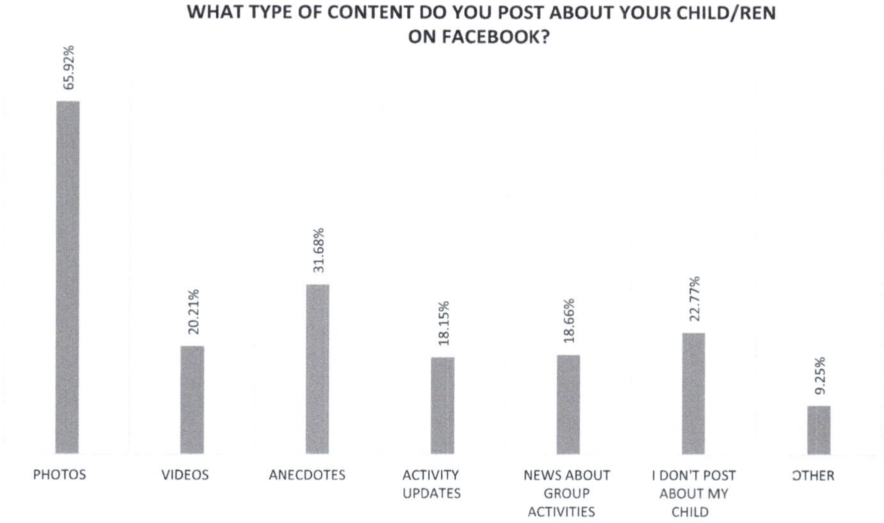

Fig. 21.2 Type of content posted by parents on Facebook about their children

reflect differing digital skill levels required to upload as a visual-based network' this could also explain the lower use of Instagram (Fig. 21.2).

RQ2 Despite high volumes of sharing, a significant proportion of respondents say sharing of content on Facebook and Instagram is risky. When asked about how risky sharing content on Facebook on a scale of 0–10 (where 10 is the riskiest) 10% rated

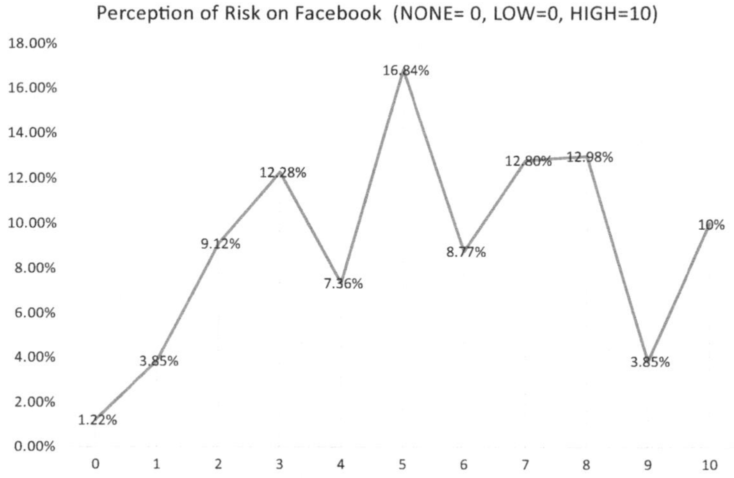

Fig. 21.3 Parents' perception of risk when posting about their children on Facebook

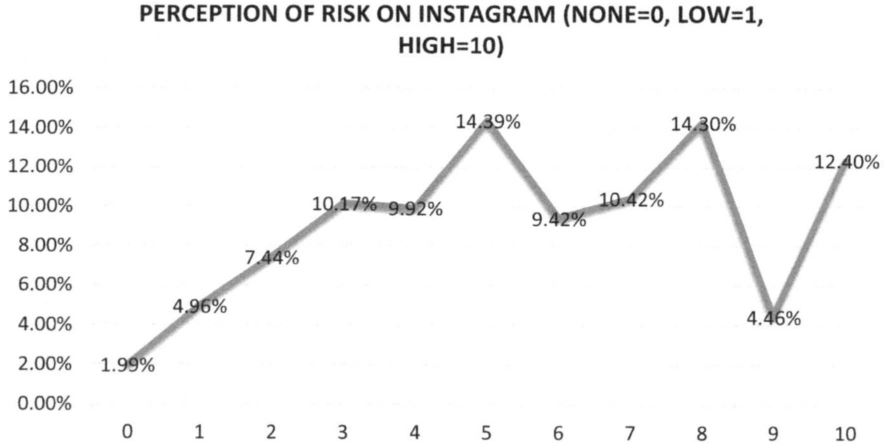

Fig. 21.4 Parents' perception of risk when posting about their children on Instagram

this as the highest of 10, while 27% of respondents believe that posting about your child on Facebook (Fig. 21.3) is very risky (rated 8–10). Respondents rated Instagram as slightly riskier with 12% rating a 10, while 31% of respondents believe that posting about your child on Instagram is very risky (Fig. 21.4).

Of note when demographics were considered, respondents aged 25–34 years were far less likely to view posting on Facebook as risky with 12% rating it as very risky, compared to 27% for 35–44 year olds and 26% for 45–54 year olds.

RQ3 Significantly, results show that 42% of respondents do not believe children always have a right to privacy (Fig. 21.5). As the graph below shows, when those who did believe in a child's right to privacy were asked about the age at which children earned this right the majority (58%) indicated at birth, however, the other responses vary wildly, with 10% identifying 5 years of age (which correlates with the beginning of primary school) and 4% nominating 12 years (the beginning of high school).

The study also considered parents' views on asking permission from children before posting. While 74% of parents believe children's permission should be asked before posting, of the parents who believe that children always have a right to privacy, 13% do not think they should ask for children's permission before posting to Facebook or Instagram. This discrepancy suggests that the concept of a right to privacy is still fraught with issues for parents. They may believe in the concept of privacy for a child, but fail to respect this in practice through, for example, seeking permission when it comes to creating digital narratives.

RQ4 Of those surveyed, 97% of Facebook users control who sees information, whereas only 67% of Instagram users do the same. Facebook users were most likely to change privacy settings from the default public setting for posts (80%), followed by using the custom-list function (56%) and asking for approval for post tags (55%). Age played a part in which measures parents took: 25–34 year olds (51%) are less likely to use the custom lists function than 45–54 year olds (64%); while 25–34 (65%) year olds are more likely to ask for permission for post tags than 45–54 year olds (33%) (Fig. 21.6).

As Fig. 21.7 below outlines, on Instagram, users were most likely to make their account private (66%), followed by removing geotags on posts (28%).

Further steps that parents were taking included turning off location data on phones (49%), ensuring identity of children was obscured (17%) and ensuring no location data was included in photos or videos (40%). However, a significant portion (32%) of respondents were not aware of, or had not considered, these options.

Discussion and Conclusion

The relationship between children and the media is being transformed by the digitisation of childhood, which has created new ways of interfering with children's privacy, in both the private and public sphere. In findings that echo Chalken and Anderson's (2017) 'privacy-openness paradox', where privacy concerns were nominated as a downside to Facebook use, but at the same time mothers were increasingly skilled at negotiating ways to protect their privacy so they can enjoy the benefits of openness, our survey data demonstrates that parents are enthusiastic sharenting users of Facebook and to a lesser extent Instagram while remaining mindful of the risks associated with posting information and images about their

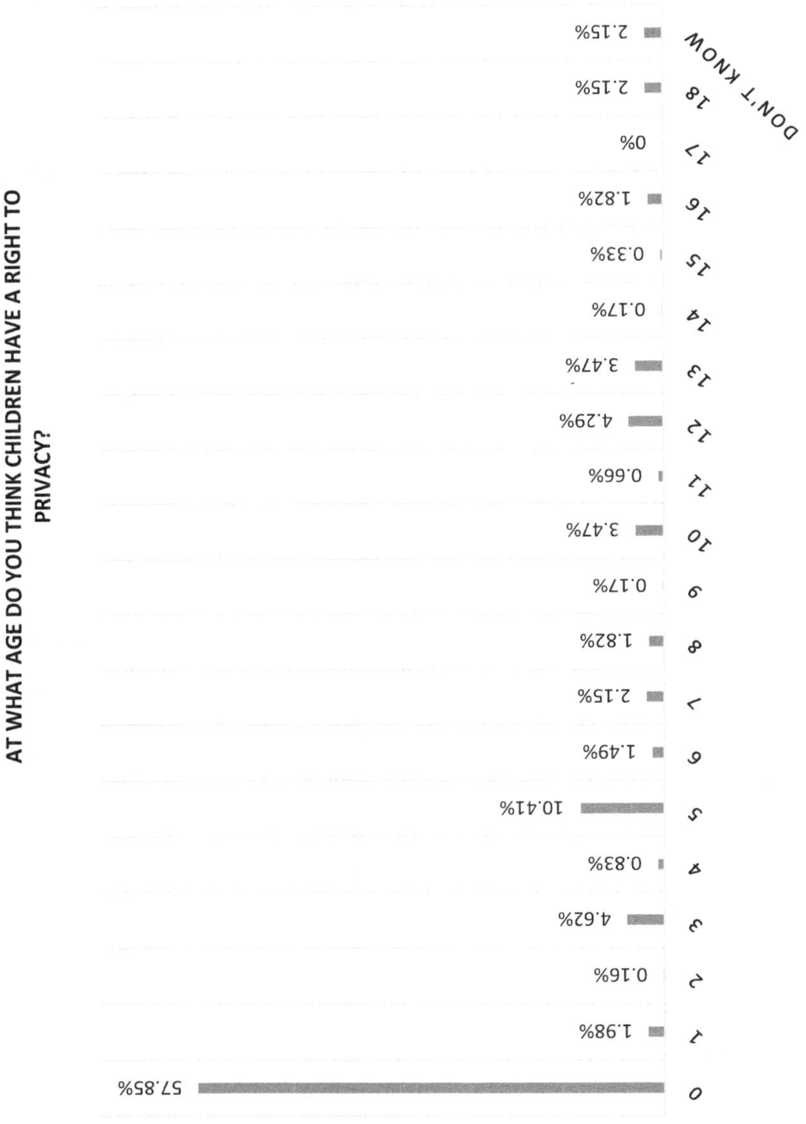

Fig. 21.5 Ages that parents thought children had a right to privacy

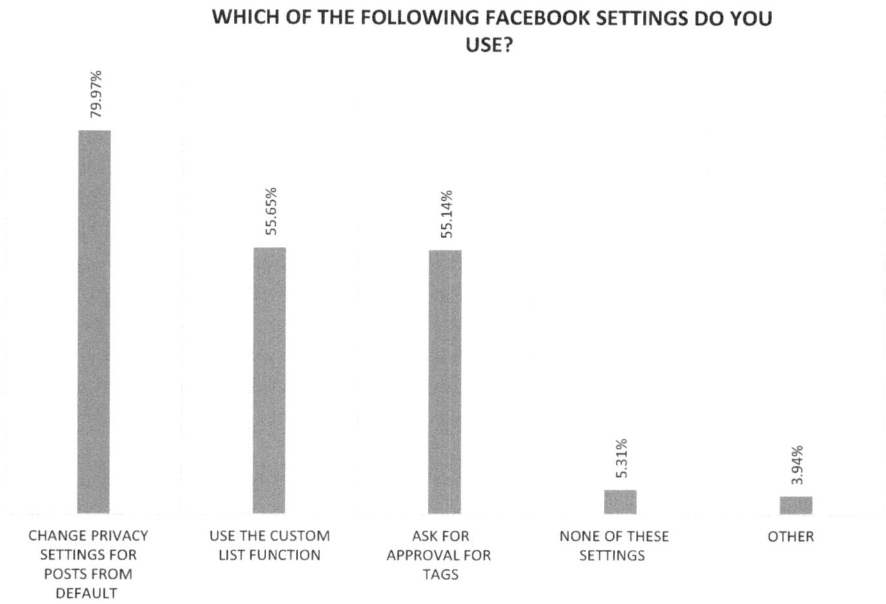

Fig. 21.6 Parents' use of privacy settings on Facebook

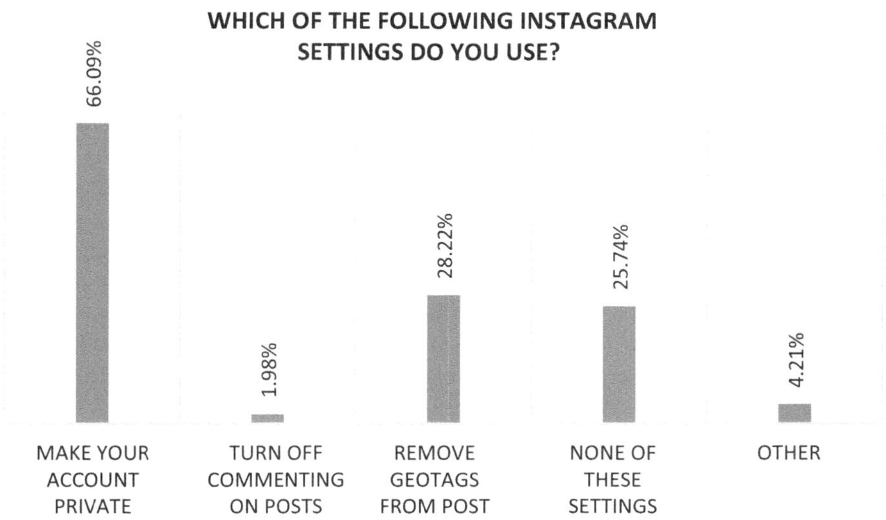

Fig. 21.7 Parents' use of privacy settings on Instagram

children. The majority are adept at using fairly sophisticated security settings on social media to minimise the distribution of their children's images, although a significant minority are not. Younger parents also seem less concerned about the risks of sharenting. But for many parents who are careful to use privacy settings on Facebook and Instagram, attitudes towards their children's privacy remain ambivalent. Indeed, more than 40% of parents did not believe their children had any right to privacy. This was despite the fact that they were highly mindful of the privacy and security risks posed by Facebook and Instagram and the need to ensure their own privacy settings were appropriately set. Of those who did consider their children had a right to privacy, the majority did not consider this right existed until the child was either capable of understanding the concept of privacy, or of requesting it themselves. The data also suggests a disconnect between a belief in the right to privacy, and actually asking children's permission to post information, photographs and videos. This disconnect not only suggests issues with the concept of privacy, but also has potential long-term consequences. Children want their parents to ask permission to post their images more often than their parents think they should (Moser et al., 2017). The long-term consequences of these breaches of children's trust by their parents are not known.

The subjective and widely different ideas about when children have the right to privacy, marry entirely with our subjective, social and geographically determined understandings of childhood, and the ages at which children are capable of understanding consent and agreeing to the sharing of their images. They are also emblematic of the conflict of interest inherent in representing one's online identity as a parent through sharenting, and parents' role as protectors of their children's privacy (Blum-Ross & Livingstone, 2017), a conflict of interest that on the whole has attracted little attention to date. Conversely, concerns about the digital behaviour of children and young people, including through activities like sexting and cyberbullying abound. But at least children and young people have some autonomy in the creation of these digital footprints, problematic as they may end up being; autonomy they certainly do not enjoy when their information and images are being distributed through sharenting (Steinberg, 2017). These privacy considerations have specific consequences when viewed with parents' digital skills, particularly their use of privacy settings and other measures to protect their children. Our research suggests parents feel confident about their digital skills, with the majority reporting they control who sees their social media updates on Facebook. However, far less were controlling their updates on Instagram. This could relate to the more open nature of Instagram (default settings allow anyone to follow updates without request). It could also reflect the younger demographic that were more likely to use Instagram for sharenting.

Overall parents' widespread lack of regard for children's right to privacy when put into practice through the act of sharenting suggest a recalibration of the relationship between children and the media. Rarely is there a perceived conflict of interest when parents are charged with the responsibility for protecting their children from inappropriate or dangerous screen content and activities. Clearly children as an audience are treated entirely differently from children as the subject of an assumed audience's gaze, an emerging disconnect which deserves greater investigation. The conflict of interest that parents undoubtedly have, as users of social media who enjoy sharenting, and as the protectors of their children's privacy also receives comparatively little attention, which is partly indicative of the pace at which the digitisation of childhood is occurring, and the comparative recency of sharenting, and associated scholarship.

As a pilot study, the research presented here has limitations. It is limited to one country, Australia. The population sample is not representative and included a disproportionate number of mothers and those with higher education. It does, however, provide a useful road map for future research about parents and consent, around policy that frames children as simply in need of protection from data harvesting from corporations, and about understandings of childhood. Further qualitative research is needed to investigate the issues related to the conflict of interest that sharenting creates. Such research should involve children and young people as well; their voices are not often heard in these debates, despite their unquestionable right to privacy, and to protection, in both the private and the public sphere.

References

Abidin, C. (2015). Micro-microcelebrity: Branding babies on the Internet. *M/C Journal, 18*(5).

Archer, C. (2019). How influencer 'mumpreneur' bloggers and 'everyday' mums frame presenting their children online. *Media International Australia, 170*(1), 47–56.

Archer, C., & Kao, K. T. (2018). Mother, baby and Facebook makes three: Does social media provide social support for new mothers? *Media International Australia, 168*(1), 122–139.

Blum-Ross, A., & Livingstone, S. (2017). "Sharenting," parent blogging, and the boundaries of the digital self. *Popular Communication, 15*(2), 110–125.

Bokhari, S. (2019). I'm 14, and I quit social media after discovering what was posted about me. *Fast Company*. Retrieved from https://www.fastcompany.com/90315706/kids-parents-social-media-sharing

boyd, D. (2006). Friends, friendsters, and myspace top 8: Writing community into being on social network sites. *First Monday, 11*(12).

Buckingham, D. (2000). *After the death of childhood: Growing up in the age of electronic media*. Malden, MA: Polity Press.

Buckingham, D. (2005). Constructing the 'media competent' child: Media literacy and regulatory policy in the UK. *MedienPädagogik: Zeitschrift für Theorie und Praxis der Medienbildung, 11*, 1–14.

Buckingham, D., Davies, H., Jones, K., & Kelley, P. (1999). Public service goes to market: British children's television in transition. *Media International Australia Incorporating Culture and Policy, 93*(1), 65–76.

Chalklen, C., & Anderson, H. (2017). Mothering on Facebook: Exploring the privacy/openness paradox. *Social Media + Society, 3*(2). https://doi.org/10.1177/2056305117707187

Clark, L. S. (2011). Parental mediation theory for the digital age. *Communication Theory, 21*(4), 323–343.

Craig, D., & Cunningham, S. (2017). Toy unboxing: Living in a (n unregulated) material world. *Media International Australia, 163*(1), 77–86.

Hargittai, E. (2002). Second-level digital divide: Differences in people's online skills. *First Monday, 7*(4).

Helsper, E. J. (2002). A corresponding fields model of digital inclusion. *Communication Theory, 22* (4), 403–426.

Holloway, D. (2019). Surveillance capitalism and children's data: The Internet of toys and things for children. *Media International Australia, 170*(1), 27–36.

Holloway, D., & Green, L. (2016). The Internet of toys. *Communication Research and Practice, 2* (4), 506–519.

Holloway, D., & Green, L. (2017). Mediated memory making: The virtual family photograph album. *Communications, 42*(3), 351–368.

Lemish, D. (2007). *Children and television: A global perspective*. Blackwell: Malden, MA.

Lipu, M., & Siibak, A. (2019). 'Take it down!': Estonian parents' and pre-teens' opinions and experiences with sharenting. *Media International Australia, 170*(1), 57–67.

Litt, E. J. (2013). Measuring users' Internet skills: A review of past assessments and a look toward the future. *New Media & Society, 15*(4), 612–630.

Livingstone, S., Blum-Ross, A., & Zhang, D. (2018). What do parents think, and do, about their children's online privacy? *Parenting for a Digital Future: Survey Report 3*, LSE.

Livingstone, S., Ólafsson, K., Helsper, E. J., Lupiáñez-Villanueva, F., Veltri, G. A., & Folkvord, F. (2017). Maximizing opportunities and minimizing risks for children online: The role of digital skills in emerging strategies of parental mediation. *Journal of Communication, 67*(1), 82–105.

Lupton, D., & Pedersen, S. (2016). An Australian survey of women's use of pregnancy and parenting apps. *Women and Birth, 29*(4), 368–375.

Lupton, D., Pedersen, S., & Thomas, G. M. (2016). Parenting and digital media: From the early web to contemporary digital society. *Sociology Compass, 10*(8), 730–743.

Mascheroni, G., & Holloway, D. (2019). *The Internet of toys practices, affordances and the political economy of children's smart play*. New York: Palgrave Macmillan.

Moser, C., Chen, T., & Schoenebeck, S. Y. (2017). Parents' and children's preferences about parents sharing about children on social media. In *Proceedings of the 2017 CHI conference on human factors in computing systems* (pp. 5221–5225). Denver, CO: ACM.

Nikken, P., & Schols, M. (2015). How and why parents guide the media use of young children. *Journal of Child and Family, 24*(11), 3423–3435.

Ouvrein, G., & Verswijvel, K. (2019). Sharenting: Parental adoration or public humiliation? A focus group study on adolescents' experiences with sharenting against the background of their own impression management. *Children and Youth Services Review, 99*, 319–327.

Pasquier, D., Simões, J. A., & Kredens, E. (2012). Agents of mediation and sources of safety awareness: A comparative overview. In S. Livingstone & L. Haddon (Eds.), *Children, risk and safety on the Internet: Research and policy challenges in comparative perspective* (pp. 219–230). Policy Press: Bristol.

Potter, A. (2017). Regulating contemporary children's television: How digitisation is re-shaping compliance norms and production practices. *Media International Australia, 163*(1), 20–30.

Potter, A., & Goldsmith, B. (2017). Reality's children: Young people and factual entertainment television. *Media International Australia, 164*(1), 44–55.

Saner, E. (2018). The 'sharent' trap – Should you ever put your children on social media? *The Guardian*. Retrieved from https://www.theguardian.com/lifeandstyle/2018/may/24/sharent-trap-should-parents-put-their-children-on-social-media-instagram

Steinberg, B. S. (2017). Sharenting: Children's privacy in the age of social media. *Emory Law Journal, 66*, 16–41.

Tait, A. (2016). Why YouTube mums are taking their kids offline. *New Statesman America*. Retrieved from https://www.newstatesman.com/culture/observations/2016/04/why-youtube-mums-are-taking-their-kids-offline

Tolliday, R. (2017). *Sensis social media report 2017-Chapter 1-Australian and social media*. Melbourne: Sensis.

UNESCO. (1989). *The UN Convention on the Rights of the Child*. Retrieved from https://www.ohchr.org/en/professionalinterest/pages/crc.aspx

Webb, A. (2013). We post nothing about our daughter online. *Slate*. Retrieved from https://slate.com/technology/2013/09/facebook-privacy-and-kids-dont-post-photos-of-your-kids-online.html

Anna Potter is an Associate Professor in the School of Business and Creative Industries at the University of the Sunshine Coast. She is the author of Creativity, Culture, and Commerce: Producing Australian Children's Television with Public Value and Creativity Rebooted: Producing Children's Television in the On-Demand Age.

Renee Barnes is a Senior Lecturer in Journalism at the University of the Sunshine Coast. Her research focuses on participatory practices online. She is the author of Uncovering Commenting Culture: Trolls, Fanboys and Lurkers, which examines the role of online commentary in society.

Chapter 22
Santa's Little Helper and Star of Instagram, Elf on the Shelf: Gendered Labour, Normalising Surveillance and Digitising a Childhood Phenomenon

Catherine Archer and Tama Leaver

Introduction

The EotS, described in all its marketing material as a Christmas tradition, despite being developed and marketed from only 2005, has become a worldwide cultural phenomenon, toy and 'interactive' Christmas decoration. According to a recent UK industry news article, 13 million of The Elf on the Shelf: A Christmas Tradition box sets have been sold worldwide since 2005, and the brand has an estimated 2.5 million followers on social media (Hutchins, 2019). Indeed, the same report suggests that 2019 would be "the biggest year in the UK market" for the brand. The toy's basic premise is that the Elf sits on a family's home shelf and then each night reports magically back to Santa whether the children in the household have been 'naughty or nice'. On the official company website, in 2019, the Elf's powers are described thus: "The magical Scout Elves help Santa manage his nice list by taking note of a family's Christmas adventures and reporting back to Santa at the North Pole nightly. Each morning, the Scout Elf returns to its family and perches in a new spot, waiting for someone to spot them. Children love to wake up and race around the house looking for their Scout Elf." While the EotS and its associated brand extension products have clearly been bestsellers, the media coverage of the toy has been ambivalent. Online news sites, including major mainstream news outlets such as the Wall Street Journal, and Australia's ABC News, suggest the Elf may be doing kids more harm than good, with his hyper-surveillance inside the home alongside the

C. Archer (✉)
Murdoch University, Perth, WA, Australia
e-mail: c.archer@murdoch.edu.au

T. Leaver
Curtin University, Perth, WA, Australia
e-mail: t.leaver@curtin.edu.au

D. Holloway et al. (eds.), *Young Children's Rights in a Digital World*, Children's Well-Being: Indicators and Research 23,
https://doi.org/10.1007/978-3-030-65916-5_22

tyranny of parents having to be creative each night with the elf's positions, creating a possible love-hate relationship with the toy. Headlines include "Elf on the Shelf trend mirrors loss of privacy" (Baranyai, 2015) and "The Tyranny of the Elf on the Shelf: Where to Put Him Tonight?" (Bindley, 2018) alongside other articles that offer 'helpful' ideas on how/where to position the Elf for maximum effect for the kids.

This chapter explores the way parental Instagrammers have approached this toy, with some seeing it as a way to garner followers and likes and present their 'ideal', creative parenting style. Others on Instagram have chosen to use humour to subvert the toy, deriding it as a waste of time or tapping into its inherent creepiness factor, with extremely adult poses and captions. The representations of the toy show the approaches parents now take in the social media age—presenting a carefully curated image and staging of their parenting style and identity. These social media traces simultaneously reinforce or subvert the inherent surveillance normalisation of this new Christmas phenomenon. In the era of 'surveillance capitalism' (Zuboff, 2015) it is argued that the Elf is reflective of our times. According to Holloway (2019: 22) children themselves are "clearly embedded actors within surveillance capitalism"—their data is of value to large organisations as is their worth as consumers.

The EotS is now a toy that has become part of many families' (and therefore young children's) Christmas rituals. The Elf has also become a digitised social media identity, with numerous 'Elfie selfies' uploaded in the lead-up to Christmas. To situate the EotS on social media, we begin with a literature review examining the rise of Instagram and selfies, the role of some mothers as first bloggers and now prominent parental Instagrammers or influencers, and the commercialisation and gendered nature of Christmas in social media and domestic labour terms. We proceed outlining our methodological approach using Instagram as a space of cultural visibility and proceed to analyse the visual presentation of EotS via the most popular images and hashtags. We then conclude by positioning the EotS not only as an avatar of Christmas commercialisation but also as an icon of the normalisation of parental 'intimate surveillance' (Leaver, 2017).

When Instagram Selfies, Christmas, Children and Mums Collide Within a Marketised Society

Instagram and Selfies

Instagram is a photo and video sharing platform social media application (app), launched in 2010 and purchased by Facebook in 2012 (Leaver, Highfield, & Abidin, 2020). In Australia, as in many parts of the world, Instagram is a hugely popular, and fast-growing social media app, with a recent survey stating that 46% of internet users in Australia used the platform (Danno, 2019). Users take a photo or video through their phone, then edit it to their liking, and post it on the app's newsfeed or Stories with or without comments. Their 'followers' can then like and comment on the post. One of the key elements of Instagram is the use of the # (hashtag) to allow users to create an intentional public display for sharing by deliberate choice. The hashtag is then searchable and marked for public consumption, and intentional public discourse (Ridgway & Clayton, 2016). Instagram is "at the heart of global digital culture, having made selfies, filters and square frames an inescapable part of everyday life since it was launched in 2010" (Leaver et al., 2020, p. 268).

As a social phenomenon Instagram has attracted academic research, not only within social media and communication fields, but also in areas of psychology, cyberpsychology and health, (see, for example, Frison & Eggermont, 2017; Lup, Trub, & Rosenthal, 2015; Slater, Varsani, & Diedrichs, 2017) with reports in the popular media centring on its use by and effect on young women and girls, on body image, potential narcissism, bullying and mood. The platform, initially a photo-sharing service, is now also an extremely important avenue for brands, not-for-profits and for-profit organisations to use as a marketing tool and tactic, including through 'influencer marketing', sponsored posts, and direct-to-consumer posts (Bergström & Bäckman, 2013; Blount, 2019; Djafarova & Rushworth, 2017). Individuals have used the platform to promote their own 'personal brand', and social media influencers have arisen who market themselves as arbiters of taste and co-creators of brand value, primarily through the platform (Abidin, 2016; De Veirman, Cauberghe, & Hudders, 2017; Marwick, 2015). The issues surrounding Instagram's increasing commercialisation and unregulated marketing practice saw Instagram introduce bans on advertising on the platform of plastic surgery and questionable weight loss products aimed at under 18s in 2019 (McCormack, 2019). Instagram also hid 'likes' for users, in some countries in 2019, including Australia, ostensibly to combat "mental health issues" (Levett, 2019).

Social media and communication scholars have examined the relatively recent 'mass phenomenon' of the selfie, as simultaneously objects but also cultural practices (Iqani & Schroeder, 2016; Senft & Baym, 2015). The rise of the selfie is deeply connected to Instagram and 'personal branding' (Eagar & Dann, 2016; Marwick, 2015). While selfies were originally thought of as having to be photos, shared on social media, of people's own face or bodies/body parts, Tiidenberg and Whelan (2017, p 146) suggested it was important to go beyond thinking of selfies as having the mandatory presence of a human face/body, "thus taking on visual self-representation beyond the dynamics of good/bad, empowering/objectifying that that selfie debate has tended to emphasise." The concept of self-presentation through visual images (and through objects/possessions) also has resonance with consumer culture/marketing academic Belk's original concept of the extended self (through what we possess) and his later discussion of the digital extended self (Belk, 1988, 2013). Depiction of EotS on Instagram could be seen as a form of selfie, as it is intended to be performative in the Goffman (1956) sense, simultaneously a staging but also a window into the 'backstage' of home life and the labour required by (mostly) mothers to make Christmas 'magical' for children.

Mothers and Children as Social Media Influencers

The visibility of women's labour as mothers has increased as women have embraced social media, with mum (or 'mommy') bloggers becoming important influencers of women's mothering practice (Archer, 2019a; Leaver, 2017; Lupton, Pedersen, & Thomas, 2016). As eyeballs have shifted from mainstream media, social media has filled the gap for both mothers as consumers and the advertisers seeking to influence mothers and children (Lopez, 2009). The mum blogger trend began more than 10 years ago and, for some bloggers, the practice has resulted in brands paying large sums to feature in 'advertorial' style posts (Hopkins, 2019). Mothers, originally motivated by the community and connection that social media offered, have, in some cases, seen the potential to earn money from their 'playbour' (Archer, 2019b). Blogging for social media influencers has now been complemented and in many cases superseded by posting shorter-forms and image-based material on popular platforms, including Facebook and Instagram (Tiidenberg and Baym 2017; Pinjamaa & Cheshire, 2016). With the rise of mum blogging and the uptake of social media by parents, the practice of 'sharenting' (parents sharing images of children on social media) has also developed (Blum-Ross & Livingstone, 2017; Chalklen & Anderson, 2017; Choi & Lewallen, 2018; Leaver, 2017). Many mum bloggers and everyday mums now share images of their children. In the case of the mum bloggers/social media influencers, children are now often seen as 'brand extensions' of their own personal brand (Archer, 2019c). In some cases, mothers (sidestepping developing their own personal brand) are going straight to market with images of their babies and children on social media, hoping to garner sponsorship and payment and fashioning their children as 'micro-microcelebrities' (Abidin, 2015; Choi & Lewallen, 2018).

Those social media influencers with children regularly use family/children's milestones and celebrations to create consumable content for their intended audiences on Instagram and other platforms (Abidin, 2017). Often these milestones and celebrations feature sponsored and/or paid content. As a major cultural event, Christmas offers plenty of opportunities for social media (and branded content) sharing. Christmas is a gendered activity and an important 'enculturation' process for children (Belk, 1989; Batinga, Pinto, & Resende, 2017; Brewis & Warren, 2011; Fischer & Arnold, 1990; Freeman & Bell, 2013; Vachhani and Pullen 2011). With the rise of 'sharenting' practices, concerns have been raised regarding the rights of the child relating to their data.

The Impact of Christmas on Mothers and Children

Mothers and women more generally are seen as the primary labourers in mainstream media when it comes to Christmas, fulfilling societal and family (and their own) expectations, performing myriad tasks in the lead up to and during Christmas day

(Brewis & Warren, 2011; Fischer & Arnold, 1990; Freeman & Bell, 2013). These tasks include planning for, buying and wrapping presents for children and other family/friends, planning the Christmas meal/s, buying the ingredients, overseeing any children's craft and baking activities relevant to Christmas, cooking much of the food and, often, cleaning up afterwards. Decorating the house and table for Christmas are also mostly seen as activities in the women's domain (Freeman & Bell, 2013).

Some scholars have noted the impact of mainstream media, including women's magazines, as a cultural 'guide' for women on what is appropriate and, in some cases, aspirational for the 'perfect family Christmas'. Content analysis of women's magazines shows the emphasis placed on the perfectly staged Christmas, with only some acknowledgement of women's dual roles as earners outside the home who may have limited time to stage a Christmas fit for a magazine photo shoot (Brewis & Warren, 2011; Freeman & Bell, 2013). Women, as the primary caregivers and purchasers of gifts for children, as well as other commodities relevant to Christmas, are the primary 'target market' for most brands. Christmas plays a large part in the process of enculturation of gender and consumption behaviour expectations in children (Batinga et al., 2017; Belk, 1989).

This chapter investigates the EotS phenomenon with an aim to view contemporary attitudes to the child and broader society, including women as gendered consumers. Children are viewed by marketers and mainstream media as one of the most important end 'consumers', even if they don't buy the products (Cook, 2008; Jenkins, 1998) and their role as important market actors cannot be overstated. Indeed, the EotS has been so deeply tied to commercialism and Christmas that in 2019 EotS partnered with Kellogg's to feature Elf giveaways with the purchase of popular breakfast cereals including Corn Flakes and Sultana Bran (see Fig. 22.1). The impact of Christmas on women has been investigated by some. Others have looked at the meaning of Christmas through children's eyes, including conducting analysis of children's letters to Santa.

The paradox of Christmas as a secular commercial extravaganza and religious Holy period has been researched and discussed for many years, with Belk and Bryce describing it as "the distilled essence of contemporary consumption" (Belk and Bryce 1993: 277). Importantly, as the most impactful of all cultural seasonal traditions in Australia, England, most of Europe, the USA, Canada, New Zealand and many other Western nations, the festival and its hyper-commercial nature (as a frenzied spending spree for many) has a profound influence on families and children in particular. In recent times that impact (in particular its commercial significance) has also been felt in other, predominantly non-Christian nations, including, but not limited to, the UAE and other parts of the Middle East, Singapore and Japan (to name some countries that have embraced it as a festival that is good for commerce even if not relevant to the dominant religion). Belk's investigation of the 'modern' US Christmas in 1989 explored mass media treatments of the meaning of Christmas and argued that Santa had many similarities to the Christ figure and was 'sacred' to America's Christmas commerce. However, unlike Christ, "Santa is a god

Fig. 22.1 Elf of the shelf competitions on Sultana Bran Kellogg's cereal. Source: Photo by authors, taken 30 November 2019

of materialism and hedonism, of modern consumer culture," Belk argued (Belk, 1989, n.p.).

We argue that the child is at the heart of consumer culture at Christmas and the EotS is a manifestation of this trend towards commodification and surveillance. Children, once valued by society for their use as labourers in the fields and factories

Fig. 22.1 (continued)

(Holloway, 2019), are now, it could be argued, valued for their data, imagined futures, and worth as consumers under the new normal of 'surveillance capitalism'. Parents, too, may use and value children as a form of self/personal brand extension. As Henry Jenkins (1998, p.22) argued "Children's culture is shaped at the global level through powerful institutions and at the local level through individual families.

Through these everyday practices the myth of the innocent child gives way to the reality of children's experiences." The 'powerful institutions' of today, of course, include social media companies and major brands (including toy marketers). We explore below the extent to which EotS on Instagram reveals the convergence of social media platforms, Christmas commercialisation, gendered labour and the normalisation of cultural expectations and practices of intimate surveillance.

Method

While analysis of Instagram is important, given its popularity as a social media site, it is also problematic, given the issues with downloading data, ethical privacy concerns and the challenges of investigating a mainly visual medium (Highfield and Leaver, 2016). Data for #elfontheshelf2018 were collected following a 7 year (2012–2019) ethnographic study of mum/mom bloggers, (some of whom are now using Instagram) by one of the authors. One of the Australian mum bloggers, interviewed in 2012, posted #elfontheshelf images to Instagram in 2018 and that piqued our interest in the phenomenon.

Analysis of this data is done with reference to visual narrative analysis (see, for example, Riessman, 2008; Rose, 2001) following Tiidenberg and Whelan's (2017) approach. Initially the content was thematically analysed for what the elf pictures were mainly used to communicate; this was followed with visual narrative analysis of specific posts to explain the layered and intertextual aspects of meaning making. The main posts for analysis were the top nine posts downloaded from Instagram that came up after searching #elfontheshelf2018 in June 2019. Image downloads, online news articles and field notes documenting routine and extensive situated immersion are also drawn upon here to describe the practices conducted by those posting with the #elfontheshelf2018 hashtag on Instagram. Nine of the top posts featuring #elfonthefuckingshelf hashtag were also downloaded for analysis, clearly emphasising a different approach to the toy.

Using Instagram's desktop client, any search of a hashtag will yield nine 'top' posts followed by the 'most recent' posts (which are presented as separate tabs in the mobile app). It should be noted that we are aware that, because of algorithms, the 'top' nine posts seen when one person searches may well be different than for other researchers. This is the 'black box' of Instagram that so frustrates many researchers of the platform. Nevertheless, the top nine posts represent a sample of some of the 'types' of posts and individuals posting using this hashtag. Our research has been informed by popular discourse on the EotS, discovered through Google searches of recent 'news' articles. We also referenced social semiotics because of the techniques it offers for the analysis of images (Kress & Van Leeuwen, 2006). Adopting social semiotic analysis, the analyst identifies graphical elements of the visual image—such as lines, shapes and colour. Social semiotic analysis has been used in the past to analyse magazine articles and television. The social semiotics analyst also examines the layout of the page/post in terms of how it may "present, realise or [...] contextualise social positions and relations" (Kress, 2009, p. 139).

Results and Discussion

Of the top nine posts downloaded with the #elfontheshelf2018 hashtag, only one was by a male, with one by a commercial (children's organic baby food in squeezable pouches) brand. The remaining posts were from women (all mothers) and one other was presented as a child/toddler influencer, presumably posted on their behalf by a parent. All of the mothers who posted had links to other social media channels, including some to blogs and others to YouTube, and clearly were posting in order to create a following, with branded content included. The following themes were noted: Elf projecting the perfect parenthood performance (and the Instagrammer as helpful guide towards this aspirational goal), Elf projecting creepy horror/adult themes, Elf projecting playfulness, Elf projecting the trials of 'work' around staging Christmas for children, and Elf as a way of staging children as objects of the gaze. These themes will be discussed in turn.

The Elf as projecting the 'performance of perfect parenting' was evident in the post of the Elf dressed in a spaceman/astronaut's clothes, fashioned from foil and strung as a decoration, and described as being presented because the Instagrammer/'s boys 'love space'. The poster references her special EoTS Facebook page that clearly has ideas for parents on how to display EoTS. The Elf's silver foil suit also ties into the colour scheme of silver and blue baubles, echoing women's magazine features that offer 'helpful' decorating tips for women at Christmas. The post has the hashtag #vlogger to foreground the mum as a video producer. Other EotS posts on Instagram project images that are creative and carefully staged by mothers in this same vein. One of the other top nine posts shows the Elf with two 'elf babies'. The post is by a mum/craft/sewing blogger who also promotes that the patterns for the babies are available on her blog, again using the Instagram post in the 'attention economy' to direct people's eyes to her blogpost.

In complete contrast, of the nine #elfontheshelf2018 posts analysed in detail, the post projecting the Elf as a creepy horror character was the only one by a man, and the Instagrammer presented himself as a make-up artist specialising in clowns/horror. The post was of the Elf strung up, upside down, mouth covered in black tape and with a clown figure about to chainsaw him between his legs. Hashtags such as #torture and #scaryclown accompanied the more benign #elfontheshelf2018 hashtag meaning children could easily stumble on this post while scrolling through their parents' phones. The 'adult' version of the Elf was also evident in other (sexualised) posts viewed outside the top nine posts for example with the Elf in a staged 'hot tub' with Barbie.

Given EotS is marketed as a toy for children's consumption, it is not surprising that many of the posts are devoted to 'playful' staging. For example, in the top nine posts, one is of the Elf contained in a jar with a note saying "I farted in here, wanna smell?" Other posts viewed outside the top nine reference the Elf 'pooing' chocolate drops or 'peeing' lemonade. One post in the top nine pictures a small 'sign' stating 'free reindeer poop' next to the Elf and chocolate drops. This post is also commercial in nature as it is by an American mum blogger offering her 'members' access to the downloadable signs, again directing eyeballs to her own blog.

Another commercial post of the top nine, from a brand for organic baby food, has a pouch of the baby food brand held by the Elf with a sign saying "That moment when you are almost asleep and realise you forgot to move the Elf." This shows a brand clearly using the EotS's relatability for parents, and cultural significance, to promote its product through Instagram. The post taps into the projection of the Elf as a chore for parents (specifically mothers).

Finally, two of the top nine posts analysed portray toddlers and babies as 'elves', dressed up as cute subjects of the adult gaze. In one post a toddler is on a toyshop shelf next to toys, dressed as an Elf. For this post, the Instagrammer uses the hashtag #Walmart and investigation of her Instagram feed shows she regularly posts content related to the large North American discount department store, Walmart. It is likely that the Instagrammer was paid by Walmart to post these images of their child in Walmart, although no #sponsored or #ad posts hashtags are evident. This account is portrayed as being of and by the toddler, rather than of the mother, showing the trend towards parents using children as social media influencers and the continued trend of (often undisclosed) advertorials.

The #elfonthefuckingshelf hashtag was a surprise find, given the EotS is supposedly a children's toy. These posts are irreverent and more 'adult' in nature. The now 'classic' and viral post from English cartoonist, blogging mum Katie Kirby's "Hurrah for Gin" is at the top of these posts. The following exchange in cartoon form is shown:

Child stick figure: Mummy, why don't we have an elf that moves around like everyone else?

Mother stick figure: Because Mummy's got enough on her fucking plate, sweetie.

The post taps into the cultural phenomenon of Elf on the Shelf, and (some) mothers' recognition that it is 'just one more chore' to perform to make the perfect children's Christmas.

Revisiting the #elfontheshelf hashtag search on Instagram in November 2019 also reveals the size of that hashtag, and the cluster of related tags that Instagram suggests when undertaking the basic search. At this time, there were over 3.5 million #elfontheshelf images on Instagram. The number of #elfonthefuckingshelf images was only 269, which suggests a number of images using this tag have been removed or deleted since it is smaller than the number of images a year earlier. Comparatively large, #elfontheshelfideas returned just over 133,000, most of which appear to, again, be posted by mothers and caregivers making helpful suggestions to others about reducing the imaginative labour required in posing the Elf each day. In contrast, the #naughtyelfontheshelf hashtag search returned just over 27,000 images, many of which were clearly adult in nature, with the top nine posts including four that featured young women wearing elf costumes in sexually suggestive poses. The range of intentions evident from these four hashtags shows the broad cultural impact of EotS as well as cementing the connection of gendered labour and the Elf on one hand and the broad use of EotS in a wide range of cultural settings on the other.

Finally, the huge presence of EotS on Instagram reveals the central place this relatively new story and toy have in the experience of Christmas. Taking the notion that Santa Claus is 'watching' the behaviour of children to determine whether they deserve toys, the Elf takes that surveillance a step further, making the Elves part of the performance of Christmas whilst also normalising the idea of surveillance, even within the home. As Pinto and Nemorin (2014) argue "The Elf on the Shelf essentially teaches the child to accept an external form of non-familial surveillance in the home when the elf becomes the source of power and judgment" and, moreover, more broadly this leads to the idea of accepting authority, even domestically, in the form of surveillance. In a cultural context where internet-enabled toys can also be undertaking a form of surveillance of children (Holloway & Green, 2016), the normalisation of surveillance can be troubling. In contrast, the subversive and overtly playful images of the EotS on Instagram point to clear limits on this surveillance which is neither always enacted, nor necessarily conceptually clear if children and families focus on the creativity of the EotS rather than the logics of being watched.

Conclusion

The Elf on the Shelf is a large and growing cultural phenomenon attached to the ever-increasing commercialisation of Christmas. The millions of EotS images on Instagram show the breadth of EotS in countries across the globe. Posts by Instagramming mothers clearly demonstrate and, at times, lament the gendered labour that places the burden of posing the Elf each night mostly on mothers. So, too, does the posts show the significant impact of the EotS narrative which, at the most basic level, normalises a form of surveillance in the domestic family space. Yet the cultural impact of EotS is by no means limited to the official commercial narrative developed by the owners of EotS. Indeed, there are many subversive, playful, adult and naughty Instagram images of EotS as well, showing that the Elf is not always faithful to its owners or the sanitised commercial message of good behaviour and banal surveillance.

References

Abidin, C. (2015). Micromicrocelebrity: Branding babies on the internet. *M/C Journal, 18*(5).

Abidin, C. (2016). "Aren't these just young, rich women doing vain things online?": Influencer selfies as subversive frivolity. *Social Media+ Society, 2*(2), 2056305116641342.

Abidin, C. (2017). # familygoals: Family influencers, calibrated amateurism, and justifying young digital labor. *Social Media+ Society, 3*(2), 2056305117707191.
Archer, C. (2019a). How influencer 'mumpreneur' bloggers and 'everyday' mums frame presenting their children online. *Media International Australia, 170*(1), 47–56.
Archer, C. (2019b). Social media influencers, post-feminism and neoliberalism: How mum bloggers' 'playbour' is reshaping public relations. *Public Relations Inquiry, 8*(2), 149–166.
Archer, C. (2019c). Pre-schooler as brand extension: A tale of Pixie's bows and birthdays. In L. Green, D. Holloway, K. Stevenson, & K. Jaunzems (Eds.), *Digitising early childhood* (p. 58). Cambridge: Cambridge Scholars Publishing.
Baranyai, R. (2015, December 11). *Elf on the Shelf trend mirrors loss of privacy*. The Sudbury Star. Retrieved from https://www.thesudburystar.com/2015/12/11/elf-on-the-shelf-trend-mirrors-loss-of-privacy/wcm/5ec030a5-3e89-328d-97bc-a472fa5b82ec
Batinga, G. L., Pinto, M. d. R., & Resende, S. P. (2017). Christmas, consumption and materialism: Discourse analysis of children's Christmas letters. *Revista Brasileira de Gestão de Negócios, 19* (66), 557–573.
Belk, R. W. (1988). Possessions and the extended self. *Journal of Consumer Research, 15*(2), 139–168.
Belk, R. W. (1989). Materialism and the modern U.S. Christmas. In E. Hirschman (Ed.), *Interpretive consumer research* (pp. 115135). Provo, UT: Association for Consumer Research. Retrieved from https://www.acrwebsite.org/search/view-conferenceproceedings.aspx?Id=12180
Belk, R. W., & Bryce, W. (1993). Christmas shopping scenes: From modern miracle to postmodern mall. *International Journal of Research in Marketing, 10*(3), 277–296.
Belk, R. W. (2013). Extended self in a digital world. *Journal of Consumer Research, 40*(3), 477–500.
Bergström, T., & Bäckman, L. (2013). *Marketing and PR in Social Media: How the utilization of Instagram builds and maintains customer relationships*. Retrieved from http://www.diva-portal.org/smash/record.jsf?pid=diva2%3A625012&dswid=-573
Bindley, K. (2018, December 12). The tyranny of the Elf on the Shelf: Where to put him tonight? *Wall Street Journal*. Retrieved from https://www.wsj.com/articles/the-tyranny-of-the-elf-on-the-shelf-where-to-put-him-tonight-11544634142
Blount, J. (2019). Want Instagram followers? *Here are 80 tools for growing your brand in 2020. Forbes* Retrieved from https://www.forbes.com/sites/joresablount/2019/10/05/want-instagram-followers-heres-80-tools-for-growing-your-brand-in-2020/#78e3c69d1bb0
Blum-Ross, A., & Livingstone, S. (2017). "Sharenting," parent blogging, and the boundaries of the digital self. *Popular Communication, 15*(2), 110–125.
Brewis, J., & Warren, S. (2011). Have yourself a merry little Christmas? Organizing Christmas in women's magazines past and present. *Organization, 18*(6), 747–762.
Chalklen, C., & Anderson, H. (2017). Mothering on Facebook: Exploring the privacy/openness paradox. *Social Media+ Society, 3*(2), 2056305117707187.
Choi, G. Y., & Lewallen, J. (2018). "Say Instagram, Kids!": Examining sharenting and children's digital representations on Instagram. *Howard Journal of Communications, 29*(2), 140–160.
Cook, D. T. (2008). The missing child in consumption theory. *Journal of Consumer Culture, 8*(2), 219–243.
Danno, M. (2019). *Digital 2019: Australia social media usage is growing*. Retrieved from: https://wearesocial.com/au/blog/2019/02/digital-report-australia
De Veirman, M., Cauberghe, V., & Hudders, L. (2017). Marketing through Instagram influencers: The impact of number of followers and product divergence on brand attitude. *International Journal of Advertising, 36*(5), 798–828.
Djafarova, E., & Rushworth, C. (2017). Exploring the credibility of online celebrities' Instagram profiles in influencing the purchase decisions of young female users. *Computers in Human Behavior, 68*, 1–7.

Eagar, T., & Dann, S. (2016). Classifying the narrated# selfie: Genre typing human-branding activity. *European Journal of Marketing, 50*(9/10), 1835–1857.

Fischer, E., & Arnold, S. J. (1990). More than a labor of love: Gender roles and Christmas gift shopping. *Journal of Consumer Research, 17*(3), 333–345.

Freeman, L., & Bell, S. (2013). Women's magazines as facilitators of Christmas rituals. *Qualitative Market Research: An International Journal, 16*(3), 336–354.

Frison, E., & Eggermont, S. (2017). Browsing, posting, and liking on Instagram: The reciprocal relationships between different types of Instagram use and adolescents' depressed mood. *Cyberpsychology, Behavior and Social Networking, 20*(10), 603–609.

Goffman, E. (1956). *The presentation of self in everyday life*. London: Penguin Books.

Highfield, T., & Leaver, T. (2016). Instagrammatics and digital methods: Studying visual social media, from selfies and GIFs to memes and emoji. *Communication Research and Practice, 2*(1), 47–62.

Holloway, D. (2019). Surveillance capitalism and children's data: The Internet of toys and things for children. *Media International Australia, 170*(1), 27–36.

Holloway, D., & Green, L. (2016). The Internet of toys. *Communication Research and Practice, 2* (4), 506–519. https://doi.org/10.1080/22041451.2016.1266124.

Hopkins, J. (2019). Communication, culture, and governance in Asia: "We connect with people through stories": Gender and affective labor in Momblogging. *International Journal of Communication, 13*(20).

Hutchins, R. (2019). Elf on the shelf: 2019 will be 'the biggest year in UK market' for the brand. *ToyNews*. Retrieved from https://www.toynews-online.biz/2019/09/17/elf-on-the-shelf-2019-will-be-the-biggest-year-in-uk-market-for-the-brand/

Iqani, M., & Schroeder, J. E. (2016). # selfie: Digital self-portraits as commodity form and consumption practice. *Consumption Markets & Culture, 19*(5), 405–415.

Jenkins, H. (1998). *The children's culture reader*. New York: NYU Press.

Kress, G. (2009). *Multimodality: A social semiotic approach to contemporary communication*. London: Routledge.

Kress, G. R., & Van Leeuwen, T. (2006). *Reading images: The grammar of visual design*. London: Routledge.

Leaver, T. (2017). Intimate surveillance: Normalizing parental monitoring and mediation of infants online. *Social Media+ Society, 3*(2), 2056305117707192.

Leaver, T., Highfield, T., & Abidin, C. (2020). *Instagram: Visual social media cultures*. Cambridge: Polity Press.

Levett, D. (2019). OPINION *Instagram removing likes is about money, not mental health*, Mumbrella. Retrieved from https://mumbrella.com.au/instagram-removing-likes-is-about-money-not-mental-health-589970

Lopez, L. K. (2009). The radical act of 'mommy blogging': Redefining motherhood through the blogosphere. *New Media & Society, 11*(5), 729–747.

Lup, K., Trub, L., & Rosenthal, L. (2015). Instagram# instasad?: Exploring associations among Instagram use, depressive symptoms, negative social comparison, and strangers followed. *Cyberpsychology, Behavior and Social Networking, 18*(5), 247–252.

Lupton, D., Pedersen, S., & Thomas, G. M. (2016). Parenting and digital media: From the early web to contemporary digital society. *Sociology Compass, 10*(8), 730–743.

Marwick, A. E. (2015). Instafame: Luxury selfies in the attention economy. *Public Culture, 27* (1 75), 137–160.

McCormack, A. (2019). *Instagram, Facebook crack down on harmful weight loss products and cosmetic procedures, ABC Triple J Hack*. Retrieved from https://www.abc.net.au/triplej/programs/hack/instagram-weight-loss-posts/11528822

Pinjamaa, N., & Cheshire, C. (2016). *Blogs in a changing social media environment: Perspectives on the future of blogging in Scandinavia* (2016). Research papers. 17. Retrieved from Http://aisel.aisnet.org/ecis2016_rp/17

Pinto, L., & Nemorin, S. (2014). *Who's the boss? "The elf on the shelf" and the normalization of surveillance*. Canadian Centre for Policy Alternatives. Retrieved from https://www.policyalternatives.ca/publications/commentary/who%E2%80%99s-boss

Ridgway, J. L., & Clayton, R. B. (2016). Instagram unfiltered: Exploring associations of body image satisfaction, Instagram# selfie posting, and negative romantic relationship outcomes. *Cyberpsychology, Behavior and Social Networking, 19*(1), 2–7.

Riessman, C. K. (2008). *Narrative methods for the human sciences*. Thousand Oaks, CA: Sage.

Rose, G. (2001). *Visual methodologies: An introduction to the interpretation of visual materials*. Thousand Oaks, CA: Sage.

Senft, T. M., & Baym, N. K. (2015). What does the selfie say? Investigating a global phenomenon. *International Journal of Communication, 9*(Feature), 1588–1606.

Slater, A., Varsani, N., & Diedrichs, P. C. (2017). # fitspo or# loveyourself? The impact of fitspiration and self-compassion Instagram images on women's body image, self-compassion, and mood. *Body Image, 22*, 87–96.

Tiidenberg, K., & Whelan, A. (2017). Sick bunnies and pocket dumps: "Not-selfies" and the genre of self-representation. *Popular Communication, 15*(2), 141–153.

Tiidenberg, K., & Baym, N. K. (2017). Learn it, buy it, work it: Intensive pregnancy on Instagram. *Social Media+ Society, 3*(1), 2056305116685108.

Vachhani, S. J., & Pullen, A. (2011). Home is where the heart is? Organizing women's work and domesticity at Christmas. *Organization, 18*(6), 807–821.

Zuboff, S. (2015). Big other: Surveillance capitalism and the prospects of an information civilization. *Journal of Information Technology, 30*(1), 75–89.

Catherine Archer is a Senior Lecturer and researcher specialising in social media and strategic communication at Murdoch University. Her current research interests include social media, focusing on families' use of social media, influencer relations and ethics, and the blurring of lines between media, marketing, public relations and communication.

Tama Leaver is a Professor of Internet Studies at Curtin University and President of the Association of Internet Researchers (AoIR). He recently co-authored Instagram: Visual Social Media Cultures (Polity, 2020) with Tim Highfield and Crystal Abidin. He is @tamaleaver on Twitter and his web presence is www.tamaleaver.net.

Chapter 23
Digital Predictions: Children's Futures, Opportunities and Obstacles

Michele Willson

Introduction

Early childhood is seen by many as the ideal time to shape, support and encourage the child in order to become fully emotionally, intellectually and socially competent adults in the future. Discussions about the degree that children can participate and have agency in these processes are ongoing (Livingstone & Third, 2017). However, what happens with these agentic capacities—of adults and children—when decisions are made based on big data analytics and predictive algorithms?

Predictive algorithms are enacted in the everyday in multiple ways: for example, autosuggested Google search terms; Amazon recommendations; Google map travel time forecasts, or more controversially in predictive (and pre-emptive) policing practices. Prediction entails forecasting possible outcomes based on modelling, pattern detection and recognition through the (supervised and/or unsupervised) analysis of large data sets using iterative machine learning algorithmic processes (McQuillan, 2016). These practices inform strategies, policies and planning.

Within the contemporary child's digital ecosystem/s, there are multiple and diverse predictive practices currently and potentially at play. In the health sector, for example, predictive machine learning algorithms anticipate the likelihood of genetically detectable disorders in IVF pre-implantation screening (Regalado, 2017) or a child's possibility of developing autism (Ananthaswamy, 2017); in the education sector, they are being applied to educational data to identify students at risk or those in need of particular types of targeted intervention (Clow, 2013; Smith, 2017); in the commercial sector they are used to nudge particular types of purchasing decisions or to prompt data disclosures.

M. Willson (✉)
Curtin University, Perth, WA, Australia
e-mail: m.willson@curtin.edu.au

D. Holloway et al. (eds.), *Young Children's Rights in a Digital World*, Children's Well-Being: Indicators and Research 23,
https://doi.org/10.1007/978-3-030-65916-5_23

This chapter explores a number of predictive practices in early childhood initiatives. In doing so, the paper raises questions about the broader ethical, and normative issues that become apparent for child-rearing practices, and the possibilities for child or parental current and future agency when predictive practices and risk aversion drive the choices that are made available, hidden or negated.

The Child as a Data Re/source

Children are increasingly positioned as data (re)sources and embedded in what I describe elsewhere (Willson, 2019) as algorithmic ecosystems. These systems intermingle, assist and disrupt. They rely heavily on various surveillance, reporting and data capture practices of the child from conception (even preconception) onwards for a range of diverse reasons and diverse stakeholders. Data about and from children are captured in multiple ways: biometric data recorded directly from their bodies through wearables and through data-enabled 'equipment' such as mattresses or child car seats; behavioural data extracted through camera surveillance, sensors and child monitor devices, or the translation of observation about these behaviours inputted into data systems by parents, carers, health or educational professionals; collected from child play activities directly through internet connected toys or through their engagement with entertainment and educational apps on tablets or smartphones are just a few of the myriad of data collection opportunities (Lupton & Williamson, 2017; Mascheroni, 2018).

For example, there has been a noticeable increase in the use of digital technologies directly by young children (0–8 years old). This uptake has been assisted by the introduction and ease of use of touchscreens and other devices such as internet connected toys (Holloway & Green, 2016). According to a 2013 EU report looking at the digital practices of children from 0 to 8 years old, at least 50% of Swedish 3–4 year old children use touchscreens; 25, 50 and 70% of American 3, 5 and 7 year olds respectively are online daily; and, 93% of 3 to 9 year old South Korean children are online for an average of between 8 and 9 h weekly (Holloway, Green, & Livingstone, 2013). Whether engaging with entertainment or educational activities offered by commercial or education providers, and with variable privacy measures and critical data literacy levels amongst children and their parents, the data collection possibilities of these types of engagement are immense.

We have been measuring, evaluating, recording and predicting children's activities and outcomes for all of modernity at least so the intent of these activities is not new. These practices form part of the underlying logic of liberal governance that informs governmental responsibility and care for populations and the individual's responsibility and care for the self. Increasing commercial encroachment into the everyday through data capture is an imperative of contemporary, or surveillance capitalism (Zuboff, 2015). The capacities for complex computing, big data analytic capacities and algorithmic machine learning push these practices into all areas of a child's life in a way previously unimaginable or physically unachievable.

As data is increasingly gathered, combined and analysed across an expanding, diverse array of everyday life activities, and as techniques and technologies become increasingly able to capture and manipulate these data, they in turn are employed as a way of managing risk, of driving agendas and shaping environments often in ways that we are not aware. What might these capacities and decisions mean for understanding agency when choices might be offered (or not) based on opaque predictions taking place unbeknownst to parents or child? What decisions might be made based on these predictions and the classificatory and correlation work that underpins it, and how might this affect a child's possible futures?

Predictive Modelling, Analytics and Action

By predictive practices, I am referring to the use of predominantly machine learning techniques using structured and unstructured data and algorithmic analysis to uncover noticeable patterns in behaviours, characteristics or relationships, to anticipate likely outcomes, to nudge behaviours and attitudes and to be able to take pre-emptive action or acts of intervention as a result. In predictive analytics, a variety of machine learning algorithms are employed depending upon the particular task, purpose, and types of data involved. As noted above, data can be drawn from and combined with almost anything: sleep patterns, movement, emotions, physiology, genetics, performance, sound...the list is endless. Different algorithm techniques can be combined into model ensembles (Burrell, 2016) and applied to innumerable data combinations to identify the likelihood of possible future outcomes; i.e. they aim to predict the likelihood of a particular event or occurrence taking place, to anticipate future scenarios or to encourage particular outcomes. Insurance companies, for example, use predictive calculations in their determination and assessment of likely risk in order to calculate premiums: house insurance premiums according to the suburb you live in, the type of building construction, what types of locks you use. These determinations are in turn built on broad analyses of instances where there has been an insurance claim in order to detect patterns and to calculate a risk score.

Predictive analytics use an actuarial form of surveillance whereby large data sets are scanned, rather than the interrogation of individual instances. As the capacity for data collection, storage, aggregation and manipulation is expanded, the possibilities for predictive analytics and the types of activities that these techniques can be applied are similarly expanded. In his discussion of predictive algorithms and their use to initiate pre-emptive action, Andrejevic notes:

> Preemptive practices do not intervene at the level of subject formation, but at that of the population. They are actuarial in the sense that they assess overall patterns of risk to determine probabilities of the emergence of particular events over time and space. The more comprehensive the data profile, the higher the likelihood of unearthing a relevant or actionable pattern. (2017: 883)

A relevant or actionable pattern for an algorithm is identified by recognising correlations amongst data sets. Note that the emphasis here is upon correlation—a seeming alignment or the co-appearance of particular types of activities such as, for example, the percentage of school absence rates and lower school achievement levels are used to suggest possible causation. While in some ways, school attendance and level of performance appears a self-evident correlation (surely if the child is not at school, they are missing out on learning activities that will hinder their overall achievement), it is a relatively blunt instrument if translated into a governance approach that directs action at the absence of the population rather than at other possible underlying contributors in individual circumstances. It can also lead to more serious or problematic correlation assumptions. The use of predictive analytics to seek patterns for identifying risk of child abuse in order to inform child welfare decisions (Willson, 2019) is an example where heavily surveilled populations—lower socio-economic families from particular ethnic or racial groups-may have higher correlation patterns with rates of abuse by default as a result of their level of surveillance; however, this does not mean that these particular social or ethnic groups are inherently more abusive or that abuses are not happening amongst other populations who are less heavily surveilled. The potential for predictive practices when applied to children and child-rearing to highlight or obscure particular characteristics of individual or groups of children intentionally but also unintentionally, therefore, warrants closer interrogation.

These predictions can have material consequences that can be advantageous or disadvantageous for the child, the family and their future pathways. As Cope and Kalantzis (2016: 13) note about predictive analytics in relation to education:

> Just as predictive analytics can be used to raise one's insurance premium or increase one's chance of arrest, so they might be used to predetermine a child's place in a learning track or a teacher's employment prospects

It is therefore also important to recognise that prediction does not happen in isolation as a simple process of input, analysis and output; it is the tying of algorithmic outputs of recognised patterns or the production of particular modelled scenarios with strategies and actions informed by particular discourses that generate outcomes (Fig. 23.1).

Such strategies or actions can be pre-emptive (in order to avert or capture an event), persuasive (to change or disrupt a predicted likely outcome) and targeted to specific individual characteristics or events. The use of predictive analytics in educational spaces can be used to identify children at risk of underperforming, overperforming or variously performing based on correlated patterns. The use of predictive analytics in health care can be used to identify possible markers in children for future diseases, or to manipulate genetic outcomes (Ananthaswamy, 2017; Regalado, 2017).

The perceived ability to anticipate and shape the future is alluring particularly in relation to children who are often positioned as vulnerable and malleable subjects. Indeed, it could be suggested there is a strong moral, social and cultural imperative that calls on society to do so. Commercial imaginaries are quick to harness this

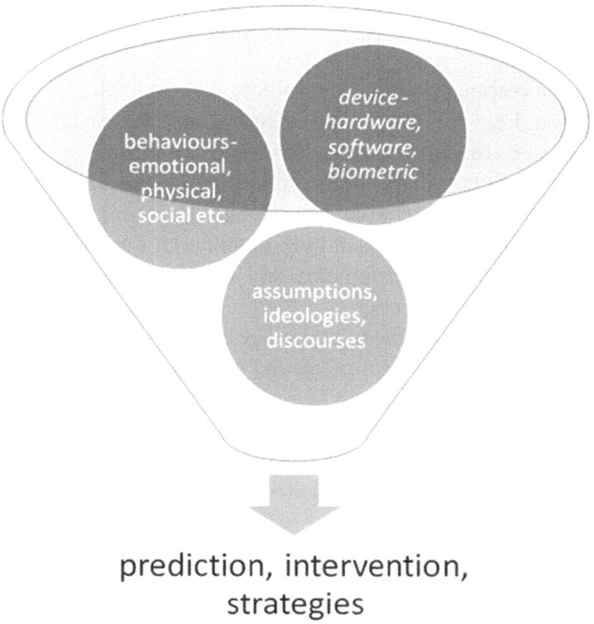

Fig. 23.1 Predictive assemblages

imperative in the types of services and offerings that are developed. Parents, educators and health professionals are all motivated to adopt technological tools and practices that will produce optimal child-rearing outcomes.

Writing about the data analytics industry, Beer (2019: 32) notes how the industry positions themselves in the marketplace:

> Data and their analyses are presented as being a powerful, ongoing and permanent presence, giving constant insights that are always there. . . . . These analytics reveal hidden value in the data, they shine a light on organisations and show things that were previously invisible. They enable the future to be seen and an imagined future to be part of the present decisions that are taken. They see everything, in detail; nothing escapes their sight.

The willingness to embrace initiatives that anticipate and shape the environment, abilities and practices of a child's future is revealed in the discussion below. Examples drawn from the education and commercial sectors reveal the range of activities upon which predictive attention is directed but also point to an increasing capacity for cross data aggregation between commercial entities, and also data collection, aggregation and analysis on the basis of data drawn from across the commercial and state sectors.

Commercial Care

In the commercial sector, products abound to allay the fears and concerns of parents. Tama Leaver (2017) has talked about the owlet: a sock the infant wears that measures oxygen intake, but there are a multitude of devices and wearables produced by private companies that capture infant behavioural and biological data, and offer benchmarking and advice in response to predicted outcomes. These devices not only capture the data for each individual child, the data can be manipulated, merged and used in other environments and in relation to other behavioural anticipation and device developments.

In order to be able to undertake relatively accurate predictive tasks, large amounts of data need to be collected and classified for that data to be able to be parsed and manipulated and 'made sense' of. Scale is important. Enough data needs to be gathered to render the outcomes generalisable and to increase accuracy in forecasting outcomes. Commercial applications and products alone may have the capacity to garner large swathes of data for such purposes, however, when linked or aggregated to data from other types of products the possibilities for predictive calculations are amplified.

Nod, a digital 'sleep coach' developed by Rest Devices, Inc. and Johnson and Johnson, is marketed as a tool to help sleep deprived parents of infants manage their offspring's sleeping patterns. According to a blog post by one of the co-founders of Mimo, Dulcie Madden:

> We realised we could deliver a personalized, self-learning sleep coach to parents, all via an app. Using huge amounts of sleep data, behavioral science, machine learning, clinical expertise, and love, we figured out that we could help identify a baby's sleep patterns, his or her parent's parenting style, key problems the family was facing, and then deliver a customized program for that family to do, night by night, to get more sleep within weeks. (https://www.mimobaby.com/single-post/2017/10/04/Solving-for-sleep)

In fact, the Nod website promoting the app claims that "within 30 days of using Nod, families experience two more hours of sleep a night, two hours fewer night wakings and four more hours added to longest overnight sleep period." (https://www.nodtosleep.com/). Nod is marketed as not only a way to manage a child's individual sleep patterns but also to be able to do this within the parameters of your parenting 'preferences'.

This has been made possible by the use of data analytics and machine learning and predictive modelling drawing from large data sets. According to a press release in 2017, "Both companies [Rest Devices, Inc. and Johnson and Johnson] have studied hundreds of thousands of baby sleep patterns, so by combining their expertise, the Nod app can provide parents with an advanced, personalised sleep coaching system—like having a baby sleep expert in your own home." (https://www.jnj.com/innovation/nod-digital-baby-sleep-coach).

These hundreds of thousands of sleep pattern data have been captured through the Johnson's baby sleep app and through infant wearables and crib monitors sold by Rest Devices' Mimo products and presumably are used to train the underlying machine algorithms. Johnson also has the capacity to aggregate data across a range of operations and contexts—Johnson and Johnson Pacific Limited encompass consumer health, medical device, pharmaceutical and biologics companies as part of their holdings.

Recommendations generated by Nod include advice on parenting practices and intervention strategies to change infant sleep patterns towards sleeping for longer with fewer interruptions: the ideal for sleep-deprived parents. Recommendations then are premised upon predicting possible outcomes based on past practices (those of the thousands of infants' sleep data that the algorithms have been trained on alongside the individual data that is captured for that particular child) in combination with data entered by sleep experts (https://www.babysleep.com/).

So, why are these types of activities worthy of consideration? Sidestepping concerns around privacy or questions about potential commercial nudging of consumers, there is a broader question about the delegation of parental consideration and judgement on the basis of abstracted data sets, and machine learning prediction.

In her discussion of a group of new mothers' use of parenting apps, Thornham (2019) notes a number of things relevant to our discussion here. First, she draws a link between the type of data these apps record and how they align with the questions these mothers are being asked by their health professionals (p. 176). This, she suggests, explains why the apps are so popular. Second, she then points to how these intermeshed relationships between data and feedback from the app are used as a way to alleviate anxiety and concern about parental approaches and infant progress while simultaneously aggravating this uncertainty through its very visibility via the prompted need to monitor and enter the requested data. As a result, not only did mothers use the information fed back to them via the app as a way to validate their experiences and performance, the health professionals and the mothers referred to this data rather than on the mothers' recollections as being more accurate and truthful (p. 177). A 'handing over' or delegation of maternal judgement, and agency in relation to infant care from mother to an app is undertaken creating a complicated relationship between infant, mother, technological device, the data obtained and interpreted and the health care professional. Leaving aside broader and important questions about how collected data is used by the app provider and how that data may, in turn, be aggregated, manipulated, and analysed to uncover further patterns with resultant observations fed back to the health care and commercial sectors, the normalisation of the use of such tools as a replacement for or better than an individual's personal observations is problematic for a range of reasons. These reasons include the fact the provision, interpretation and predictive outcomes are always open to inaccuracies whether due to messy or inaccurate initial data, due to underlying programming assumptions and parameters that might amplify particular discourses over others or require categorisation actions that render some groups as invisible or less powerful, due to opaque machine learning formulations and possibly erroneous rules, or simply that nuance or subjective or alternate interpretations are not made available when the prediction is made on the basis of rigid data collection categories. However, these possibilities are not open to scrutiny or broader interrogation: in many cases, they are accepted uncritically and then acted upon.

> Big data enables a universalisable strategy of preemptive social decision making. Such a strategy renders individuals unable to observe, understand, participate in, or respond to information gathered or assumptions made about them. When one considers that big data can be used to make important decisions that implicate us without our even knowing it, preemptive social decision making is antithetical to privacy and due process values. (Kerr & Earle, 2013, 71)

These pre-emptive decision making possibilities can have important ramifications for the child's developmental, educational and relational future opportunities and pathways. Yet these ramifications are not transparent or easily critiqued.

Captured at School (Educational Data Mining)

Data analytics using predictive algorithms and modelling are also well entrenched in the education systems and educational discourses at all levels—from childcare (Willson, 2019) through to the university sector (Clow, 2013; Knox, 2017). These analytics extend from the analysis of student text, student progress, peer interaction, personalisation of learning tasks and assessment, broadening out to include cognitive, behavioural and emotional analysis. Indeed:

> Educational data scientists are becoming new kinds of scientific experts of learning with increasing legitimate authority to produce systems of knowledge about children and to define them as subjects and objects of intervention. (Williamson, 2016: 401)

Ben Williamson (2016) explores the multitude of, what he refers to as, biopolitical pedagogies increasingly employed within the education sector. These pedagogies situate data extracted through biometric devices and techniques within interpretative frameworks drawn from psychology, physiology and neuroscience to explain, predict and anticipate learning and developmental outcomes. By extending the sphere of educational influence beyond simple learning analytics derived through online assessment, monitoring and delivery through personalisation practices, the potential sphere for possible identification, prediction and intervention into the child's development is expanded to bring bodies, emotions and minds into a data enhanced educational approach. These interpretations can be at a remove from individual educator's or carer's own observations and interventions in relation to a particular child's learning, instead undertaken in a pre-emptive, presumptive and anticipatory manner as a result of correlation with an identified pattern or behaviour derived from large aggregated data sets: a just-in-case scenario.

Education providers are increasingly reliant on data-capturing commercial devices and providers of services for the provision and analysis of their educational activities across all age groups from childcare, kindergarten through to secondary and tertiary education although the nature of the activity undertaken and resultant analysis may differ. The increasing demand for interactive and engaging content underpinned by the argument about the importance of engaging different learners through fun, interactive and personalised learning activities compels educators to acquire and use online educational games and activities provided by third-party commercial providers. These third-party providers collect and may disseminate or share data on children's activities with little control over this disclosure by parents or children or indeed the education provider. Relatedly, the education providers themselves seem to be relatively opaque as far as their information data collection activities of their learners' activities (and that of parents) and the use of that data. For example, a search undertaken by this researcher in late 2018 of a number of West Australian public school websites and the state's education department web profile could not locate any evidence of a privacy or information disclosure policy to assist in interpreting the use of data collected through commercial learning management systems or applications and any data analytic or predictive activity that might be undertaken by either public or commercial entities. There were certainly no disclosures to that effect displayed.

The conflation and intermingling of spheres of activity—commercial and educational, commercial and health, public and private-not only offer opportunities for deeper and more complex data collection, aggregation and analysis, including prediction, they also extend the coverage and potential impact of the predictive application to children's futures. According to a 2012 Federal Trade Commission report, the range of data collection practices by commercial apps targeted at children is extensive and the level of disclosure of this data collection and distribution activity available to parents is inadequate. The report noted that "nearly 60% (235) of the apps reviewed transmitted device ID to the developer or, more commonly, an advertising network, analytics company, or other third party" yet "...., only 20% (81) of the apps reviewed disclosed any information about the app's privacy practices (Mohapatra & Hasty, 2012: 6).

Issues with Predictive Techniques

The range of issues with the application of predictive techniques have been touched on in the above discussion and in various literature elsewhere (see for example, Andrejevic, 2017; Crawford & Schultz, 2014; Dencik, Hintz, & Carey, 2018; Willson, 2019). There are clearly benefits to predictive techniques where, for example, attention is drawn to the likelihood of an adverse event such as a detrimental health outcome that might be prevented or alleviated as a result of either intervention or by increasing vigilance and resultant monitoring. However, there are a number of things to be mindful, and that require critical awareness in terms of the aims, context, and process when evaluating predictive approaches.

Aims and Context

Aims denotes the underlying rationale for the predictive analysis being undertaken. This may be intentional—deliberately intending to find a way to determine particular outcomes—or accidental by machine identification or unanticipated or unforeseen patterns but seen as useful and actioned on that basis.

In critically assessing predictive analytics and their aims, context becomes important as the analytic aim or intent is entwined with cultural and social expectations and power differentials. A health context with the aim of enhancing child health outcomes (Ananthaswamy, 2017) is innately different to the use of child health data by an insurance company interested in identifying future risk and possible premiums or a commercial company interested in selling child health monitoring products. Yet the possibilities for delineating the boundaries between these two contexts—health and commercial—in their collection and use of data for predictive analysis becomes increasingly unclear where not regulated.

Different child and family cohorts may be subject to different types and levels of surveillance, with different types of predictive intents and outcomes likely. These are coupled with cultural assumptions, and disciplinary discourses as to what is normal, desirable or commodifiable. For example, the surveillance and predictive intent targeted at welfare recipients in order to identify children at risk will differ in nature but also in consequence, to the types of predictive analytics targeted at the purchasing practices of wealthy parents who can be influenced by their children to purchase particular products. This targeting highlights some cohorts and occludes others from consideration or participation.

Process

The processes employed in predictive analytics are important also because if the underlying data (often messy and drawn from different domains with different underlying parameters) is incorrect, inadequate, partial or biased then the outcome will be also. However, the capacity to check these processes is limited: machine learning techniques are opaque, the rationale for decisions and outcomes frequently unclear and unable to be interrogated. A health professional can be questioned about a treatment plan and advise on the research that informs that decision, a machine learning recommendation is not open to this level of interrogation or interaction—at least not yet. This opacity will become even more comprehensive with the increasing adoption of machine learning and artificial intelligence. The immense scale of data sets employed means a reliance on technological calculation is automatic as human calculation can be too time consuming and costly (if even possible).

Predictive techniques encourage the conflation of correlation with causation: however, the identification of patterns does not automatically denote any particular causal relationship exists as a result. They also have the capacity to replicate or amplify particular assumptions simply by the programming and attention to particular characteristics or data sets based on flawed assumptions. As Dencik et al. (2018: 12) note,

> Algorithms may create self-fulfilling prophecies whereby the targeting of certain groups in the initial analysis raises their visibility in all future calculations while obscuring other forces at play.

This obscuration potentially distorts predictive accuracy and applicability impacting upon the decision making and application of the outcomes.

Moving Forward

For contemporary children whose lives are increasingly datafied from the outset (i.e. from birth or, increasingly, even before birth), predictive potentials and consequences are amplified exponentially. Questions about what choices become available to them or not, for what reasons and what recourse they may have to change these opportunities and pathways become an increasingly pressing consideration. In a comment about the general population, Andrejevic (2013: 297) notes:

> Every message we write, every video we post, every item we buy or view, our time-space paths and patterns of social interaction all become data points in algorithms for sorting, predicting, and managing our behavior. Some of these data points are spontaneous, the result of the intentional action of consumers; others are induced, the result of ongoing, randomized experiments. The complexity of the algorithm and the opacity of correlation render it all but impossible for those without access to the databases to determine why they may have been denied a loan, targeted for a particular political campaign message, or saturated with ads at a particular time and place when they have been revealed to be most vulnerable to marketing.

These observations are even more relevant in the case of children who have less capacity to control the data collection and the predictively motivated decision-making that is undertaken yet has real impact on their lives. It is also more relevant given the capacity to collect data from conception onwards offering future potential for extensive profiles to be generated. Moving forward, mechanisms to interrogate, to make transparent and to contest data predictions or interventions and to highlight opportunities denied or offered as a result will need to be developed and critical literacy in relation to data collection and predictive practices will need to be acquired by all. Some nations and governance entities are putting in place overall regulatory measures to address individual data privacy management, data collection and data analytic practices but these are partial, situated as responsive to current technical possibilities and do not accommodate techniques and data predictive capacities yet to be developed. Until that time, ongoing questioning of children and their parents' agency in these predictive environments requires critical attention.

References

Ananthaswamy, A. (2017, June 7). Baby brain scans can predict who is likely to develop autism. *New Scientist*. Retrieved from https://www.newscientist.com/article/2133941-baby-brain-scans-can-predict-who-is-likely-to-develop-autism/

Andrejevic, M. (2013). Estranged free labor. In T. Scholz (Ed.), *Digital labor: The Internet as playground and factory* (pp. 279–309). London: Routledge.

Andrejevic, M. (2017). To preempt a thief. *International Journal of Communication, 11*, 879–896.

Beer, D. (2019). *The data gaze: Capitalism, power and perception*. London: Sage.

Burrell, J. (2016, January–June 1–12). How the machine 'thinks': Understanding opacity in machine learning algorithms. *Big Data & Society*.

Clow, D. (2013). An overview of learning analytics. *Teaching in Higher Education, 18*(6), 683–695.

Cope, B., & Kalantzis, M. (2016). Big data comes to school: Implications for learning, assessment and research. *AERA Open, 2*(2), 1–19.

Crawford, K., & Schultz, J. (2014). Big data and due process: Toward a framework to redress predictive privacy harms. *BCL Rev, 55*, 93. Retrieved from http://lawdigitalcommons.bc.edu/bclr/vol55/iss1/4

Dencik, L., Hintz, A., & Carey, Z. (2018). Prediction, pre-emption and limits to dissent: Social media and big data uses for policing protests in the United Kingdom. *New Media & Society, 20* (4), 1433–1450. https://doi.org/10.1177/1461444817697722

Holloway, D., & Green, L. (2016). The Internet of toys. *Communication Research and Practice, 2* (4), 506–519. https://doi.org/10.1080/22041451.2016.1266124

Holloway, D., Green, L., & Livingstone, S. (2013). *Zero to eight. Young children and their internet use*: London: EU Kids Online, LSE.

Kerr, I., & Earle, J. (2013). Prediction, preemption, presumption: How big data threatens big picture privacy. *Stanford Law Review Online, 66*(65).

Knox, J. (2017). Data power in educations: Exploring critical awareness with the "learning analytics report card". *Television & New Media, 18*(8), 734–752.

Leaver, T. (2017). Intimate surveillance: Normalizing parental monitoring and mediation of infants online. *Social Media + Society, 3*(2). https://doi.org/10.1177/2056305117707192

Livingstone, S., & Third, A. (2017). Children and young people's rights in the digital age: An emerging agenda. *New Media & Society, 19*(5), 657–670.

Lupton, D., & Williamson, B. (2017). The datafied child: The dataveillance of children and implications for their rights. *New Media & Society, 19*(5), 780–794.

Mascheroni, G. (2018). Datafied childhoods: Contextualising datafication in everyday life. *Current Sociology*, 1–16. https://doi.org/10.1177/0011392118807534

McQuillan, D. (2016, July–December). Algorithmic paranoia and the convivial alternative. *Big Data & Society*, 1–12.

Mohapatra, M., & Hasty, A. (2012). *Mobile apps for kids: Disclosures still not making the grade.* Federal Trade Commission staff report. Retrieved from https://www.ftc.gov/sites/default/files/documents/reports/mobile-apps-kids-disclosures-still-not-making-grade/121210mobilekidsappreport.pdf

Regalado, A. (2017). Eugenics 2.0: We're at the dawn of choosing embryos by health, height, and more. *MIT Technology Review*, 1 November. Retrieved from https://www.technologyreview.com/s/609204/eugenics-20-were-at-the-dawn-of-choosing-embryos-by-health-height-and-more/

Smith, R. (2017). The emergence of the quantified child. *Discourse: Studies in the Cultural Politics of Education, 38*(5), 701–712.

Thornham, H. (2019). Algorithmic vulnerabilities and the datalogical: Early motherhood and tracking-as-care regimes. *Convergence: The International Journal of Research into New Media Technologies, 25*(2), 171–185.
Williamson, B. (2016). Coding the biodigital child: The biopolitics and pedagogic strategies of educational data science. *Pedagogy, Culture and Society, 24*(3), 401–416.
Willson, M. (2019). Raising the ideal child? Algorithms, quantification and prediction. *Media Culture & Society, 41*(5), 620–636.
Zuboff, S. (2015). Big other: Surveillance capitalism and the prospects of an information civilization. *Journal of Information Technology, 30*, 75–89.

Michele Willson is Professor of Internet Studies at Curtin University, Perth in Western Australia. Her research explores the intersection of technology and the everyday with a particular focus on algorithms, digital childhood, social games and online sociality more generally.

Chapter 24
Research Ethics and Digitising Early Childhood

Madeleine Dobson, Karen Murcia, Kim Gifkins, and Donell Holloway

Introduction

Ethical issues arise in all aspects of social research and are arguably magnified when the participants are young children living, learning, and developing in environments that integrate digital technologies. Research is responding to the need for a better understanding of the challenges presented to families and educators who are supporting young children growing up in a rapidly-evolving digital age. Established research guidelines provide a loose ethical framework but often don't answer the questions and dilemmas facing researchers who are working to understand the impact of digital technologies on the lives of young children. These questions and dilemmas can present complexities which, in this emerging space, are challenging to navigate and resolve.

In this chapter, we share researchers' experiences and perspectives on this topic and explore potential pragmatic anchors for the ethical decisions made in response to issues emerging in the field. We present vignettes from researchers' practice and analyse these through the Digital Child Ethical Research Framework developed from the current National Statement on Ethical Conduct in Human Research and the EECERA Code for Ethical Conduct. Our analysis and discussion include topics such as the negotiation of young children's informed consent; their participation in digital data collection; respecting children's rights; honouring their voices in

M. Dobson · K. Murcia (✉) · K. Gifkins
Curtin University, Perth, WA, Australia
e-mail: Madeleine.dobson@curtin.edu.au; Karen.murcia@curtin.edu.au; Kim.gifkins@curtin.edu.au

D. Holloway
Edith Cowan University, Perth, WA, Australia
e-mail: Donell.holloway@ecu.edu.au

D. Holloway et al. (eds.), *Young Children's Rights in a Digital World*, Children's Well-Being: Indicators and Research 23,
https://doi.org/10.1007/978-3-030-65916-5_24

research; data security; and, managing anonymity when disseminating visual data. Our objective is to provide greater ethical clarity and to support the design of sensitive research which embraces respectful inclusion of young children living in the digital age.

Pragmatism as a Research Paradigm

Pragmatism as a paradigm opens the lens of research inquiry and recognises multiple world perspectives and potential realities. A paradigm, defined as an accepted model or pattern (Kuhn, 1962), describes the parameters of research and makes explicit the often-intangible values and beliefs that cannot be directly observed, yet impact and shape social phenomenon and peoples' behaviour. Arguably, valuable analysis and interpretation of ethical principles and implications arises from immersion into the complexity of the real world which includes some more subtle actions and practices that are driven by researchers' values and beliefs.

Informed by the work of Dewey, pragmatism values the joining of beliefs and actions which underlie the quest for knowledge and understanding through social research. Dewey (1958, p. 387) states, "We must begin with things in their complex entanglements rather than with simplifications made for the purpose of effective judgment and action". Pragmatic inquiry challenges us to question how we know and how we make representations of the social world. Morgan (2014, p. 8) explains how pragmatism draws focus towards questions such as: "How do researchers make choices about the way they do research? Why do they make the choices they do? And, what is the impact of making one set of choices rather than another?" Understanding social experiences requires deep immersion and questioning of both actions and beliefs and how each informs the other in an ongoing cyclic fashion. Pragmatism can be thought of as a philosophical approach, a stance, or a working point of view. It can be generative in nature by considering multiple peoples' experiences of knowledge or events and ethical pathways through the many possibilities for actions (Badley, 2003). Experiences that people have may reveal possibilities and potential actions or responses but it is the ever-evolving nature of how people interpret what is experienced that points us towards ethical research.

As a research paradigm, pragmatism supports modes of analyses based on cycles of abductive reasoning (Feilzer, 2010). We used this approach to frame our interrogation and response to the complexities and dilemmas that confront social researchers endeavoring to practice ethically. Our aim was to transform the challenges of conducting research in the digitising of early childhood to comprehensible situations which are illuminated by critical questioning. Researchers must consider the ever-increasing ways child's identity can be accessed and personal information shared due to the digital nature of communication and data. We must be diligent and always questioning in order to protect the rights of children who participate in research.

Digital Child Ethical Research Framework

In developing our Digital Child Ethical Research Framework, the National Statement on the Ethical Conduct in Human Research (National Health and Medical Research Council, 2007—updated in 2018) was reviewed and considered in the context of early childhood and the digital age. First, it is noted that the values of respect, research merit and integrity, justice, and beneficence are emphasised and reinforced as underpinning the relationship between researchers and participants. The updated National Statement includes a section that aims to cover the elements that occur across the lifespan of a research project, including design, review, conduct, and outcomes. The guidance provided is promoted as being applicable to all fields of research and some advice is provided for research involving digital technologies. Discussion of the elements of research includes how to explore/justify/manage risks and build ethically good relationships with participants—but with a focus on those participants being autonomous adults.

The background section of the National Statement notes that difficult ethical questions in research can arise due to consideration of trust—i.e. research participants may form a relationship with a researcher they do not know but will need to trust. When participants are young children, further consideration and explanation of how a trusting relationship will be built and maintained becomes critical to a successful research project.

The document includes a section specifically referring to children and young people that assists in guiding interactions with schools and considers the uniqueness of research with young children. Although the National Statement recognises that children may have a "developing capacity ... to be involved in decisions about participation in research" (National Health and Medical Research Council, 2007/2018, p. 65), young children are generally regarded as having limited understanding and there is no requirement for their consent. Instead, it is required that consent be obtained from parents/guardians. Nonetheless, researchers are encouraged to engage with young children in discussing the research and its possible outcomes. Researchers therefore have a dual responsibility to build trust with both parents/guardians and the young children engaged in their research. Where research suggests that the experience of young children's interaction with digital technology is qualitatively different to that of their parents, this presents a challenge when seeking to conduct research *with* (rather than *on*) young children.

Recognition of the positive and important role that digital technologies can play in early childhood education is becoming established. For example, Early Childhood Australia's (ECA) statement on young children and digital technologies takes a nuanced and multidimensional view by looking at the participation of children and families in digital contexts and the purposes and value of digital technologies across multiple areas. The statement is aligned to the United Nations Convention on the Rights of the Child and ECA's Code of Ethics, thus taking a rights-based perspective (UN General Assembly, 1989; ECA, 2018). Throughout this chapter, we will explore these guidelines and what they can offer to researchers as they approach work involving young children and digital technologies. Although ethics guidelines have been criticised for presenting a deficit model in relation to young children, researchers can and should employ the guidelines as a tool to enable the construction of trust in the research relationship.

The European Early Childhood Education Research Association (EECERA) Ethical Code (EECERA, 2014) provides the following ethical principles, which center around an ethic of respect:

1. The child, family, community, and society;
2. Democratic values;
3. Justice and equity;
4. Knowing from multiple perspectives;
5. Integrity, transparency, and respectful interactions;
6. Quality and rigour;
7. Academic scholarship; and,
8. Social contribution.

These principles ask that researchers embrace ethics throughout the entirety of their project, in a conceptual and methodological sense, and also with a future focus. Possessing a short, medium, and long-term focus with regards to research ethics is of great value for researchers and the individuals and communities with whom they work.

There is also benefit for researchers in examining the ECA Code of Ethics (ECA, 2016). While this document is pedagogically focused, a number of principles also relate to research—particularly in terms of the conceptualisation of children and families participating in research. These principles recognise:

- Each child has unique interests and strengths and the capacity to contribute to their communities;
- Children are citizens from birth with civil, cultural, linguistic, social and economic rights;
- Partnerships with families and communities support shared responsibility for children's learning, development and wellbeing; and,
- Democratic, fair and inclusive practices promote equity and a strong sense of belonging.

Through consideration of these principles, researchers can consider ethical issues around children's participation, their role as researcher, and how to create a democratic and equitable research process for all involved.

The above-mentioned principles have informed the development of the Digital Child Ethical Research Framework (see Fig. 24.1). We approached the development of this framework as a team, engaging in critically reflective discussions and drawing from our research experience and our knowledge of challenges, tensions, and opportunities where children and digital technologies are the focus of research. Our intention was to formulate and pose specific questions that are useful for researchers to consider when planning and undertaking work with young children. We arrived at a shared understanding that the 'digital child' is one who is engaged

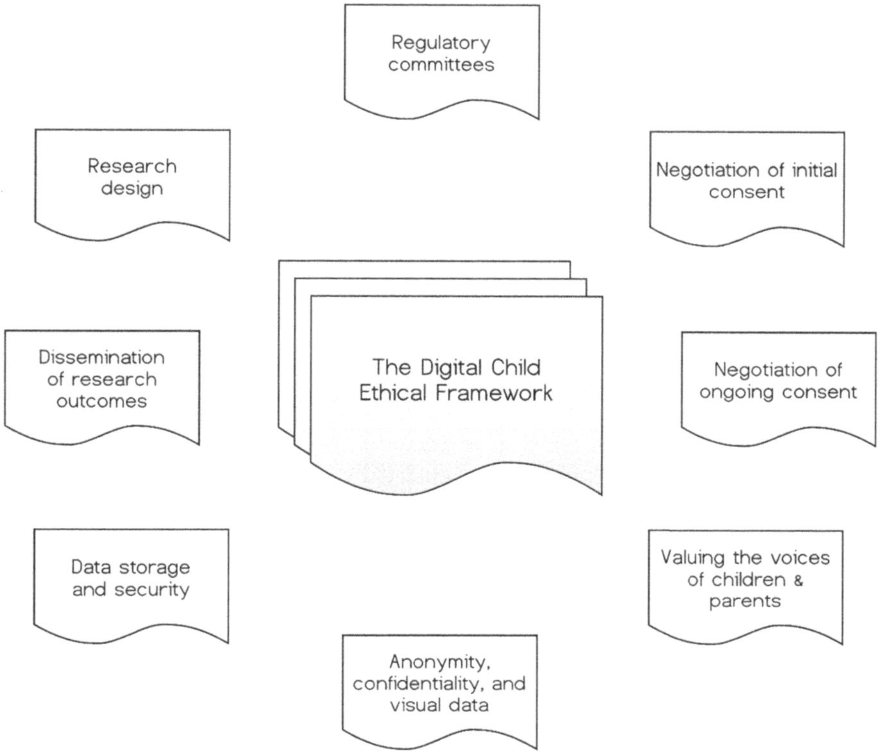

Fig. 24.1 The digital child ethical research framework

with digital technologies and impacted by the use of digital technologies by their family members, educators, and communities. The framework is presented and used in this chapter to unpack the researchers' vignettes. The analysis is further informed by international work where the ethics of research with young children is viewed as situated, dialogical, and relational (European Co-operation in Science and Technology, 2018; Flewitt, 2005).

Our Approach

Pragmatism as a research paradigm underpinned our interrogation of ethical challenges and dilemmas in research into the digitisation of early childhood. We considered and debated different researchers' perspectives regarding their work with children and digital technologies in order to provide an enriched understanding of ethical implications within the field. For the current framing and interrogation of ethics in the digitisation of early childhood in this chapter, four researchers shared vignettes that captured the ethics of their research. A University Research Ethics Officer whose work involved reviewing and overseeing the management of academic research ethic applications, joined with the research team and contributed to the joint critical analysis of each vignette.

The range of problematic ethical implications arising from the examples of research with young children—such as approaches to privacy and confidentiality and ways of honouring the rights of children—required thoughtful reflection and inquiry. We considered inquiry as a process focused on asking and answering questions where "beliefs that have become problematic are examined and resolved through action" (Morgan, 2014, p. 3). We reflected on the actions described to consider the beliefs of the researchers which in turn led us to consider how beliefs influence the choice of action. This analysis process was reflective and cyclic; moving back and forth between beliefs and actions. It represented the dynamic nature of ethics in research and the way early childhood philosophies and national ethics guidelines are interwoven with practice.

This inductive analysis process was guided by the literature-informed Digital Child Ethical Research Framework, which was specifically developed for this purpose and with the early childhood research context in mind. Significant questions raised and discussed by the research team in each element of the framework were captured and are shared here as an outcome of the inquiry (Table 24.1).

This framework and guiding questions are intended to support other researchers' ethical deliberations throughout the life of a project where the focus is on children and digital technologies. We illustrate the use of the framework and guiding questions in the discussion of each vignette below.

Vignette One

There were ethical issues aplenty during the development of my emerging research project which seeks to explore how children and childhood are constructed in parents' Instagram posts.

The development of this project took some time as myself and my research partner were preoccupied by ethical questions. Our background is in Early Childhood Education and Care and we embrace a strong respect for children and families. While many Instagram users have publicly available accounts and a high follower count, we wondered whether it was really right to simply collect data from parents' Instagram accounts and to analyse these without their consent. Some would say this is acceptable given the visibility of such accounts, however we were unsure—in the end, we arrived at a decision that prioritised our respect for children and families and asked for consent. Asking these families for consent may pose limitations in terms of opportunity for data collection, but it seemed the most ethically sound approach from the perspective of engaging respectfully with families.

We focused on recruiting parents with public Instagram accounts where the focus was the family and the children. Our communication with prospective participants

Table 24.1 Ethics elements and critical reflection questions

Research stage	Questions for reflection
Research design	How do we ensure there is rigour in our design and the process is worthwhile and meaningful for the participating children and families? What is the ask on children's time; are we adding something to normal activities or want to observe and understand what they are already doing? How do we frame our study and make explicit our paradigm and conceptual lens for examining the digitising of childhood? How do we design our research with children and families in a way that gives voice and agency to them in the context of the project?
Regulatory committees (including University and Industry-based Ethics committees)	How can the competing needs of risk management at a site level be managed with ethics of research? How feasible is it to gain access to the context required for conducting the research? How much time should be allowed for the required ethics review of the proposed research? How can we work constructively with feedback that challenges your preferred research design? How can we communicate effectively to regulatory committees about innovative digital technologies and or practices in research?
Negotiation of initial consent	What strategies can we employ so children understand our research? How can we adapt the initial information provided in order to meet the different language development levels of children? How can negotiation of initial consent take into account both parents' consent and children's assent? How can children indicate their willingness or not to participate? How can we be confident children are giving consent in an informed and authentic way?
Negotiation of ongoing consent	What strategies can we employ to 'check-in' with children as the research progresses? How can we take a multi-sensory approach to understanding children's on-going consent to participate? How do we respect children's agency whilst also managing children's off task behavior and limited attention levels to research tasks? How can we offer parents, site managers and or education leaders updates about progress so they maintain informed ongoing consent?
Valuing the voices of children and parents	What digital technologies can be used to accurately capture children's actions and voices? How can digital technologies be put into the hands

(continued)

Table 24.1 (continued)

Research stage	Questions for reflection
	of children so they capture and record their perspective of their world? How can information and images shared on social media be used ethically in research?
Anonymity, confidentiality and visual data	What strategies can we use to protect the identity of the children and ensure no harm is done as a result of the research? What techniques can be used to protect the identity of children in digital photos and video date? How can online images of children shared in public domains be used in research while ensuring anonymity and confidentiality? How can we share a child's image with their parent, while still protecting the identity of other children collaborating in the learning environment?
Data storage and security	How can images of children be de-identified and safely stored online? How can video data showing children's faces be stored during the research process? How should digital data such as transcribed interviews be de-identified and stored during the research process? How should photos and videos of children be managed after the completion of a research project?
Dissemination of research outcomes	How can you communicate visual data in a creative and innovative way to maintain participants anonymity? How can we navigate the complexities of reporting data in an accurate and representative way while also maintaining a protective stance towards children, families, and settings?

provided information which was detailed and accessible, with an appreciative tone and an open stance—we wanted to make it clear to parents that we valued their time and their consideration of the project, and that we were available to engage in discussion with them should they have any questions. This was a dimension of creating positive working relationships with participants from the outset.

We also explored questions around the future presentation of data (e.g. anonymising posts by blurring facial features or rendering in sketch style) and the possible implications for participants' privacy. We arrived at a decision to crystallise any identifying features while adding captions regarding any demonstrated emotion—for example, "child is gazing directly at the camera and smiling".

Discussion

The above vignette primarily presents ideas and issues linked to the following dimensions of the framework: Negotiation of initial consent; Anonymity, confidentiality, and visual data; Data storage and security; and, Dissemination of research outcomes. The researchers draw attention to their emphasis on the collection of data from publicly available social media accounts and the complexities at play here. They acknowledge the challenge presented by seeking informed consent, but have persisted with this approach out of respect for children and families. This aligns to the framework and its emphasis on relating respectfully to participants throughout the project's lifespan. Their point about creating positive relationships from the earliest stage of negotiating initial informed consent is an important one. Researchers should carefully consider the manner in which they engage and communicate with participants throughout the research to ensure partnerships are ethical, supportive, and productive.

From here, the researchers should consider the negotiation of ongoing consent and how this can be managed effectively. For instance, what progress updates can be offered to participants as a dimension of maintaining informed ongoing consent? Consideration should be given to the regularity and framing of progress updates, to ensure they effectively reinforce open communication and provide transparency about the project. Researchers could consider developing a detailed timeline of communications to map their interactions with participants, and should be continually attentive to how communications are conveyed so that participants remain informed and engaged.

The researchers also demonstrate awareness of issues around the ethical use of social media in research including anonymity and confidentiality, data storage and security, and how to engage in ethical dissemination. Mannay (2015) provides useful guidance around creative and innovative forms of dissemination with a mind towards sustaining both impact and an ethical approach. For example, using forms such as poetry can create engaging representations that maintain anonymity. Creative writing strategies could be used to disseminate research findings and impact readers by balancing contextual details and elaborated personal responses, emotions or beliefs while protecting the identity of research participants as they become fictional characters. Further provocations around the ethical dissemination of visual research are raised by Waycott et al. (2015) where, in particular, they call attention to careful planning regarding the context of dissemination and implications for participants, researchers, and the audience. There are tensions here that can prove challenging, but are worth engaging with productively to ensure an ethical approach that serves the research and the researched.

Vignette Two

My action research project, *Early Childhood STEM: Digital Technologies, provided Educators from an Early Years Learning Centre's Kindergarten (3 and 4-year-olds)* with the opportunity to express their views on digital technologies for learning and develop inquiry teaching and learning strategies. I decided to work in an independent Early Years Centre rather than a Kindergarten program within the State's Education Department. This was due to the increased need to apply for both University ethics and approval from the Education Department. The time delays associated with gaining approval from the Education Department could not be accommodated in the project's 6-month timeline. Subsequently, with informed consent from the participating Centre's Educators, the researchers were invited to design, implement, observe, and critically reflect on children's learning with three different types of coding technologies. The Educators were asked to keep a photographic digital diary of the demands of the learning experiences incorporating digital technologies, children' inquiry activities, and evidence of their computational thinking or coding. The learning stories about children by the Educators were a normal part of their practice but also contributed valuable information to the research. Throughout the project, Educators were positioned as collaborators and practitioner researchers. As such, they provided data about the children and their engagement with the coding toys. However, they were also research participants and provided reflections on their own beliefs and practices with digital technologies. When presenting this research at conferences, the Educators co-presented so were not anonymous. However, all children were photographed strategically to focus on the activity and not their faces. If present in photos, children's faces were blurred to protect their identities. This also extended to a child's full name written on their hat!

Discussion

The research context described was limited to independent kindergarten sites that sit outside of the state education system due to unmanageable timelines for the ethics review and site access approval process. Internationally, there is an increasing tendency for institutional review committees and boards to be legalistic and focused on management of perceived 'risk'. Consequently, the requested changes to ethics applications are "not grounded in knowledge of a researcher's field and may even prevent research entirely" (Harger & Quintela, 2017, p. 12). In this case, restricting research to independent sites creates an ethical challenge as state educators are not able to participate and receive the professional learning benefits associated with action research initiatives of this nature.

Contributing to professional learning from action research, is the positioning of educators as collaborators who try out strategies and provided data about children's learning. Yet the educators were also participants who provided reflections on their views and practices. During research of this nature, ethics can be considered relational as knowledge is gained through the relationship between researchers and participants. The relationship both shapes and informs ongoing, dialogic negotiation of ethical decision-making (Flewitt, 2018). This was evident by the joint decisions made by the research team for managing visual data capturing children's playful learning and how this was presented at conferences.

Inviting the educators to co-present at conferences did create an ethical dilemma. It was impossible to conceal their identity but it did respect the educators as collaborators and practitioner researchers. The educators worked with set protocols and protected the children's identity in the photos shown and the conversations shared. Another ongoing issue throughout the project was allowing for all children to play with the introduced technologies while ensuring only the children whose parents had provided signed consent were photographed or discussed in research de-briefings and learning stories.

Vignette Three

In my doctoral research, I embraced a feminist view of ethics. This encompassed an ethic of care premised on compassion towards participants, recognition of different ways of knowing, respect for participants, and research integrity (Brabreck & Brabreck, 2009). This research took place in a primary school with girls from Year Two–Year Seven. The focus of the project was on the children's relationships with digital and traditional media. Throughout the project, I prioritised establishing an equitable, accessible, transparent, safe, and supportive research context, where I would act as an ethically literate researcher by reflecting on and engaging with ethics throughout the research process. In the early stages of planning my project, I took a rights-based approach aligned to relevant articles of the United Nations Convention on the Rights of the Child. When seeking informed consent, I engaged in dialogue with participants and invited their questions and feedback. Checks for understanding were employed to ensure the girls were clear on the nature of the project and its requirements. This approach continued throughout the research—I often checked in with the girls to confirm they were comfortable with the research, which was premised on individual, in-depth interviews, and to invite them to lead our conversations based on their interests, experiences, and perspectives. This was critical in honouring their voices and views. Member-checking was used to confirm their contentedness with my write-up of the research—this was presented to them in PowerPoint format and in age-appropriate language. The girls were welcome to reflect on this and initiate revisions, additions, or omissions. They were encouraged to view their contribution to the research as 'their story', over which they had agency. They were also invited to select their own pseudonym, which enhanced their ownership over the narrative they were contributing.

Discussion

This vignette reflects a commitment to embedding ethics in the research design and raises important ideas around negotiation of consent, both initial and ongoing, and how children's voices are valued. The researcher mentions committing to ethics throughout the project and focusing on ethical literacy—this indicates a holistic and comprehensive approach, which is in keeping with the Digital Child Ethical Research Framework.

The researcher has been respectful and responsive in seeking informed consent initially and continually throughout the project, by entering into dialogue with children, checking for understanding, and checking in regularly. Strategies like these can assist with ensuring that children are consenting in an authentic way. In terms of checking in, there are a range of approaches that researchers may employ—these may be adapted depending on the context of the project (e.g. the children's age (s), the nature of the researcher's work with the children) but can include dedicating time to reflecting on the project, its progress, the child's place in it, and asking questions that centre the child's feelings about being engaged in research. Multi-sensory approaches can be utilised to ensure responsiveness to all children—for example, observing children's body language, listening for shifts in their tone, and staying tuned in to their level of engagement in the research tasks at hand.

The member-checking strategies employed by the researcher are appropriately child-centric and respectful. The researcher mentions making this accessible to the children and encouraging them to take ownership of their contributions. Member-checking is a valuable strategy that can contribute to how researchers in this space are designing their projects to give participants voice and agency, and in terms of maintaining informed consent. It can also contribute to the accurate and representative dissemination of data, in that the participants will have had an opportunity to confirm the legitimacy of what will be reported.

Vignette Four

Our research project *Toddlers and tablets: Exploring the risks and benefits 0–5 s face online* involved qualitative ethnographic research in the family homes of children aged five and under. In the design phase much thought was placed on ethical issues, especially gaining children's consent and trust throughout the research process. The parents of all children were asked to give informed consent and to confirm that they had received and understood the information about the study. Children aged 4–5 years old were all asked to signify consent through the use of pictograms and images (i.e. smiley face to indicate 'yes', sad face to indicate 'no'). Interviewers experimented with prompt cards, questions, drawings, digital tours, and play to solicit verbatim accounts of the pre-schoolers' responses to digital technologies, and to prompt narrative stories about their use. Ongoing consent was assumed until a child showed disinterest or refusal to play or talk. This was interpreted as their right to withdraw consent and stop participating.

Research in the context of the family home provides rich, context-laden data. However, these research moments are entangled with family routines, unexpected interruptions, and are held within the context of multiple relationships within the home including those between the child, parents, siblings, friends and visitors. We found that older siblings tended to join in on the 0–5 child interviews and that these siblings often contributed context and depth to the interview. However, they also presented an ethical dilemma. We solved this problem by asking the parent and sibling to give signed consent post-interview. This possible scenario was included within our ethics application with the other option being to treat the sibling's data as erroneous and unusable.

Discussion

This vignette focused on negotiation of initial and ongoing consent from young children. Flewitt (2005, p. 556) described this as provisional rather than informed consent. She explained, "the participants agreement was understood to be provisional upon the research being conducted within a negotiated, broadly outlined framework and continuing to develop within the participants expectations". Firstly, the communication about the research was adjusted to be age-appropriate and included symbols for children to register their willingness or not to participate. Indicators of ongoing willingness to participate were age-appropriate, including signs of disinterest or refusal to engage.

The researcher's ongoing awareness regarding negotiating ethical consent for participation was evident in their response to older siblings' contributions. They valued the voices of these children and the contributions they made but without consent they were unable to use the data. Responding with a post-interview consent form to be signed by the siblings was appropriate, but in so doing, changed the nature of the research and the nominated participants. This required an amendment to the original ethics application and approval.

Conclusion

This chapter has explored ethical issues pertaining to research involving children and digital technologies. The Digital Child Ethical Research Framework that highlights some of the considerations and areas/questions to be addressed when researching children's use of digital technologies has been presented and used to analyse researcher vignettes. This framework made explicit a range of issues impacting ethical decision-making in the digital era and as such could provide support to early career researchers as they make ethics applications and ensure ongoing ethical practices throughout research programs. The vignettes included provided insight into the ethical complexities and considerations that researchers in this space may experience. We look forward to further developing the Digital Child Ethical Research Framework as this area of inquiry continues to progress and evolve and recognise that with further examples and illustrative vignettes could increase the trustworthiness and transferability of the framework. While this chapter cannot provide resolute answers to ethical questions and quandaries, the framework and vignettes presented intend to provide provocations and guidance which will support researchers in the ethics of their work.

References

Badley, G. (2003). The crisis in educational research: A pragmatic approach. *European Educational Research Journal, 2*(2), 296–308.

Brabreck, M., & Brabreck, K. (2009). Feminist perspectives on research ethics. In D. Mertens & P. Ginsberg (Eds.), *The handbook of social research ethics* (pp. 39–54). Thousand Oaks, CA: Sage. https://doi.org/10.4135/9781483348971.n3.

Dewey, J. (1958). *Experience and nature*. New York: Dover.

Early Childhood Australia. (2016). *ECA code of ethics – Core principles*. Retrieved from http://www.earlychildhoodaustralia.org.au/our-publications/eca-code-ethics/code-of-ethics-core-principles/

Early Childhood Australia. (2018). *Statement on young children and digital technologies*. Retrieved from http://www.earlychildhoodaustralia.org.au/our-work/submissions-statements/eca-statement-young-children-digital-technologies/

EECERA. (2014). *Ethical code for early childhood researchers*. Retrieved from http://www.eecera.org/custom/uploads/2016/07/EECERA-Ethical-Code.pdf

European Co-operation in Science & Technology (COST). (2018). *The digital literacy and multimodal practices of young children*. Retrieved from http://www.digilitey.eu

Feilzer, M. (2010). Doing mixed methods research pragmatically: Implications for the rediscovery of pragmatism as a research paradigm. *Journal of Mixed Methods Research, 4*(1), 6–16.

Flewitt, R. (2005). Conducting research with young children: Some ethical considerations. *Early Child Development and Care, 175*(6), 553–565.

Flewitt, R. (2018). Working group 5: Ethics and researching digital literacy. DigiLitEY. European Cooperation in Science & Technology. 6th Project Meeting. Riga. Retrieved from http://digilitey.eu/sixth-project-meeting-riga-latvia-21st-and-22nd-june-2018/

Harger, B., & Quintela, M. (2017). The IRB as gatekeeper: Effects on research with children and youth. In I. E. Castro, M. Swauger, & B. Harger (Eds.), *Researching children and youth: Methodological issues, strategies, and innovations* (pp. 11–33).

Kuhn, T. S. (1962). *The structure of scientific revolutions*. Chicago: University of Chicago Press.

Mannay, D. (2015). Making the visual invisible: Exploring creative forms of dissemination that respect anonymity but retain impact. *Visual Methodologies, 3*(2), 67–76.

Morgan, D. (2014). Pragmatism as a paradigm for social research. *Qualitative Inquiry, 20*, 1045–1053.

National Health and Medical Research Council. (2018). *National Statement on Ethical Conduct in Human Research*. Retrieved from https://nhmrc.gov.au/about-us/publications/national-statement-ethical-conduct-human-research-2007-updated-2018#best_interests_of_the_child

UN General Assembly. (1989). *Convention on the rights of the child*. Retrieved from https://www.unicef.org/crc/

Waycott, J., Guillemin, M., Cox, S. M., Warr, D., Drew, S., & Howell, C. (2015). Re/formulating ethical issues for visual research methods from the ground up. *Visual Methodologies, 3*(2), 4–15.

Madeleine Dobson is a Senior Lecturer and Director of Student Experience and Community Engagement in Curtin University's School of Education. Her research interests include media, digital technologies, social justice, children's rights, and resilience and wellbeing within educational contexts.

Karen Murcia is an Associate Professor in the School of Education, Curtin University. Her research interests include scientific literacy for citizenship, creativity with digital technologies and STEM education. She is a researcher with the Australian National Centre of Excellence for the Digital Child, Chair of Child Australia and a Scitech Discovery Centre Board Member.

Kim Gifkins is a recognised specialist in human research ethics and has provided many years of service to the field as a University Senior Ethics Advisor. She is currently affiliated with Curtin University where she is supporting University Governance and committees of Academic Board.

Donell Holloway is Senior Research Fellow at Edith Cowan University, Australia. She is a Chief Investigator on three Australian Research Council grants, the ARC Centre of Excellence for the Digital Child, The Internet of Toys: Benefits and risks of connected toys for children, and Toddlers and Tablets: Exploring the risks and benefits 0–5 s face online.

The manufacturer's authorised representative in the EU is Springer Nature Customer Service Centre GmbH, Europaplatz 3, 69115 Heidelberg, Germany. If you have any concerns regarding our products, please contact ProductSafety@springernature.com

Printed and bound by CPI Group (UK) Ltd, Croydon, CR0 4YY
30/09/2026
02225692-0010